ENCYCLOPEDIA OF AGRICULTURE

ENCYCLOPEDIA OF AGRICULTURE

Vol. 3

AGRICULTURAL BIODIVERSITY

By

Dr. Renuka Sharma

DISCOVERY PUBLISHING HOUSE PVT. LTD.
NEW DELHI-110 002

Published by:
Tilak Wasan
DISCOVERY PUBLISHING HOUSE PVT. LTD.
4383/4A, Ansari Road, Darya Ganj
New Delhi-110 002 (India)
Phone : +91-11-23279245, 43596064-65
Fax : +91-11-23253475
E-mail : parul.wasan@gmail.com
discoverypublishinghouse@gmail.com
web : www.discoverypublishinggroup.com

***First Edition:* 2012**

ISBN: 978-93-5056-020-4 (Set)
978-93-5056-021-1 (Vol. 1)
978-93-5056-022-8 (Vol. 2)
978-93-5056-023-5 (Vol. 3)
978-93-5056-024-2 (Vol. 4)
978-93-5056-025-9 (Vol. 5)

Encyclopedia of Agriculture

Printed at:
Shree Balaji Art Press
Delhi

Preface

The breakthrough in science that permitted genes to be identified and manipulated as molecular ushered in the agricultural technology era, which is now more than a decade old. The new tools of agricultural technology are changing the way scientists can address problems in the life sciences; agriculture is one area facing major changes as a result of this new technology. The unanticipated rapid ratio at which discoveries and their applications in technology have unfolded has stored the capacity of society – more specifically, our agricultural research and educational institutions to absorb and adjust to change. We are challenged by pressing decisions opportunities, and problems that we face now and will continue to face in the future. Competition from abroad impels us to devise and use new technologies that can improve the efficiency and quality of agricultural production. These concerns led to this study – an overview of how the agricultural research system is responding to the latest technology and how it might prepare for future opportunities. Agricultural technology is moving in many directions with positive results – crop improvement, production of transgenic plants, vaccine development and diagnostic methods are some impending applications – but the development of genetic engineering's tools can be found in almost every agricultural discipline. The exception is that, through lessions learned, adequate food will be made available to the whole world in the future. The present title **"Endcyclopedia of Agriculture"** has been planned, written, and edited with the intention of being useful for the beginners, researchers and scientists involved in the field of agricultural genetic transformation.

I would like to express my sincere thanks to Dr. M.P. Arora, whose continuous inspiration and encouragement initiated me in bringing out this title.

I am specially indebted to my husband Mr. Rajeev Sharma for his enthusiastic support, constant helpfulness and good spirit during the writing of this title.

Quite frankly, this book would not have been written without the aid of my daughter Shreya, who co-operated patiently during periods of neglect.

Special and sincere thanks to my parents and in laws for their blessings and continuous encouragement in bringing out this title.

To make the work more comprehensive and informative, I have consulted many authoritative books, research journals, abstracts, monographs etc., so there can be no claim to originality except in the manner of treatment.

I also express thanks to my friends and colleagues whose continuous inspirations have initiated me to bring out this title.

I express my gratitude to Mr. Wasan and staff of M/s Discovery Publishing House Pvt. Ltd. for their whole hearted co-operation in the publication of this title.

I acknowledge the fact that the development and publication of this book would not have been possible without the assistance and encouragement of colleagues, research scholars and students. They are so numerous to mention. I thank all of them most warmly for helping to create the present title.

Any worthwhile criticism and suggestions for improvement would be thankfully acknowledged.

Renuka Sharma

CONTENTS

Chapter 1

INTRODUCTION

The emergence of has expanded the human capacity to take advantage of genetic resources and manipulate biological material to obtain food, medicine, and other valuable substances. Biotechnology products generally contain a large intellectual component requiring significant up-front investment, and they have a highly valuable commercial potential, which has created an impetus for the privatization of the knowledge input to their production.

The *establishment* of intellectual property rights (IPRs) for knowledge about biological processes and properties has increased the value and importance of *maintaining* biodiversity, as genetic resources are a key input to biotechnology production.

The management of genetic resources and biotechnology has created new policy *challenges* in the attempt to attain a *socially* optimal allocation of costs and benefits between the public and private sectors. The gene revolution *originated* in the developed world, but much of its promise may lie in addressing production and consumption problems in developing countries, with *significant* potential for *alleviating* hunger and poverty.

However, there is a great deal of criticism and uncertainty about the capacity of the current *institutional* framework governing access to *biotechnology* to facilitate the *transfer* of technologies controlled by the private sector in developed countries to benefit a broad range of producers and *consumers* in developing countries.

At the same time, most of the world's biodiversity *resources* are located in developing countries. Thus, strategies for their conservation and *utilization* in sustainable economic development need to be considered in the context of generating equitable access to the benefits from the *management* and development of genetic *resources*, as well as the need for efficient approaches to their conservation under conditions where economic development is imperative.

The study of the policy nexus of *managing* biodiversity and biotechnology, *especially* within the context of the developing world, is an intellectual challenge which is highly relevant to current policy debates. This books aims to provide a state-of-the-art *summary* of knowledge and policy debate in this *critical* area.

The book *presents* the results of three years of collaborative research in which the *authors* aimed to develop a coherent and economics-based approach to policymaking in the *management* of biotechnology and *biodiversity*.

Leading experts in various aspects of this policy debate were asked to contribute chapters on specific issues to which they could apply their unique expertise. By *integrating* the continuous effort of the editors with the insight of the other authors, we hope to have a fluid *augmentation* that is rich with insight and unique knowledge.

Target audiences for this book include agricultural *economists* who are working on technology and resource management issues, and *especially* on biotechnology and biodiversity, development economists addressing issues of resources and agricultural sector in developing countries, and *environmental* and resource *economists*.

These *individuals* may be in academia, in government, in *nongovernmental organizations*, and in private companies. Another target audience is policy scholars in government, schools of public policy or schools of *environment* that are interested in issues of biotechnology policy, IPRs, and biodiversity, as well as the *interaction* between developed and developing nations *regarding* these issues.

A third target audience is scholars in both development studies and resource management studies. By *largely de-emphasizing* technical *presentation* in the main text and *emphasizing* conceptual and policy issues, we believe that we will reach scholars whose aim is to *analyze* these major issues of development and resource management without heavy emphasis on economics.

Finally, interest in these topics presented in the book is strong among scholars and policymakers both in the developed and developing world, and in international organizations such as The World Bank and *United Nations* agencies.

Thus, policymakers throughout the world who are addressing the issues of this book are an important target audience. While most of the chapters rely upon economic analysis and tools, most of the book is written in a manner that aims to reach a broad range of experts interested in the topic, including *noneconomists*.

It also contains contributions by noneconomists who are experts in biotechnology and biodiversity. It aims to *familiarize* the reader with some of the major debates, policy options associated with management of biotechnology and *biodiversity* in the developing world, and conceptual approaches that aim to identify policies and management *schemes* that will lead to strategies that will improve social welfare and reduce poverty.

This book begins with a section containing chapters overviewing the global setting in which the management of biotechnology and biodiversity are taking place, *including* an analysis of major *socioeconomic* trends and *institutional* developments and their potential impacts.

This section is followed by one containing chapters *summarizing* the major issues in the management of *agricultural* and wild biodiversity, *including* valuation and incentives for conservation. Equity concerns and their implications for the distribution of costs and benefits *associated* with the conservation and use of *genetic resources* are the subject of the chapters in the following section.

The next section provides an analysis of the current and potential value of biotechnology in developing countries and the types of institutional reforms needed to realize this *potential.* The book is then concluded with a summary chapter, which integrates the policy implications drawn from earlier sections on *biodiversity* and biotechnology in the context of development.

In the remainder of this chapter, we provide a general introduction to the links between *agricultural* biodiversity and biotechnology, based upon issues and *findings* of the chapters as summarized below.

LINKAGES BETWEEN AGRICULTUAL BIODIVERSITY AND BIOTECHNOLOGY

Between 1961 and 1999, global per capita cereal production increased by 22% while total acreage devoted to cereals increased by only 4.9%. This increase in productivity is *partially attributable* to an increase in fertilizer, pesticide, and water use.

However, in a recent study, Evenson and Gollin (2003) show that the development and adoption of improved genetic materials were a *significant* and large part of the increase in agricultural productivity over this period.

They estimate that between 1961 and 1980, 21% of the growth in yields in food production in developing countries was attributable to the adoption of modern *varieties* among farmers, as was 50% of the yield growth *experienced* between 1981 and 2000.

Modern production systems are frequently characterized by their domination by monoculture, the adoption of which can lead to decreased genetic diversity, at least by some measures of diversity. The loss of genetic diversity generates costs in terms of reduced *resilience* of farming systems and *reduced* options for future crop and variety development.

The concern over the *erosion* of genetic *resources* may be linked in part to the increasing globalization of the economy, which has created pressures and conditions for the increasing intensification of agriculture, leading to the adoption of modern plant varieties around the world and, in turn, possible loss of traditional plant varieties.

At the same time, some *developing* countries perceive that major international *corporations* primarily from developed countries are likely to earn much income through the utilization of genetic materials that have been *conserved* mainly by farmers in developing countries.

The desire to *maintain* national sovereignty over their genetic resources has led to at least a dozen *countries* establishing controls over access to their genetic *resources,* and an equal number of nations developing such *controls.*

Enough international concern has developed over the need to conserve *agricultural* genetic resources to lead to the establishment of a multilateral system of access and benefit-sharing for key crops, the *International* Treaty on Plant Genetic Resources for Food and *Agriculture-hereafter* denoted as the *International* Treaty.

The International Treaty is considered a major step towards guaranteeing the future availability of the diversity of plant genetic resources for food and agriculture (PGRFAs) on which farmers and breeders depend, as well as a fair and equitable sharing of benefits.

This treaty entered into force on June 29, 2004. Of course, PGRFAs are the basic biological input into the breeding of new crop *varieties.* Molecular *biotechnology* is *increasingly* at the forefront of modern crop breeding *techniques.*

However, as applying such techniques is costly, modern crop varieties tend to be produced with developed country conditions and markets in mind, thereby limiting the extent of their relevance to developing country conditions and likely adoption rates.

This situation is *unfortunate* as biotechnology can *potentially* be of great use to developing countries in helping them meet the demands of feeding their *populations.* At the same time, to the extent that farmers in developing countries are adopting modern crop varieties, their *adoption* may be coming at the expense of traditional farmers' *varieties,* or landraces, concentrations of which tend to be in developing countries.

In the *process,* it is possible that PGRFAs of potential future value in crop breeding may be lost. However, as the cost of biotechnology applications fall, and *consequently,* biotechnology transfers to developing *countries* increase, agricultural biodiversity in these countries could be *increasingly* threatened.

In sum, PGRFA conservation and the promotion of biotechnology applications in developing countries may be strongly linked. If so, policy mechanisms addressing each would be more efficient if they were linked.

This book explores the economics of both the conservation of PGRFAs and adoption of molecular biotechnology and the economics of whether or not their *respective* policies should be linked and, if so, how.

OVERVIEW OF THIS BOOK

This section provides a summary of the contents of the rest of the chapters in this book. Chapter elsewhere in this chapter is an overview of the processes of globalization (*particularly trade liberalization*), *environmentalism,* consumerism, and the rise of the information economy, all of which are key factors that shape the evolution of agriculture, biotechnology, and biodiversity.

Chapter elsewhere in this chapter covers the evolution of plant *improvement* research, focusing on changes in the research process from the "green revolution" to the "gene revolution." Among the chapter's key points are that the green revolution was largely based in the public sector and involved crops and varieties that were suited for developing countries with highly *productive* fanning areas.

Varietal adoption patterns were also very much conditioned upon the presence of local breeding capacity. With the *subsequent* gene revolution, *agricultural* research and development (R&D) are now largely based in the private sector. The result of a shift in center of research innovation from the public to the private sector is that the focus of R&D will be on seeds or varieties with significant commercial value, which tend not to be seeds or varieties adopted for specific developing country *conditions*.

Given the focus of *biotechnology* applications on varieties intended for *profitable* developed country conditions, the ability of developing countries to benefit from *biotechnology* will depend on their local *breeding* capacity.

Of three breeding options examined in the chapter-local breeding with local varieties, regional breeding with adapted varieties, and adoption of seeds produced *elsewhere-the first* option is most expensive and last is the least. However, the first option is more likely to produce *higher* benefits in terms of biotechnology adoption as well as *biodiversity conservation*, as local *varieties* will be used in breeding.

Anywhere else in this book covers *valuation* and conservation issues for genetic *resources* and biodiversity. Chapter elsewhere in this chapter discusses the economic value of maintaining crop diversity as insurance against *vulnerability* to disease and pests.

Based on an empirical assessment of the change in welfare resulting from a marginal change in number of potential parents, the author finds little value overall in *maintaining* a large number of potential parents in breeding lines.

On the other hand, while noting that this chapter does not cover exactly the same subjects as in other chapters, these latter chapters argue that *in situ* conservation is an important means of *conserving* a *valuable* aspect of plant genetic diversity: The *evolutionary* process which occurs as a result of both human and natural selection pressures.

Else where in this chapter tackles the economic incentives for conserving crop genetic diversity on farms. The chapter starts off with an *assessment* of the market failure that arises from the public good nature of *in situ* conservation, in which farmers bear the cost of conservation but perhaps a small share of the benefits to *society* of such *conservation*.

The chapter argues that there is a greater harmony between public and private values in terms of managing biodiversity for reduction in vulnerability to pests and diseases (i.e., a form of portfolio *diversification* at the farm level), but not for *reducing* genetic erosion, which has public good aspects.

Rural populations depend to some extent on diversity in the genetic base, particularly in areas with *isolated* markets, as a form of insurance. However, such a *dependence* is not necessarily sufficient to promoting socially optimal levels of *in situ* conservation. Policies to promote conservation *in situ* include *promotion* of demand for products of diverse (*landrace*) varieties, e.g., *building* niche markets, labeling, and raising public awareness.

Other methods include changes in plant-breeding methods, such as participatory plant breeding, *community* seed banks, seed registers, and *protection* of farmers' *varieties* through "farmers' rights." The argument made in this chapter-that *in situ* conservation is cheapest where opportunity costs associated with the adoption of modern varieties are highest-also comes out in other conservation chapters in this section.

In other chapter focuses on *in situ* conservation methods and their costs, but in a less micro-oriented fashion than the previous chapter. Like Chapters elsewhere in this book asserts that the cheapest means of *promoting in situ* conservation is to look for such *conservation situations* with the lowest opportunity costs for *maintaining* diversity *in situ*.

However, the chapter provides more analysis of the processes required to keep conservation incentives in place even while promoting economic development, given that the cheapest conservation possibilities tend to be in areas with relatively low levels of *economic* development.

The chapter differs from all other chapters on *in situ* conservation in that it addresses the consequences of having more PGRFA conservation than is optimal for society. Chapter elsewhere in this chapter uses *empirical* evidence from Mexico to *investigate* the factors driving on-farm diversity in PGRFAs.

It argues that we need better information on what factors determine the selection of particular varieties for adoption by farmers, what impacts the process of selection has on genetic populations and what (e.g., trading networks, markets, seed *exchange* networks) *determines* genetic flows in and out of PGRFA populations.

Of all the chapters that address conservation issues, this chapter goes into the most detail about how human selection of PGRFAs interacts with natural selection in determining patterns of diversity. Chapter elsewhere in this book talks about the backbone of all regional and international collaboration for PGRFA conservation, namely, the presence of reliable national conservation programs.

International funding does not remove the need for domestic funding. Emphasis must be on measures that improve the efficiency of conservation, and measures to be targeted for improvement include regional and international collaboration, data and information management, and over-duplication of samples.

For example, conservation efficiency can be raised through the creation of a multilaterally accessible database with information on the *ex situ* and *in situ* germplasms that are available in the regions from which the germplasm is drawn. The author of the chapter asserts that a final prerequisite for any collaboration on the regional or international level is the maintenance of national sovereignty of those countries involved.

Namely, only with their sovereign rights *maintained* over materials such as germplasm are countries willing to place such materials in secure storage facilities outside their borders. *Elsewhere* in this book covers distributional issues in the *management* of genetic *resources*.

Chapter elsewhere in this book discusses the sharing of benefits derived from the utilization of PGRFAs in the breeding of new varieties. From an economic efficiency as well as equity standpoint, it seems reasonable to tie a country's contribution to a benefit-sharing fund (such as that envisioned under the auspices of the new International Treaty) to the benefits it receives from its use of PGRFAs.

Every country benefits from utilization of PGRFAs in the *production* of new goods, but some countries may benefit more than others. *Unfortunately*, as discussed in this chapter, these benefits cannot be *quantified*, except *perhaps* in limited case studies.

Hence, given that political considerations dictate that a benefit-sharing fund be created, an alternative can be to appeal to indicators that take equity and development considerations into account in *determining contributions*, and that acknowledge at least some of the characteristics of the benefits of PGRFAs. Thirteen *potentially* feasible *indicators* are examined in this chapter.

All the feasible *indicators* are deficient in some way. While Chapter elsewhere in this book discusses who should contribute to a benefit-sharing fund, and how much, Chapter anywhere else in this book *examines* potential economic criteria for *distributing* money from the conservation fund for the *conservation* and *sustainable* development of plant genetic resources. However, benefits accruing from the distribution of these funds for conservation *activities* are almost impossible to ascertain.

The question then becomes what is the most *economically* efficient method of *distributing* the funds among countries or throughout the world, given the available data. This chapter describes a proxy indicator for the *importance* of a region as a primary center of diversity.

It then goes on to rank their importance to the global *community* and to OECD countries based upon the consumption of crops *originating* from various centers of diversity. Chapter elsewhere in this book extends the institutional discussion in the previous chapter with a demonstration of how *cooperative* game theory can be applied to determining the "fair and *equitable* sharing" of the benefits arising from the use of PGRFAs, using as a starting point the regional allocations from the previous chapter.

Using this *approach*, the impacts of the players' (e.g., countries') *bargaining* power on the resulting allocations can be empirically assessed. Furthermore, the approach allows us to explicitly account for potentially competing interests of the players, thereby introducing some equity to the allocation.

The *implications* of three different allocation regimes are modeled. One of these assumes that funds will be distributed by the *International* Treaty only to world regions, which then will be responsible for allocating the funds within their regions. This scenario was found to be particularly appropriate as part of a flexible *mechanism* for biodiversity *conservation* as it allows the use of different types of control *mechanisms* at different levels of negotiation processes.

In the other part of this book includes chapters that address *biotechnology* concepts, economic *valuation* of *biotechnology*, and *management* of *biotechnology production* and processes. Chapter elsewhere in this book gives an overview of evolution of agricultural biotechnology concepts and applications.

While this chapter provides a conceptual overview of the present state of biotechnology *applications* to agriculture, the first section of Chapter elsewhere in this book provides a technical overview that includes examples of specific products.

Specifically, Chapter elsewhere in this book provides a relatively detailed summary of existing *biotechnology* applications and provides details on *secondgeneration* biotechnologies that are being developed and that may be of relevance to developing countries. The chapter also presents data on the adoption of *genetically* modified organisms (GMOs) in developing countries.

It also discusses GM products that are further down the production pipeline, and does so by country, crop, and trait and, for livestock, by country, species, and trait. Chapter elsewhere in this book examines how differing IPR regimes, states of *development* of the seed industry, and *agricultural* R&D capacities will affect the nature of biotechnology adoption in developing countries.

For example, with strong IPRs, a strong breeding sector, but high transaction costs in trading IPRs, the most likely outcome is that the biotechnology company will directly introduce GM *varieties* that are not locally adapted, resulting in a loss of *in situ* diversity in PGRFAs.

With weak IPRs and a strong breeding sector, every breeder or seed company can use *commercialized* GM varieties in order to cross-breed the technology into their own *germplasm*. Thus, many *different* GM varieties will be available on the market, although in the long run there may be less access to technology, due to developer's inability to capture rents.

Chapter in this book utilizes the example of biotechnological innovation in the global canola sector to identify some lessons for how developing countries might participate and benefit from this *innovation*. Developing countries are facing *ever-rising technical,* economic, and political barriers that limit their capacity to use *biotechnology* in their fight against hunger.

Developing countries require functioning economic markets, physical and scientific *infrastructure*, and political and legal capacities. In many cases, these countries will need to create the appropriate input and output market *conditions* for the new technology to be disbursed.

Firms will only go where there is supporting *infrastructure*, research collaborators, functioning labor markets, competent regulators, and markets that are accepting of GM products. Some developing countries, such as China, India, and Brazil, have the prospect of assembling *institutions* adequate to promoting adoption of GM crops.

Like the previous chapter, Chapter 16 addresses the economics of the adoption of biotechnology and the constraints to its adoption, but does so from the farm level. The chapter argues that divisible technologies that are simple to use and that have limited fixed costs (e.g., GM seed *varieties* and tissue culture *technologies*) hold the most promise for adoption by small, poor farmers.

The fact that biotechnology varieties do not require high inputs of human capital-in fact, they often result in reduced management *requirements*-also means they may be well suited for adoption among lowincome farmers. Nevertheless, adoption may be *constrained* by several farm-level factors including farm size, *agroecological* conditions, availability of credit, and risk.

These factors have been shown to be important in the adoption decisions among smallholders in developing countries. Adoption levels also depend on macro-level factors including a country's research capacity and characteristics of its input and output markets.

China has shown the greatest success with GM crop adoption-where farmers have benefited instead of foreign firms due to a combination of weak IPRs and significant government *involvement* in biotechnology research.In Latin America, a *growing* gap between small, poor farmers and large multinational cooperation, as well as negative public perception of GM crops, continue to be significant constraints to farmer adoption of GM crops.

In other part of this book, the final section, draws policy implications by identifying and expounding on the themes that cut across the biodiversity, biotechnology, and development issues raised in this book. Chapter elsewhere in this book examines the potential of biotechnology for poverty alleviation and is the only chapter to consider indirect impacts (via labor markets and food prices) of biotechnology adoption on the poor.

The chapter raises *questions* about *biotechnology* as a means of poverty alleviation: (1) Do faster and cheaper means of economic development exist than through agricultural technology change; (2) are there faster and cheaper means of agricultural technology change than through biotechnology; (3) do many market failures (e.g., in the provision of credit) exist that may prevent agricultural biotechnology from being *effective*; and (4) do other basic needs of the poor need to be addressed before biotechnology adoption would be effective?

After raising these questions, the chapter provides many policy recommendations for how to get biotechnology to work as a tool for poverty alleviation. Chapter elsewhere in this book describes a possible *mechanism* for reconciling the economic tension that exists between the public and private economic forces that drive agricultural research.

That mechanism is the establishment of an intellectual property clearinghouse for agricultural biotechnology. This *clearinghouse* would provide three essential functions: (1) *identification* of all relevant intellectual property that exists over a given technology and what properties are available and how they could be accessed; (2) the establishment of a pricing scheme and terms of contract that depend on the identity of the buyer; and (3) the establishment of an arbitration mechanism for monitoring and enforcement of the contracts made through the clearinghouse.

The purpose of the clearinghouse would be to reduce market failures in agricultural biotechnology markets. It would also increase access to agricultural biotechnology in the National Agricultural Research Systems (NARS) in the developing countries, the Consultative Group on International Agricultural Research (CGIAR) system, universities, and, ultimately, farmers in developing countries. Chapter elsewhere in this book picks up on and amplifies policy themes in agricultural biodiversity conservation and sustainable use that were raised in earlier chapters. The chapter discusses the *effectiveness* of various types of payment *mechanisms* for conservation.

It identifies the wide range of actors who are, or potentially could become, involved in *conservation* through the use of a wide range of mechanisms that go well beyond the traditional concepts of *conservation* activities.

A key theme throughout the discussion in the chapter is the importance of *recognizing human* knowledge as a key component of agricultural *biodiversity* and the necessity of incorporating means for knowledge preservation as much as the physical conservation of agricultural biodiversity.

In other chapter of this book provides a detailed history and description of the *International* Treaty on Plant Genetic Resources for Food and Agriculture. Finally, chapter elsewhere in this chapter is a synthesis that attempts to identify and *reconcile* the common themes across the chapters and draws some major *economic* conclusions and policy *recommendations* from this synthesis.

2

Chapter

EVOLUTION OF DIVERSITY

Over the past 20 years, several global trends have been *unfolding* which have *implications* for the evolution of agricultural biotechnlogy and the *conservation* and sustainable use of agricultural *biodiversity*. These trends are *interlinked* and in some cases have opposing effects, and their final outcomes are yet to be *determined*. In this chapter we provide a short survey of these developments together with an analysis of their potential implications for the use of agricultural *biotechnology* and the *management* of agricultural *biodivesity*.

The trends covered include trade and capital market *liberalization*, the rise of the *environmental* movement, consumerism, the *privatization* and devolution of *government* services, and the emergence of the information age. Both biotechnology and the concept of biodiversity are fairly recent arrivals onto the human scene, and their management has raised several controversies.

For example, biotechnology is a product that is comprised of a large intellectual *component*, e.g., it represents the culmination of a process of research. This *research* has mostly been carried out in the private sector, *although* it also often involves the use of genetic *resources* which originated in the public domain.

There is *considerable disagreement* on the best means of *protecting* the property rights to the intellectual component embodied in biotechnology, while recognizing both the private and public *contributions* to the end product. In addition, agricultural biotechnology products are the result of a major scientific advance and have only very recently become *available*.

Due to their novelty, there is only *limited* information on the long-run impacts they might have on *environmental* and food safety. A great deal of *uncertainty* exists on how much risk such products entail, as well as much controversy on how it should be *measured* and how much is socially acceptable. Considerable *uncertainty* and conflict exist over the conservation of *agricultural* and wild biodiversity as well. Assigning values to *biodiversity conservation* is fraught with *uncertainty.*

One of the most *significant* values associated with biodiversity is *preserving* potential future options for the use of the genetic resources maintained-and this is very difficult to assign value to. There is even considerable uncertainty with *determining* the use values of agricultural biodiversity, which ostensibly is easier to measure. Uncertainty over values leads to controversy over conservation strategies: how much and what should be preserved.

Controversy is particularly sharp when conservation conflicts with economic development. These controversies are currently under discussion and *negotiation* in a *variety* of formal and informal forums, and they are being shaped by the global trends, which we identified in the first paragraph. In the discussion which follows below, we discuss how these global *processes* are shaping the ongoing debates in various contexts and draw conclusions as to their potential implications for the *management* and use of *agricultural biotechnology* and *biodiversity* in developing countries.

Our discussion is kept to a fairly general level, which does not fully capture the *tremendous* variation that exists among developing countries in terms of their *endowments* and capacities. More specific analyses related to the *management* of biotechnology and agricultural biodiversity in the varied context of developing countries are given in later chapters of this book.

GLOBALIZATION OF TRADE AND CAPITAL MARKETS

Over the last 20 years, the volume of trade between countries has expanded remarkably as a result of the reduction of trade barriers, as well as decreasing costs in transport and *communications* and the increased mobility of capital across *international boundaries.*

International and regional trade *agreements* have been the primary mechanism by which trade barriers have been lowered, such as the General Agreement on Trade and Tariffs (GATT), and subsequently the World Trade Organization (WTO) at the global level, and North American Free Trade Agreement (NAFTA), the *European Community,* and MERCOSUR as examples of *regional* blocs.

Liberalization has also occurred in agricultural trade markets, although this is one of the most contentious areas of international trade policy and one where significant distortions still exist, particularly among developed countries. Indeed it was deadlock over agricultural trade which caused the breakdown of negotiations at the 2003 WTO meeting in Cancun.

Nonetheless, there has been *considerable* movement towards the *liberalization* of agricultural trade markets, and more is expected in the future. In the United States, there is a move towards *converting* commodity support programs towards *"green payments,"* e.g., paying for *environmental* services. In Europe, the expansion of the *European Union* is creating pressures to reform the Common Agricultural Policy (CAP) and reduce production supports.

Farmers are increasingly expected to rely on insurance instruments provided by the private sector and *sometimes* subsidized by the government for the management of production and *revenue* risk. Future markets and forward contracts are also likely to play a major role in reducing risk in agriculture.

In basic grain markets, the impact has been a shift in production from high cost to a few lower cost *producers* such as the United States, *Argentina*, and *Australia*, as well as Thailand and Vietnam. At the same time many developing countries as well as the transition economies of Eastern Europe have become net importers of grain, and this trend is expected to continue with *liberalization*.

If indeed agricultural support prices in developed countries are reduced, producer prices for some agricultural commodities are likely to increase in developing countries and new market *opportunities* created. One impact of these changes may be *increased* incentives for the adoption of new *yield-increasing* biotechnologies.

Agricultural trade *liberalization* increases competitive *pressures* among producers and creates incentives for increasing yields and reducing costs in agriculture. It also exposes producers to the demand *requirements* of a larger group of consumers. This may expand the demand for both yield-increasing and *pest-controlling biotechnology* products.

For example, the ability to export to markets in Japan, Canada, and other countries may be *determined* by the ability to control pest problems with minimal or n o chemical residues. Concern about ozone depletion is leading to regulations banning the use of methyl bromide and other chemicals. These *measures* provide increased incentives for the *adoption* of *pestcontrolling* biotechnology products.

By reducing investment barriers, trade *liberalization* creates the *potential* for investors such as multinational companies to invest in both production and marketing *infrastructure* in developing countries with *promising* commercial market potential, or which establish incentives to attract mobile capital. Profound changes are occurring in the *organization* of the food sector in developing countries due to globalization, as well as *urbanization*, *increasing* incomes, and the *opportunity* costs of food *purchasing* and preparation.

The rise of multinational retail chains, supermarkets, fast food chains, and other forms of pre-prepared foods are manifestations of this change. The developments in the structure of food markets raise challenges and opportunities for local and global suppliers, and have *implications* for both agricultural biotechnology and biodiversity. On the one hand, food producers can potentially take advantage of the *income-earning opportunities* created in a dynamic and rapidly expanding market.

This could increase the demands for agricultural biotechnology and incentives to adopt among producers in order to remain competitive. On the other hand, small producers unable to adapt to the required *institutional* and organizational changes, and the technology and *management requirements* that they entail, risk *marginalization* in terms of market participation.

Some evidence of this trend is available with concentration in the food supply chain linked to increased farm consolidation and reduced market participation among small producers. It is not clear what impact this will have on either agricultural diversity or *biotechnology*, although it is likely to lead to a higher demand for biotechnology products from both the *commercial* farm sector and the food processing industries, but will reduce demand from small farmers.

While agricultural trade *liberalization* may result in increased incentives for producers in developing countries to adopt agricultural *biotechnology,* the extent to which adoption actually will occur depends on the types of innovations biotechnology delivers, and the degree to which these substitute for scarce factors of production and address key *production* and *consumption* constraints.

At present, agricultural biotechnology innovations are being developed primarily to reduce production costs or increase yields under conditions present in developed countries, which constitute the main market for these products. In many developing countries, production *constraints* are of a different nature than those in developed countries; barriers to *productivity* increases are often more related to *controlling* for the incidence of drought, poor soil quality, and high rates of pest and disease, whereas in developed countries reducing *management* costs and pesticide use are more important concerns.

Trade liberalization may exert some positive influence on the commercial *attractiveness* of developing innovations to address these needs through its impact on the global demand for inputs; however, this will only apply to technologies that have the potential for a significant *commercial* market. Even where technologies are suitable for the production conditions in developing countries, it will be necessary for countries to have in place an adequate level of research, extension, and regulation to achieve *dissemination* and adoption of such technologies.

The *institutional* requirements are significantly higher and more sophisticated than has been the case in the past for the adoption of improved agricultural *technologies.* Issues such as biosafety regulation, the negotiation of *intellectual* property rights (IPRs), and the technological capacity to modify technologies to suit local conditions place fairly significant burdens on the research and development (R&D) infrastructure of developing countries, and the capacity to meet such demands varies widely among them. An *important* effect which the liberalization of trade may have on both agricultural biotechnology adoption and the *management* of *biodiversity* is the degree to which *consumer* concerns for the *environment* and food safety are allowed to be manifested through trade regulations and labeling.

The key principle of the WTO is *nondiscrimination* among member states, e.g., a standardization of product definition and treatment. However, consumer preferences for the *environmental* and health attributes of *agricultural* products are *heterogeneous* across national boundaries and could potentially be manifested in trade regulations.

The ability of countries to regulate trade based on *environmental* and food safety concerns and *specifically* the degree to which countries may apply their own standards to reflect such concerns are governed by two agreements made under the WTO: the Agreement on Sanitary and *Phytosanitary Measures* (SPS) and the Agreement on *Technical Barriers* to Trade (TBT).

These *agreements* allow members to impose *restrictions* on trade based on *environmental* and food safety concerns, but they also seek to ensure that such regulations are no more trade restrictive than necessary by imposing restrictions on the use of such "nontariff barriers to trade." In addition, they do not apply to the processes by which agricultural and other goods are produced, but only to the products themselves, which limits the degree to which environmental and food safety concerns can be used to establish trade barriers.

Nonetheless, consumer concerns over the environment and food safety could potentially impact the *production* practices in *exporting* countries. *Ultimately,* this impact will depend on the type of specific *attributes* that are demanded, the willingness to pay among consumers for such attributes, the capacity to distinguish such *characteristics* in products (e.g., labeling), and the degree to which the expression of such *preferences* is allowed under the WTO regulations.

The WTO includes another important agreement that has major implications for the dissemination of biotechnology and the management of biodiversity: the Agreement on Trade-Related Aspects of Intellectual Property Rights (TRIPS). The main thrust of this agreement is to *facilitate* trade in products that have a high *intellectual* property content.

The agreement mandates a *minimum* standard for IPRs among member states, but leaves them free to determine the appropriate method of implementing them under their own legal system. Article (27.3(b)) of the agreement explicitly refers to the protection of plant varieties and stipulates that new varieties need to be protected either by patents or an "effective sui generis" system such as that of the *International* Union for the *Protection* of New Varieties of Plants (UPOV).

Under the UPOV system of plant protection, plant breeders' and farmers' rights may be recognized; e.g., breeders have the right to use protected genetic materials in the development of new varieties, and farmers may have the right to save and re-use seeds from protected varieties for their own use. The TRIPS agreement also allows members to exclude from patentability inventions whose use would seriously prejudice the *environment*.

Implementation of this agreement is likely to increase the incentives for the developers of biotechnology innovations to expand into new markets, due to the increased protection it provides for their *investment* into the technology. Since agricultural *biotechnology* innovations are being produced mostly by the private sector, this protection is critically important for creating incentives among the suppliers of the technology for its *dissemination*. The TRIPS agreement also has *implications* for the *conservation* and sustainable use of agricultural biodiversity.

Agricultural biodiversity is maintained through systems of access and exchange from the farm to the international level, and property rights to plant genetic resources are likely to effect current patterns of exchange. There are several options for property rights over plant genetic resources, and their impacts on diversity are expected to be varied.

Property rights and their degree of enforcement are also likely to impact the extent and nature of transgenic crop adoption, which will have *implications* for both spatial and temporal agricultural diversity. The increased value of plant genetic *resources* as an input to breeding under private breeding programs may lead to increased demand for diversity.

Concern that the establishment of property rights will lead to reduced levels of access and *exchange* of plant genetic resources and thus reduced levels of agricultural *biodiversity* have also been raised. This includes concerns about the *potentially* negative impacts on access imposed by farmers' rights mechanisms.

The agreements made under the WTO are not the only international agreements which drive the way the globalization of trade networks proceeds and impacts on biotechnology and biodiversity; there are several *environmentally* related conventions and agreements which are discussed in the following section and which may have an impact on the ways the WTO agreements are *interpreted* and *implemented*.

However, the framework laid out under the WTO is the most important in determining what the potential impacts of trade liberalization on biodiversity and biotechnology will be, as this agreement has wide and *expanding* membership and its signatories include some of the key national players in this arena, which is not the case with many of the environmentally related agreements discussed below.

ENVIRONMENTALISM

Environmentalism has arisen from two main motivations: (1) the interest in *preserving* species, environmental quality, and ecosystems and (2) the concern about environmental and health side effects of agricultural practices. The 1957 *publication* of Rachel Carson's book, *Silent Spring, was* a major benchmark in the evolution of the environmental movement.

It raised awareness about the negative side effects of pesticides and other agricultural practices. Over the last 30years, with a growing availability of information on the incidence and costs of environmental degradation, concerns over the necessity and means for controlling and reversing the process have become manifested in governmental policies from the international to the local level, as well as through activities in civil society.

A key thrust of the *environmental* movement is the promotion of awareness of the nonmarket as well as market values of environmental goods and services and pressures to account for this value through government regulations as well as consumer behaviour. Specific *manifestations* of the impacts of the *environmental* movement are *considered* in the next few paragraphs.

Establishment of Environmental Protection Legislation and Agencies

Since the late 1960s, most countries have established national agencies of environmental *protection* that are at the ministerial level and an increasing body of *environmental* regulations at all levels of government. However, in many cases the implementation of environmental regulation has been hampered because of political economic constraints of *information* about the *processes* that drive environmental *degradation* and the means to control them.

There is a large body of evidence showing that higher income countries attain higher standards of environmental quality and that corruption and flawed *governance* reduce the *effectiveness* of environmental policy. The primary means of environmental *regulation* have been through the implementation of "command and control" measures, which are fairly blunt and achieve environmental objectives at excessive costs.

However, at present, there is gradual transition to financial incentives (payment for environmental services) and market-based mechanisms (trading in water rights or pollution permits). The *regulation* of chemical pesticides and drugs consists of strict *preregistration* testing and "learning by doing" once a product is released.

The regulating authorities establish applications, standards, and tests for efficacy and side effects before *registration*. Products are recalled once a sever defect (carcinogenicity) is detected. The high cost of registration may be a barrier to entry, but it serves to address *concerns* about product safety and environmental *impacts*.

Cropper et al. (1992), in an analysis of the regulations of pesticides in the United States, suggest that the Environmental Protection Agency is capable of weighing benefits and costs when regulating environmental hazards; however, the implicit value placed on health risks-$35 million per applicator cancer case *avoided-may* be considered high by some people.

The same regulatory approach is used for genetically modified (GM) varieties. The effectiveness of this regulatory approach depends on *quantitative understanding* of the processes through which *biotechnology* affects the environment. For example, concerns about the buildup of pest resistance have led to the establishment of *refugia* requirements (demanding allocation of some land to nonmodified varieties) with *Bacillus thuringiensis (Bt)* cotton.

The *challenges* of establishing and implementing these regulations are apparent from a growing body of literature on their *evaluation*. Performance *measures* are also very *difficult* to establish for the conservation and sustainable use of agricultural biodiversity. There is uncertainty on the status, measurement, and value of *biological* diversity, both for wild and agricultural biodiversity. Chapter elsewhere in this book the *irreversibility* of the loss of genetic resources also creates difficulties in *assigning performance* measures. Until recently, agricultural biodiversity conservation policies have focused primarily on the *ex situ* preservation of genetic resources associated with economically important crops.

At present, the portfolio of policies includes *ex situ* gene banks, the establishment of botanical gardens and experiment stations, and various forms of incentive measures to promote *in situ* conservation. The former are *mechanisms* for preserving genetic resources, while the latter conserve *evolutionary* processes and human knowledge in addition to genetic resources.

International Agreements on Global Environmental Problems

Increasing concerns about global environmental problems and the need for *international coordination* in addressing them have given rise to a *proliferation* of international agreements. At the U. N. Conference on Environment and Development (UNCED) held in Rio de Janeiro in 1992, a basis was laid for several international agreements in the areas of biodiversity preservation, climate change, desertification control, and others.

Of direct relevance to the management of biodiversity and biotechnology is the Convention on Biological Diversity (CBD). The CBD is an *intergovernmental* convention that entered into force in 1993, which has now been ratified by 180 parties with the aim to achieve three main goals:

(1) the conservation of biodiversity;

(2) sustainable use of the components of biodiversity, and

(3) sharing the benefits arising from the commercial and other utilization of genetic resources in a fair and equitable way.

In January, 2000, a *supplementary* agreement to the CBD, known as the Cartagena Protocol on Biosafety, was adopted. This agreement seeks to protect *biological* diversity from the potential risks posed by living modified *organisms* resulting from modern *biotechnology*. The two cornerstones of the Protocol are the concepts of Advance Informed Agreement (AIA) and the *Precautionary* Approach.

The AIA enables importing countries to subject all imports of Living Modified *Organisms* (LMO's) to a risk assessment before allowing its entry, and such risk assessments may be made using a precautionary approach. This could have *implications* for the adoption of agricultural biotechnology, as this agreement could allow countries to block imports of seeds of GM plant varieties in the absence of sufficient scientific evidence about their safety.

The agreement entered into force in September, 2003. It is important to note that the members of the CBD and the Cartagena Protocol differ from the members of the WTO. Notably, the United States has not ratified the CBD (although it is a signatory) and is not a signatory to the CP and, as the primary developer and exporter of GM products, this is likely to have major *implications* for how these agreements are implemented.

How these differences in legally binding commitments among countries will be resolved in international fora is still not clear, and there are attempts to try to harmonize any conflicting provisions. It is also not clear how varying standards for risk assessment allowed under the WTO and multilateral environmental agreements will be resolved. This will most likely emerge through dispute resolution and arbitration in *international* bodies.

Proliferation of Environmental Groups in Civil Society

Public support for the *environmental* movement has been manifested by the *establishment* of *nongovernmental* organizations that emphasize various aspects of environmentalism. Some, like the Nature Conservancy, are engaged in the purchase of valuable environmental resources (mostly land and water), and others (e.g., Greenpeace) are engaged in political activism.

Other key players include the World Conservation Union (IUCN) and the World Resources Institute (WRI), which play a role of information provision and policy support, and the World Wildlife Fund (WWF), which is engaged in the *implementation* of conservation-related projects. A key activity of many environmental groups is educating consumers on the environmental implications of various goods and services offered in the marketplace and the *mobilization* of pressure from consumers on producers through their *purchasing* decisions. Several studies have found that the demand for environmental quality is related to income.

The demand for environmental services and goods varies across income groups, with higher income categories being more likely to focus on conservation, while for lower income groups the *sustainable utilization* of natural resources is a more pressing concern. In developing countries major environmental concerns are related to problems of water quality, waste management, and *sanitation, particularly* in urban areas, as well as the sustainable use of natural resources in the process of economic development.

Countries with higher income levels are more concerned with natural resource preservation, such as the preservation of open space, and the *protection* of endangered species. In general, concerns about global *environmental* goods and services, such as biodiversity and climate change, have been driven by developed countries, although there is *increasing* awareness and concern of the importance of these issues among developing countries.

CONSUMERISM

As income increases, consumer rights and *preferences* for *improved* quality have become the major determinant of economic activities. In most developed countries, the primary potential for revenue generation is through *enhancing* the value-added of food products. Indeed, in developed countries, sectors in the agricultural economy (e.g., poultry) that have been able to provide a wider variety of quality choices and extend their product mix have been very successful. Becker (1965) provided a conceptual framework to analyze consumer choices for improved product quality.

They suggest that consumers derive enjoyment from the characteristics of market goods that they consume, and that consumption activities may entail some effort. For example, the value of a meal to a consumer may be comprised of the value of its *nutritional* content, its taste, its safety to consume, and the degree of effort that its preparation requires.

Economic factors are a major determinant of food quality *preferences.* Some *characteristics,* such as convenience in preparation, exhibit higher *elasticities* of income. Cultural factors may also influence the values assigned to various food characteristics. Thus, one of the challenges of *agricultural* industries is to *economically* produce products that contain the food *characteristics* desirable in their target markets.

As income in developing countries rises, the demand for improved food quality is likely to increase significantly. In the next 50 years, we expect that vast populations in Asia and South America will reach income levels that will enable them to pursue improved food quality. Projections made by the FAO indicate that by 2015 rises in income will translate into consumption of an average level of over 3000 kcals/day/person by 54% of the world's population.

This increase in caloric intake will stimulate a transition in food consumption patterns as well, from starchy staples toward "*luxury*" goods such as dairy products, fish, and meat. The demand for food characteristics associated with a high elasticity of income, such as food safety, nutritional content, and convenience is thus also likely to increase. According to Welch and Graham (2002), "Micronutrient malnutrition (e.g., Fe, Zn and vitamin A deficiencies) now afflicts over 40% of the world's *population* and is increasing *especially* in many developing nations.

Green revolution cropping systems may have inadvertently contributed to the growth in *micronutrient* deficiencies in resource-poor populations. Current interventions to eliminate these deficiencies that rely on supplementation and food *fortification* programs do not reach all those affected and have not proven to be sustainable."

They argue that one approach to the micronutrient deficiency problem is enhancing the nutritional content of staple food products. One of the major promises of *biotechnology* is its potential to enhance food characteristics. Biotechnology may be used to extend shelf life, modify size and shape, and enhance flavors and nutritional content.

Parker and Zilberman (1993) have shown that improved food quality may more than double the retail price of peaches, and quality-enhancing biotechnology may be a major source of income for agriculture in the long run. *Environmental* preferences are also manifested through consumer behaviour. One dimension that may enhance the demand for *biotechnology* products is the desire to consume pesticide-free food. At the same time, consumer concerns over the health and environmental impacts of biotechnology products is resulting in a slower rate of their adoption in *agricultural production.*

On the health side, concerns over the potential for increased levels of allergic reactions from consuming foods generated through biotechnology have been raised. Environmental concerns have also been raised regarding the potential for genetically modified organisms (GMOs) to escape into the larger gene pool, resulting in an irreversible change in the composition of genetic resources and the potential for the spread of *undesirable* organisms such as "*super weeds*". A critical *determinant* of the future use of agricultural biotechnology products lies in the attributes consumers will demand of products and to what extent they will pay for these.

At present this response is unclear and will be driven by conflicting concerns on *environmental* and food safety and the perception of biotechnology's impact on these, as opposed to the desire for quality *characteristics* biotechnology can deliver, such as improved taste and nutrition, enhanced shelf life, and also improved environmental performance associated with a reduction in pesticide use. Considerable variations in consumer attitudes towards agricultural biotechnology products, particularly GMOs, are found in the *potential* markets for the products.

Attitudes are often linked to income, with people from poorer countries having more positive attitudes than those from richer countries, although there are exceptions to the pattern. A survey conducted by *Environics International* in 34 *countries* revealed that, in general, people in developing countries are more likely to value the benefits of biotechnology over the *potential* risks, as compared with those in developed countries, *particularly* Europe.

Consumer attitudes were also found to vary depending on the type of benefits biotechnology conveyed: Applications that address human health or environmental concerns were viewed more favourably than those that increase agricultural productivity. Consumer rejection of GMOs has two major implications for the *dissemination* and adoption of agricultural biotechnology.

Threat of loss of market share has caused exporting countries to ban the use of biotechnology in production, and this factor is now included in the riskassessment procedures of some countries. For example, one of the largest soya-producing regions of Brazil banned the planting of GM soya and India stopped trials of BT cotton. Consumer demand for differentiated products has implications for the structure of the food processing industry as well.

We have already seen the emergence of *differentiated* products in poultry and fresh fruits and vegetables in developed countries. Producers of these differentiated products are *frequently* either *vertically* integrated firms or a chain of firms that is linked through contracts. It is likely that some dominant firms in these industries (Proctor and Gamble, Gerber, etc.) will become actively involved in *utilizing* biotechnology to produce differentiated products.

Both the marketing techniques and production structures that are associated with these *industries* are likely to transform the agricultural sectors that adopt biotechnology to meet differentiated consumer preferences. Increases in vertical integration and contracting in agriculture are likely to accompany the development of biotechnology to *respond* to these consumer demands.

DEMOGRAPHICS

Population growth and *mortality* rates will be key determinants of the *composition* and size of demand for agricultural production, and also the technology under which it is supplied. Increased populations generate increased demand for agricultural products, which must be supplied through an expansion or *intensification* of agricultural *production*.

Demographic change is the key determinant of population pressures on the land, and thus important determinants of the rate and nature of agricultural intensification, with major implications for both biodiversity and biotechnology. We are living in times of rapid and radical changes in *population* size and *distributions*. *Global population* growth rates are declining swiftlyfrom a peak of 2.04% per annum in the late 1960s to 1.35% per annum by the late 1990s.

It is projected to fall even further, to 1.1% per annum by 2015. Although rates are dropping, the absolute numbers of people added to the world's population each year are still quite large, *particularly* in developing countries.

South Asia, East Asia, and Sub-Saharan Africa are the three areas where annual *incremental population* increases have been the highest over the past two decades and, thus, where a rapid growth in the working-age population is now occurring.

Continuing large annual increases are *projected* to occur in South Asia and East Asia up to 2015. For Sub-Saharan Africa, however, the pandemic of HIV AIDS has resulted in a major shift in population projections and annual incremental increases, due to its impact on mortality rates among working age populations.

In most of eastern and southern Africa the prevalence of HIV is over 10%. For some *countries*, negative population growth rates are projected by 2010 as the *mortality* from HIV outstrips new births. Overall, the absolute numbers of adults projected to be alive in countries of Sub-Saharan Africa with HIV prevalence rates over 10% is roughly similar to what it is today.

According to the projections, between 2000 and 2025 there will be a slight increase in the number of men between 2 0 and 59 years of age, but no change in the number of women. However, HIV will also likely affect the productivity of the labor force, due to increased incidences of illness and lower capacity to perform work among afflicted laborers, as well as the need to divert labor to child care, funerals, and tending the sick among the population in general.

The impact of demographic change on agricultural technology choice and ultimately on biotechnology and agricultural biodiversity depends on the supply of factors of production aside from labor, such as land, capital, and technology. The distribution of these factors and policies that affect their relative prices will determine the degree to which an expansion in agricultural output will be met through increases in the extensive or intensive margin of agricultural production.

FAO projects, which approximately 80% of the *required* growth in crop *production* will come from, increase in the intensive margin (i.e., increases in yields per hectare per year). Arable land expansion as a source of growth (the extensive margin) will be important in some *Sub-Saharan* and Latin American countries, although much less so than in the past.

In the past, and with the green revolution in particular, the intensification of agriculture and yield increases were accomplished partially through the adoption of improved varieties, which has also been associated with changes in crop genetic diversity, although there is some controversy over whether the direction has been negative or positive.

The impacts of intensification on increasing crop genetic erosion and vulnerability are a serious concern. However, much of the areas where agricultural intensification through the adoption of monocultural systems has not yet taken place are characterized by a high degree of agroecological heterogeneity and poorly functioning markets, resulting in a higher value to maintaining diversity in the farming system. Intensification in these areas may require higher reliance on agricultural biodiversity due to the barriers to adoption of monocultural agricultural production systems.

 PRIVATIZATION AND DEVOLUTION

Many of the powers that governments wielded in the past have been transferred to the private sector or local governments in recent years. These processes of privatization and devolution are occurring parallel to the process of globalization; thus, we see a shift of power from national governments towards bigger *international* organizations as well as smaller, local governments and private firms.

The logic of this devolution is an assignment of responsibilities that are scale appropriate and correspond to core competencies of *organizations*. There are several dimensions of privatization and devolution with *implications* for biotechnology and biodiversity, which will be discussed below.

The Privatization of Agricultural and Life Science Research

One of the most striking areas where privatization has occurred is in the agricultural R&D industry, particularly those related to biotechnology. In the 1970s and 1980s developed countries experienced a major reduction in the amount of public funds devoted to agricultural research, accompanied by significant increases in private-sector spending. In developing countries private sector-funded research is still a much smaller share of total research, but increases are occurring there as well.

Declining public budgets, poor performance record of publicly funded research, increased appropriability of the returns to privately funded research due to IPRs, and the increased use of purchased inputs in agriculture as a result of increasing competition all contribute towards an increased role of private-sector funding in agricultural research. Private firms in developed countries largely dominate the R&D of agricultural biotechnology with an estimated $2.6 billion invested in 1998.

Only a small share of this investment is directed towards developing countries. There are significant market failures in harnessing the benefits of biotechnology for the benefits of poor producers and *consumers* in developing countries, which are discussed in detail in other chapter of this book. Several chapters note that a key determinant of the degree to which biotechnology R&D can be harnessed for *addressing* the needs of developing *countries* is the capacity of the public sector research institutions to access the technologies generated in the private sector of developed countries.

There is tremendous variation in this capacity among developing countries, both in terms of handling the science and the institutional issues involved. Forging innovative links between private and public R&D systems is an important way to create better access to biotechnology in developing countries and one which is taken up in other chapters of this book.

Privatization of Natural Resource Property Rights and Expansion of Trading Schemes

Land reform and the decollectivization of commonly held properties have been major trends in transition economies and developing countries in recent years. Lands that previously belonged to the state or other forms of communal ownership have been allocated to individual owners who obtain property rights for utilization of the land and its resources. These measures are intended to eliminate inefficiencies that existed under centrally planned economies and inequities in distribution in others.

At the same time a move to privatize natural resources and environmental services together with the *introduction* of market trading to improve *environmental* management has *arisen-although* on a much smaller scale. For example, individual rights to water and water trading are being introduced in countries such as Chile and the United States and are being considered in several countries in *South America* and *South Asia*.

Carbon emission reduction credits is another area where trading regimes have been established and which have the potential for considerable *expansion* depending on the nature of future *international* agreements to control climate change. In the area of agricultural biodiversity, international agreements on the potential for establishing transfer *mechanisms* to pay for the conservation of resources, such as the International Treaty on the *Conservation and Utilization* of *Plant Genetic Resources* and the CBD have been established, although considerable work still needs to be done on the design and implementation. The *privatization* of land and land reform could provide producer incentives for the adoption of biotechnology-to the extent that it contributes towards productivity gains, but impacts on biodiversity are less clear.

Where land reform programs involve use of forested lands or previously uncultivated lands for agriculture, then impacts on wild biodiversity are likely to be negative. The privatization and commoditization of other natural resources and environmental services provide farmers and natural resource owners with more flexibility and may provide them with incentives to provide *environmental* goods and services, such as *biodiversity*.

Privatization of Extension and Emergence of Private Agricultural Consultants

Many countries are *experimenting* in privatizing some of the services that public sector agricultural extension has provided.' These reforms reflect both increased scarcity in public funds and the new reality where agriculture becomes more knowledge intensive, and farmers operating in the commercial sector are looking for more detailed and specialized knowledge and are ready to pay for it.

In the United States and other industrialized countries, dealers of input manufacturers (irrigation equipment and seed and chemical companies) have increased the amount of management information that they provide to farmers. The complexity of pest control decisions and the need to comply with environmental regulations have led to the emergence of independent pest control consultants.

In specialty crops where contracting is prevalent, the buyers may dictate some production practices and provide technical assistance to the contractors. As farms grow in size, they may hire their own specialists in pest control and other aspects of production and design their own production systems.

In many regions, state extension specialists now provide advice and training to independent consultants, provide general retraining to farmers and farm workers, address some of the needs of smaller farms, specialize in treating regional problems (conflict resolution among farmers, *environmentalists*, and the urban sector), and provide *information* on the requirements and means to meet environmental regulations. Extension centers are also used to adapt and test new *technologies* under local conditions.

With a decrease in the role of the public sector in providing *information* to farmers, a need for an overall increase in the resources allocated to education and the transfer of information has arisen. In developed countries there has been some response to this need, but in many developing countries there is still a considerable lack of resources devoted to education and information transfer with a consequent negative impact on the ability of farmers to assess and adopt new technologies.

Reduction in Size and Increased Specialization of Central Governments

The reduction in the responsibilities of state governments is also associated with a reduction in taxation to support state governments (or at least a reduction in the rate of growth of taxation). Moreover, a larger share of the tax revenues of the central governments is returned to local governments that actually provide services.

There are several government agencies now attempting to subcontract provision of key services (*waste management* and *education*) to private companies, thus *significantly* reducing the size of the public sector. Governments are attempting to concentrate on the areas that they do best, such as provision of public goods such as national defense, support for basic research, and monitoring and *enforcement* of environmental *protection* and economic competitiveness.

The declining role of central governments and the transfer of responsibilities to the private and *nongovernmental* sectors may lead to increased efficiency but may also lead to gaps in unsatisfied needs, and new *arrangements* need to be established to fill these gaps. In some cases, the reduced role of the central *governments* may negatively affect the poor, at least in the short run.

On the other hand, the realignment of responsibilities will provide more resource mobility and flexible institutional infrastructure that will enable faster adoption of biotechnology innovations and better conservation of biodiversity. Devolution changes the scale at which transfers are made and may create conflicts between local needs and the provision of goods and services that are national or global in scope.

Biodiversity conservation clearly benefits a wide group but requires cost bearing at a local level, so there is a need for some sort of mechanism to address this. In terms of agricultural biodiversity, the relevant scale for management is often broader than the local level, which also creates some *coordination* problems.

Thus, devolution may have opposing effects on the management of *biodiversity*, and increasing *flexibility* in management at the local level may be positive but can be offset by a decrease in the potential for coordination at higher levels.

THE EMERGENCE OF INFORMATION AND KNOWLEDGE ECONOMICS

Arguably, the *dominant* form of technological change in the last 2 5 years has been in the area of information, communications, and data processing. Over the last 25 years, we have witnessed drastic reductions in the cost of data processing and the proliferation of computer use among families and small firms, emergence of global *communications* networks that enable instantaneous financial transactions and fast, massive transfer of data across locations, and establishment of a network of satellites that facilitate *monitoring* of resource management with a high degree of accuracy.

The emergence of the information economy has important implications for both biotechnology and biodiversity in terms of its impact on the capacity to develop new technologies and the *institutions* that are needed to promote such development, the introduction of modern production methods which are responsive to environmental heterogeneity, the analysis and monitoring of agricultural production impacts on *environmental* conditions, and the ability to inform and mobilize large groups of people over large geographic distributions. The development of *biotechnology* has benefited largely from the increase in *computational* abilities.

Biotechnology is data intensive, and mapping of genes would not have been feasible without advanced computer technologies. With *information-intensive* technologies such as biotechnology, most of the economic value is not attributed to equipment (hardware) but, rather, to *management* knowledge and information (which are in many cases embodied in software). Thus, with the evolution of information *technologies,* we have seen much more emphasis on establishing definitions and enforcement criteria for IPRs.

Without the ability to capture accrued rents using software or new knowledge of information, private parties would not have the incentive to develop these items. Therefore, both patent and copyright laws have been modified to protect IPRs, and the extent of their coverage is being expanded through international trade agreements such as the TRIPS agreement under the WTO. *Establishing* and protecting international IPRs for *biotechnology* innovations is a major challenge.

A narrow definition of IPRs for biotechnology *innovations* may not provide sufficient incentives to cover R&D costs. On the other hand, a definition that is too broad may give owners of these rights excessive monopolistic power and deter access and further innovations by others.

IPRs for the knowledge embodied in biotechnology need to take into account the *contribution* that indigenous knowledge has played in the development of an innovation and assign value to these rights accordingly. However, *assigning* property rights to goods that were previously freely available and exchanged among farmers could also reduce the accessibility to those resources and actually reduce diversity.

Adoption of Precision Agriculture

Precision *agriculture* can be defined generically as a bundle of technologies that adjusts input use to variations in environmental and climatic situations over space and time and reduces residues associated with input use. Many of these technologies rely on space age *communication* technologies and incorporate the use of geographic *positioning* systems (GPS).

Modem irrigation technologies that adjust input use according to variability in soil and weather conditions relying on weather stations and *moisture-monitoring equipment* are also examples of precision technologies. Precision technologies have the potential to increase input-use efficiency, increase yields, and reduce residues of chemicals that may *contaminate* the environment.

In many cases it may lead to input saving, but in others the yield effect may also entail increased input use. Thus far, there have been significant variations in adoption rates of technology that can be *generically* defined as precision technologies. Some modem irrigation technologies have high rates of adoption in high value crops.Some components of what is promoted as "precision agriculture" such as yield monitors are gaining significant acceptance. But, overall, adoption rates of many *components* of precision agriculture have not been very high even in developed countries. Adoption of precision farming technologies may be *hampered* by the cost of investment. *Furthermore*, the management software needed to take advantage of the information has not been fully developed.

Adoption of precision farming technologies will likely increase in the future as their cost declines, as productivity increases, and as new management software becomes available. Precision technologies may both complement and substitute for biotechnologies. Precision technologies that enable the planting of a field with several varieties of seeds will increase the demand for *diversified* genetic stock that can be adjusted to slight variations in soil conditions.

Precision agriculture may also improve sorting and harvesting methods, making the production of high-quality produce more economical and improve incentives to develop higher quality varieties. On the other hand, precision farming may reduce significantly the environmental side effects of pesticides and provide more refined mechanical ways to address weed problems, thus, reducing the demand for some of the pest control *applications* of *biotechnologies* and reducing the loss of *biodiversity* stemming from inadvertent *contamination*.

Introduction of Precision and Information Technologies for the Management of Biodiversity

Some of the major problems with biodiversity conservation and management may be better addressed with applications of precision technologies. *By-catch*, the destruction of nontarget species by fishermen, is a major *environmental* side effect in fisheries. Similarly, forest *clearcutting* is a major cause of biodiversity loss. Adoption of more refined harvesting *technologies* may reduce these side effects and, thus, result in higher levels of biodiversity preservation.

However, both the development and adoption of such technologies may not occur, at least in a socially optimal manner, unless financial *incentives* are introduced. These may include *subsidization* of research and *technology* adoption as well as penalties and *regulations* on harvesting technologies that damage the *environment*.

Monitoring and enforcement of such incentives provide a *significant* challenge, but taking advantage of emerging *technologies* in remote sensing can solve some of the technological aspects.

Improved Marketing and Product Flows

Computer technologies enable the documentation and monitoring of sales in real time and instantly provide useful information on inventory conditions and producers' preferences. Thus, marketers and distributors can obtain a faster response and reduce inventory costs. Also, marketers may be able to better identify quality *preferences* at specific locations and respond to them more promptly.

Indeed, some of the recent product diversification in agriculture, especially in the poultry and produce sectors, took *advantage* of new information *technologies*, resulting in a higher quality and more diversified product. The efficiency gains that modern information technologies provide in marketing *qualitydifferentiated* products is likely to enhance the *introduction* and adoption of biotechnology. Information *technologies* reduce the cost of product *differentiation* in agriculture, but also increase the relative advantage of contracting and vertical integration.

The introduction of a new brand of *differentiated* products requires precise coordination among retailers (who provide the shelf space), distributors, and producers. It is subject to a strict timetable. The *organization* responsible for providing a new *differentiated* product to a retail chain will prefer to contract with farmers to produce a new product or control the production itself Thus, the introduction of *differentiated* biotechnology products will be associated with "*industrialization*" of agriculture, including increased contracting and vertical integration.

CONCLUSIONS

There are several forceful and rapidly moving processes occurring globally that will affect the management, and *ultimately* the status, of biotechnology and biodiversity. In this chapter we have given an overview of some of the major social, political, and economic forces that we believe will shape the way in with the two "bios" will co-evolve with *humankind.*

Of course, the processes we have focused upon here are not the only ones which will affect how biotechnology and biodiversity issues are resolved, and their relative importance will vary among countries. Climate change could have a major impact on the demand for agricultural technology, as well as international *agricultural* supply and production, and thus affect both *biodiversity* and *biotechnology*.

Chapter 3

GENETIC DIVERSITY OF CROPS

One of the *arguments* often made for the *maintenance* of genetic diversity among agricultural crops and their wild relatives is that such diversity acts as a sort of "insurance policy" against the effects of *unanticipated* risks. New pest infestations or changes in climate, for example, can result in large crop losses or other reductions in yield unless resistance or adaptability can be bred in.

In addition to this insurance function, there is also the more mundane matter of agricultural improvement. The broader is the set of materials from which breeders can draw to improve cultivated varieties, the more valuable those varieties will be. The basic economic principle involved in the valuation of genetic material is *straightforward,* but reducing this principle to actual *implementation* is far from simple.

Rather than launching *immediately* into a discussion as to why this is "far from simple," we will defer these matters to the final section of this chapter. We will begin instead by discussing some highly simplified, but illustrative models. We will present two basic models because, as we noted above, two considerations motivate concern for maintaining genetic diversity in agriculture.

First, dramatic events could largely or perhaps totally wipe out a crop. Pest infestations or diseases are examples. Resistance to such threats is a *qualitative* characteristic. *Qualitative* characteristics often are related to the presence or absence of a single gene. Second, crops can be improved or adapted to small changes in circumstances by "*optimizing*" the *combination* of genes they contain.

Characteristics involved in such incremental improvement or *adaptation* are generally linked to large numbers of genes. Such characteristics are *categorized* as quantitative characteristics, and involve *attributes* such as height and weight. In what follows we will develop two very simple models of the valuation of genetic resources. The first will treat considerations with respect to the potential to provide qualitative attributes, and the second, *quantitative* attributes.

Some basic principles are the same between each. First, the social value of genetic improvement, either qualitative or quantitative, is related to the change in benefits arising from the particular improvement. Second, the *contribution* of a larger pool of genetic resources is, in *expectation* at least, to enhance the *characteristics* of those individual *organisms* chosen for cultivation.

The principle of marginal analysis that underlies economic valuation, then, implies that the social value of genetic resources is related to the expected difference in attributes between the best individuals drawn from larger as opposed to smaller sets. We will illustrate these points in the sections that follow. My intention here is more *illustrative* than *descriptive.*

Thus, while we will provide an example drawn from some empirical work with which we have been involved, we will not attempt to provide an actual estimate of value. Again, we will defer a discussion of "how things really work" to a later section, and confine myself to a stylized example. That *example* is as follows. Consider a situation in which a pool of individual organisms of size N is being *evaluated* for their potential to develop a superior variety for *commercial* cultivation. We will suppose that:

1. A single parent is identified, and any number of offsprings can be developed from this single parent organism. Moreover, the attributes of the parent are replicated exactly in the commercial offspring. Note that I am abstracting from both sexual reproduction and *environmental* variation in making these *assumptions.*
2. The parent is selected for a single attribute (*although* this might be a complex index *incorporating* a number of *dimensions*). In other words, I will not be considering a situation in which different organisms are selected for different purposes or growers care about *diversifying* their risks by planting different varieties.
3. We will suppose that the selection is made for purposes of planting a single generation of offspring. As we will argue later, one of the most *complicated* aspects of valuing genetic resources concerns the weight to be assigned to the contribution of one generation to the propagation of others. For the purposes of *expositional* clarity, WE am going to abstract from this consideration for now.

Much of what we write here may be interpreted as applying to situations in which crops are propagated by conventional methods (although *assumption* 1 above might only literally be *accomplished* via large-scale cloning). With the advent of the gene-splicing methods of biotechnology, new methods of inserting valuable attributes into commercial *varieties* might be adopted.

Rather than simply *propagating* the "best" of a set of parent plants, one might pick and choose among the genes of many members of the set, inserting only the "best of the best" in the *commercial* cultivars. This may presume, however, greater knowledge of genetic function than is now, or will likely soon be, available.

More generally, the effects of developments in biotechnology on the economic value of genetic diversity may be difficult to predict. On one hand, the potential scope over which the useful genes of one *organism* might be applied is greatly increased when genes can be inserted, rather than having to be bred into only those *organisms* that are *sexually* compatible.

On the other hand, however, the same principle works in reverse: The number of *potential* sources of genetic material for use in the *improvement* in any one particular crop is greatly expanded. In the next section we discuss the basic economic value of genetic resources in crop improvement.

The second section that follows illustrates the application of this valuation framework to *qualitative* improvements. The third section illustrates its application to quantitative *improvements*. Following that, we present an *application* using data from "provenance trials" of teak trees in Thailand. The fifth section discusses the impediments to *applying* such approaches to "real world" valuation problems, and the sixth briefly concludes.

THE ECONOMIC VALUE OF GENETIC IMPROVEMENTS

Let us define an expected welfare function W(N), i. e., the welfare, *W*, expected to be derived from conducting genetic improvements starting from a gene pool of *N* potential parent organisms. We will suppose that the welfare *realized* by consumers can be measured by the surplus they realize from the *consumption* of the *commercial* product, which is the area under the demand curve between *zero* and the quantity of actual consumption.

Let inverse demand be *p(q)* where *q* is quantity consumed, and suppose that actual consumption is q_1. Then consumer surplus is q_1. Then consumer surplus is

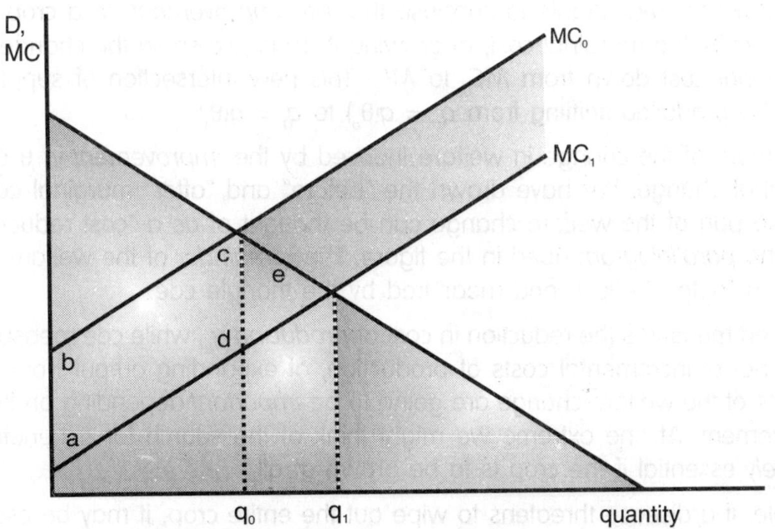

Figure 3.1: Welfare changes induced by genetic improvement.

$$CS = \int_0^{q1} p(q)dq \tag{1}$$

The cost of production may depend on both the total quantity produced and the attribute(s) for which selection occurs. For example, the identification of genes that convey pest resistance will save on the costs of pesticides employed.

Perhaps the most straightforward example of costsaving genetic improvements is an increase in yield per plant. If a genetically improved plant can produce more grain, for example, per unit area planted than would a traditional variety, less land area will be required to produce the same total volume of grain.

Let us suppose now that the measure of the attribute of interest in the variety chosen for commercial propagation is θ. Let the cost of growing a quantity q_1 of a "type θ, variety" be C(q_1, θ_1). Then the welfare to be realized from growing improved crops would be

$$W = \int_0^{q1} p(q)dq - C(q, \theta_1) \tag{2}$$

Now how does welfare change as a result of finding a "better-θ" variety of the crop? Suppose that the value of the attribute for which selection is *undertaken* absent any efforts at improvement would be θ_0. Then if a "type θ_1" variety is planted, the change in welfare will be

$$\Delta W = \int_0^{q(\theta 1)} p(q)dq - C(q(\theta_1), \theta_1) - \left[\int_0^{q1(\theta 0)} [p(q)\,dq] - C(q(\theta_0), \theta_0) \right].$$

This relationship is *illustrated* in Figure elsewhere in this chapter. The output of this agricultural crop is initially determined by the intersection of the demand curve, $D = p(q)$ with marginal cost, $MC_0 = \delta C/\delta q$. It seems reasonable to suppose that an "*improvement*" in a crop will reduce the cost, and, specifically, the marginal cost, of growing it. Thus, we show the change in θ from θ_0 to θ_1 *shifting* marginal cost down from MC_0 to MC_1. This new intersection of supply and demand results in quantity produced shifting from $q_o = q(\theta_o)$ to $q_1 = q(\theta_1)$.

Now the nature of the change in welfare induced by the *improvement* in θ depends on the magnitude of that change. We have drawn the "before" and "after" marginal cost functions as straight lines, so part of the welfare change can be thought of as a "cost reduction" effect, and measured by the *parallelogram* abed in the figure. The *remainder* of the welfare change can be thought of as an "output" effect, and measured by the triangle cde.

In short, abed measures the reduction in cost of *producing* q_0, while cde measures the benefits to consumers, net of incremental costs of production, of expanding output from q_0 to q_1.Clearly, different aspects of the welfare change are going to be *important* depending on the nature of the genetic improvement. At one extreme we might think of the search for a genetic *improvement* that is *absolutely* essential if the crop is to be grown at all.

For example, if a disease threatens to wipe out the entire crop, it may be essential to find a gene that confers resistance. In this case the welfare gain arising from the *discovery* of the gene would be the entire area between *D* and MC_1. On the other hand, if we're thinking of a situation in which plant breeders have already developed a relatively high-yielding variety, the welfare gain from finding a *marginally* better one will consist largely of the cost reduction effect, with the output effect being of the second order of importance. In the examples considered in the next two sections, it seems that qualitative genetic attributes are likely to give rise to discrete improvements, and hence to situations in which welfare gains would combine cost reduction and output effects.

Conversely, quantitative genetic *improvements* are more likely to give rise to *incremental* improvements, so welfare changes would be dominated by cost reduction effects. In short, then, we are likely to encounter a *common* but vexing problem in environmental valuation: *comparing* very unlikely, but potentially large effects with more likely but probably small effects.

Qualitative Characteristics

The economic value of a genetic resource is related to the value of the expected outcome of a process of search for improved attributes with and without that particular genetic resource included in the set over which search is conducted. When we are *considering single-gene qualitative characteristics*, the probability distribution of outcomes is simple: a gene providing the required service is either available or it is not.

More formally, the probability distribution function simplifies to a Bernoulli trial. The desired trait either is or is not present in the genome of a *particular* organism. Let us denote the *probability* with which the gene is found in a particular organism sampled asp, so it is not present with probability 1 – p.

To keep the analysis tractable, suppose that each of N organisms in a *population* may contain a crucial genetic attribute with the same, independent, probability *p*. If an organism contains the crucial gene, a payoff of *R* is realized.

This payoff, *R*, could be related to social welfare as in the previous section. For some crops, *R* might be *astronomical*: Consider, for example, the costs society would bear were it *impossible* to grow wheat, maize, rice, or potatoes. If the desired gene is not found, let us normalize the payoff to zero. Let us suppose that there is a cost c of evaluating any particular *organism* to determine if it *exhibits* genetic *resistance* to a particular pest.

Combining the *probability*, payoff, and cost considerations, the value of the "*marginal organism*" with respect to its expected *contribution* to the *development* of genetic resistance to a particular pest is

$$v = (pR - c)(1 - p)^N \tag{3}$$

Heuristically, the value of the "marginal organism" is its expected value net of the cost of testing, *pR* – *c*, times the probability that the desired gene is not found among any of the *N* other organisms.

For fixed values of *R* and *c*, the *maximum* value *v* can take on is

$$v^* = \frac{R - c}{(N+1)e}\left(\frac{R-c}{R}\right)^N \tag{4}$$

Clearly, this becomes a small number as *N* gets large. Eq. (4) results when it is assumed that the *probability* with which any particular organism contains the *necessary* gene just happens to be that which *maximizes* the value of the marginal organism. For values of p other than those that approximately maximize the value of the marginal organism, v would be smaller yet.

The issue then concerns the *magnitudes* of the payoff to successful discovery, R, the cost of testing, c, the probability, and uncertainty concerning the probability, p, with which success occurs, and the number of organisms over which testing can occur. It is difficult to say exactly how these considerations interact with one another. Any conjectures are necessarily controversial given the *magnitudes* of the *uncertainties* involved.

Perhaps if, as suggested above, *humanity* were truly confronted with the loss of a widely grown staple crop such as wheat, maize, rice, or potatoes, the prospect of so calamitous a loss would translate into a substantial value for the "marginal" element of genetic diversity. Given the existing quantity of potential "solutions" to such challenges, however, the value of *diversity* may still not be great with respect to the potential to provide *qualitative* genetic traits even in important crops.

Quantitative Characteristics

Let us now consider quantitative characteristics. These characteristics, recall, depend on the combined effects of a number of genes. While the analysis could be presented at a higher level of generality by *supposing* arbitrary probability *distribution* functions defined over *quantitative* genetic attributes observed in *populations*, both theory and *empirical* evidence suggest that many *quantitative* attributes are *distributed* normally.

Quantitative attributes can be regarded as arising from the approximately additive effects of a large number of individual genes. It is a fundamental principle of genetics—the "law of *independent* assortment"—that the random combination of genetic attributes is statistically independent. Hence, a large number of statistically *independent* random variables are added together, and the conditions for the central limit theorem are satisfied: The quantitative attribute is normally distributed .

Returning to Eq. (2), the value of production of a variety of type θ is

$$W = \int_0^{q(\theta)} p(q)\,dq - C(q(\theta),\theta)$$

Assume that the values of θ encountered in N trials are independently and identically normally *distributed* with probability $\phi(\theta/\mu, \sigma^2)$. The distribution of the *greatest* value of θ encountered in these N trials is, then, the distribution of the greatest order statistic from N draws. *Suppressing* the mean and variance *arguments* of the normal *distribution*, then, the probability density of this greatest order *statistic* is

$$N\phi(\theta)\Phi(\theta)^{N-1}$$

where $\Phi(\theta)$ is the cumulative normal density. Thus, we can write the *expectation* of welfare resulting from choosing the best among N potential parents as

$$E(W) = \int_{-\infty}^{\infty}\left(\int_0^{q(\theta)} p(q)\,dq - C(q(\theta),\theta)\right) N\phi(\theta)\Phi(\theta)^{N}\,d\theta$$

Technically speaking, in order to find the effects of a change in N on expected welfare, one would need to consider the derivative of the above expression with respect to N. The problem can be made simpler, however, by taking a reasonably close *approximation*. This *approximation* is derived by ignoring Jensen's inequality substituting the "function of the *expectation*" for the "*expectation of the function*," i.e.,

$$E(W|N) \approx \int_0^{q[E(\theta|N)]} p(q)\,dq - C(q[E(\theta|N)], E(\theta|N)) \tag{5}$$

Note that we have stated the expectation of welfare conditioned on the size of the set from which samples are drawn. This arises because the greatest order statistic is also stated as a function conditioned on N, $E(\theta/N)$.

Differentiating with respect to N, we have

$$\frac{\partial E(W)}{\partial N} \approx \left(p - \frac{\partial C}{\partial q}\right)\frac{\partial q}{\partial \theta}\frac{\partial E(\theta|N)}{\partial N} - \frac{\partial C}{\partial \theta}\frac{\partial E(\theta|N)}{\partial N} = -\frac{\partial C}{\partial \theta}\frac{\partial E(\theta|N)}{\partial N}$$

where the second equality results because price is equal to marginal cost in competitive equilibrium.

Tractable results are further facilitated by taking advantage of another *approximation*. It is intuitively *straightforward* that the expectation of the greatest order statistic in a sample of size N is *approximately* that value of the random variable for which a fraction $N/(N + 1)$ of the sample is less than that value. Thus,

$$\Phi[E(\theta|N)] \approx \frac{N}{N+1}$$

or

$$1 - \Phi[E(\theta|N)] \approx \frac{1}{N+1} \tag{6}$$

Differentiating with respect to N and rearranging, we have

$$\frac{\partial E(\theta|N)}{\partial N} \approx \frac{1}{\phi(N+1)^2}$$

or using (6)

$$\frac{\partial E(\theta|N)}{\partial N}; \frac{1}{N+1}\frac{1-\Phi}{\phi} \tag{7}$$

where the cumulative and probability density functions are evaluated at $E(\theta/N)$. The expression $\phi/(1 - \Phi)$, whose inverse appears in Eq. (7), is encountered frequently in statistical and econometric applications. It is known as the hazard rate, defined as the probability with which some event occurs for values of 6 in excess of $E(\theta/N)$, conditional on it not having occurred yet for any value of θ less than $E(\theta/N)$. The hazard rate is often abbreviated as λ. Using this *shorthand*, we can combine *expressions* to restate the change in welfare with respect to a change in the genetic

Table 3.1: Elasticity of expected welfare as a function of number of potential parents and elasticity of demand.

Number of potential parents	*Elasticity of demand*				
	0.10	0.25	0.50	0.75	0.90
10	1.181	0.394	0.131	0.044	0.015
100	0.645	0.215	0.072	0.024	0.008
1,000	0.446	0.149	0.050	0.017	0.006
10,000	0.344	0.115	0.038	0.013	0.004
100,000	0.281	0.094	0.031	0.010	0.003
1,000,000	0.237	0.079	0.026	0.009	0.003

diversity from which superior varieties can be drawn as

$$\frac{\partial E(W)}{\partial N} \approx \frac{\partial C}{\partial \theta}\frac{1}{(N+1)\lambda} \tag{8}$$

The remaining conceptual task is to consider $\partial C/\partial \theta$. The form of this expression will depend in general on the nature of the genetic improvement introduced. While this could take many forms, let me offer one general *consideration* and then examine one special case. The general consideration is the following. Eq. *(7)* can be rearranged as

$$\frac{\partial E(W)/E(W)}{\partial N/N} \approx \frac{\dfrac{\partial C/C}{\partial \theta/\theta}}{E(\theta|N)\lambda}\frac{C}{E(W)} \tag{9}$$

That is, the elasticity of expected welfare with respect to N is approximately equal to the elasticity of cost with respect to θ times the ratio of cost to expected welfare, all divided by the hazard rate times the expectation of θ conditioned on *N*.

There are a number of instances—the special case we are about to discuss being a prominent example—in which one would expect the elasticity of cost with respect to θ to be relatively small. We might also expect the ratio of costs to welfare to be small in many instances of interest.

This leaves the expression $E(\theta/N)\lambda$, in the denominator, which is the elasticity of the probability of finding an individual with a 8 value greater than $E(\theta/N)$, $1 - \Phi[E(\theta/N)]$, with respect to $E(\theta/N)$. It can be shown that this *expression* grows without bound in the limit as $E(\theta/N)$ grows large, so, not *surprisingly*, the marginal value of *additional* resources must eventually be negligible.

The special case arises when θ denotes yield per unit area planted and land is the only *purchased* input in production. If land devoted to a particular crop is *T* and yield per unit land is θ, total production is

$$q = \theta T \tag{10}$$

Let the rent on land be r, so the cost of production is

$$C(q, r, \theta) = \frac{rq}{\theta} \tag{11}$$

Thus,

$$\frac{\partial C}{\partial \theta} = -\frac{rq}{\theta^2} = -\frac{C}{\theta} \tag{12}$$

or

$$\frac{\partial C / C}{\partial \theta / \theta} = -1 \tag{13}$$

Returning to Eq. (9), let us think of how the fraction $C/E(W)$ might be evaluated. Suppose for the purposes of illustration that demand is of the form $p(q) = \beta q^{-\eta}$ for some constant β. Since $p = \partial C/\partial q$, and we are assuming marginal cost is constant given θ, we have

$$\frac{C}{E(W)} \leq \frac{E(q)^{1-\eta}}{\int_0^{E(q)} q^{-\eta} dq - E(q)^{1-\eta}} = \frac{1-\eta}{\eta} \tag{14}$$

Finally, in order to evaluate $E(\theta/N)$, we can revert to Eq. (6). Inverting the cumulative density function—which is monotonic, of course, and thus invertible—we have

$$E(\theta | N) \approx \Phi^{-1}\left(\frac{N}{N+1}\right)$$

Thus, combining the results emerging from our specific assumptions, we have

$$\frac{\partial E(W)/E(W)}{\partial N/N} \approx \frac{1}{\Phi^{-1}\left(\frac{N}{N+1}\right)\lambda\left(\Phi^{-1}\left(\frac{N}{N+1}\right)\right)} \frac{1-\eta}{\eta} \tag{15}$$

AN APPLICATION TO TEAK IMPROVEMNT

Teak *(tectona grandis)* is native to Thailand, Burma (Myanmar), Laos, and India. *Plantations* were established in Indonesia as early as the 14th century, and more recently in Bangladesh, Sri Lanka, China, Vietnam, and New Guinea. In addition, the tree has been grown in *Mexico, Puerto Rico, Cote d'Ivoire, Ghana, Togo, Tanzania,* and *Nigeria.*

There are today some 2.2 million hectares under cultivation in teak. Teak is the subject of selective breeding experiments in Thailand. It has been estimated that selective breeding and the establishment of plantations incorporating genetically superior trees could result in improvements in yield of 17% or more.

Areas of natural teak forest are being felled when land is converted to agriculture or other purposes. There is concern expressed then that genetic reservoirs that could be used to improve subsequent generations of commercial teak are being eliminated. Teak breeders have found it useful to establish "*provenance trials*" in which they grow trees under controlled circumstances in order better to distinguish between genetic (and *consequently*, heritable) and environmental factors in performance.

A provenance trial is an experiment in which teak trees (or other types of organisms) from different regions ("*provenances*") are grown under controlled circumstances in order to identify the genetic contribution to the appearance of attributes of commercial importance. A number of such trials have been *conducted* in Thailand and elsewhere.

I will employ data from a long-running experiment on several *international provenances* conducted in Thailand. Such trials collect data on a large number of quantitative and qualitative characteristics of the trees in the sample. As my purpose at present is merely illustrative, we will concentrate only on "commercial height." Commercial height is defined to be the maximum height at which diameter is at least 10 centimeters.

More generally, of course, one would care not only about height, but also diameter and, more generally, wood volume, as well as quality of wood. In addition, owners of *commercial* plantations would want to plant trees that are known to be resistant to *infestations* and have other desirable survival and input cost *minimization* properties.

Again, however, let us, for simplicity, simply *concentrate* on *commercial* height as a measure of yield-per-hectare. The sample mean and standard deviation are sufficient statistics for normally distributed variables. The mean *commercial* height among 578 trees from the provenance trial we are considering was 11.6 meters.

The standard deviation was 4.0 meters. From these sufficient statistics, we can derive the cumulative normal distribution, its inverse, and the *corresponding* hazard rate. These figures can then be employed in Eq. (15) above. The remaining *variables* are *N,* the number of potential parent organisms, and 77, the elasticity of demand. The number, *N,* is varied in the *left-most* column of Table elsewhere in this chapter, and the elasticity, η, is varied in the top row of the table.

Entries in each cell of the table show, for that number and elasticity, the elasticity of *expected* social welfare with respect to the size of the size of the genetic base from which selection can occur.

DISCUSSION

The *calculations* reported in Table elsewhere in this chapter indicate that, as expected, the elasticity of expected welfare in the number of potential parent *organisms* declines in the number of potential parents, keeping the *elasticity* of demand fixed, and in the elasticity of demand, keeping the number of potential parents fixed.

As the elasticity I have reported is the percentage change in welfare resulting from a 1% change in the number of potential parents, the change in welfare resulting from a small change in the absolute number of potential parents would be small indeed in most of the instances I have reported.

While it would be difficult to estimate the total welfare derived from the *consumption* of something like teak wood, the results I have reported suggest that the *incremental* value of *additional* genetic resources would be modest. This conclusion should be qualified in a number of ways. First, I have been assuming that consumer surplus is measured under *uncompensated* demand curves. As is well known, this approach is valid when income effects are not *important.*

While this may not be too *unreasonable* of an assumption in discussing the demand for teak wood, it would be more *questionable* when applied to, for example, rice or wheat. Staple crops may claim large shares of income, at least when they become rare. Clearly, the stakes rise as the scope of use increases.

Martin Weitzman (2000) has recently done interesting work in which he considers the tradeoffs between, on the one hand, *maintaining* a diversity of organisms to protect against the failure of each type of crop and, on the other, the opportunity costs of growing varieties *anticipated* to be less productive.

The general issue he identifies-the desire to prevent very low probability but also very catastrophic outcomes-is important, but difficult to resolve. Another major omission of the approach I have taken here concerns the treatment of dynamic *considerations.*

One could make the approach I have illustrated dynamic by *supposing* that, in every period, breeders must identify a variety that best suits a set of conditions that have completely changed since the previous round of *selection* occurred. It may, in fact, be the case that this complete-change-in-circumstances scenario leads to the highest estimate of value for the *"marginal potential parent"*: it would seem that, to the extent that selection continues to occur for the same traits over time, a finite upper bound would *eventually* be *approached.*

More generally, however, each new generation of *propagated organisms* would comprise a new random draw from the gene pool. Thus, one might, in a more complex analysis, want to consider not only the direct benefits of genetic diversity in terms of producing superior varieties for immediate cultivation, but also for producing varieties that might in turn produce still other varieties. This leads me to the final *consideration* that I will address here. I have supposed that the process of selection involves the *identification* of the single "best" individual, and that unlimited numbers of identical individuals can be replicated exactly from this source.

In practice, selective breeding typically involves the identification of a group of individuals and the attributes of their offspring reflect only imperfectly those of the parents. The analysis I have presented here can be extended to consider selection of a set of parent *organisms* and to incorporate imperfect heritability of attributes. Results are *compromised* somewhat by the practical necessity of abstracting from Jensen's inequality in order to *maintain tractability* (results are much more easily derived by *supposing* each offspring *organism* exhibits the attribute at the mean level).

Given the other imprecisions inherent in the analysis, however, this does not seem to be a major impediment to deriving illustrative results of the type I have illustrated here. I might note in closing that the model I have sketched is becoming dated by advances in biotechnology. I have supposed that the only way to generate improved *commercial* varieties is to identify promising "*packages* of *genes*" in the form of parent organisms.

As the process of agricultural *improvement* comes to rely more and more on the insertion of favourable individual genes, the search for superior *quantitative characteristics* may come more and more to resemble that for favourable qualitative *characteristics.*

Moreover, the ability to transplant genes even between different species (and sometimes higher phylogenetic taxa) may imply that genetic resources are becoming less and less scarce with respect to particular crop *improvement applications* and, hence, of lower economic value.

Chapter 4

FARM DIVERSITY

Empirical research has documented that in harsh, isolated *environments* where climatic and soil conditions are variable, farmers may depend on the cultivation of multiple crops and varieties to meet their food and cash needs. For farm households to be food secure, they require stable supplies for consumption from either their own production or market *purchases*.

As markets develop, farm households generally specialize in fewer products oriented toward the demands of distant consumers, relying less on a portfolio of crop varieties and more on a portfolio of income sources to smooth their consumption.

Yet, those in isolated areas continue to face heavy transactions costs because they have limited and uncertain options for buying and selling in markets. They have a "demand" for crop biological diversity' that is derived from the range of *production* traits and *consumption* attributes they require.

Cultural *autonomy* may reinforce this demand through shaping the preferences of rural people for the food they consume, their perceptions of crop biological diversity, and its importance. Managing the biological diversity of crop genetic resources on farms is of economic importance in part because it is a survival strategy for some of the world's rural poor.

Conserving these resources on farms also reduces the loss of potentially valuable alleles in genetic stocks still held by farmers. Geneticists often hypothesize that rare, locally adapted genotypes may be found among the landraces *cultivated* by farmers in such extreme or *heterogeneous environments*.

Some genotypes are thought to contain tolerance or resistance traits that are not only valuable to the farmers who grow them but also to the global genetic resource *endowment* on which future crop *improvement* depends.

Rare alleles are often discovered in centers of origin, though depending on the crop, valuable diversity can often be found elsewhere. Genetic resources are renewable assets, but are renewed in farmers' fields only as long as farmers continue to sow the seed.

When farmers replace "*landraces*" with "*modern varieties*" that may be more attractive to them as their economies industrialize, the gene combinations in landraces may be "lost" to future *generations* of farmers and consumers unless special efforts are made to collect them or encourage their continued cultivation. Since the 1970s, large numbers of landraces and wild relatives of cultivated crops have been sampled and stored in *ex situ* gene banks. An alternative form of conservation *in situ* has also received some *scientific* attention.

For cultivated crops, conservation of genetic resources *in situ* refers to the continued *cultivation* and *management* by farmers of crop *populations* in the *agroecosystems* where the crop has evolved. Storing genetic resources in collections as backup seed stocks in *ex situ* collections therefore substitutes imperfectly for the evolution of crop plants in the fields of farmers.

Not only do genetic resources evolve differently when conserved *ex situ* and *in situ,* but the distributions of their economic benefits and costs also differ fundamentally. The costs of genetic resource conservation in gene banks are now borne largely by public *investments,* and consumers (as well as farmers who are consumers) benefit *indirectly* from the genetic resources incorporated into improved crop varieties when output expands and prices decline.

In contrast, both the costs and benefits of conserving genetic resources *in situ* are felt directly (and in a very immediate sense) by the farmers who grow them. To suggest that some of the poorest farmers of the world should shoulder the full burden for conservation of crop *biodiversity* seems inappropriate.

For this reason, *international* agreements such as the Convention on Biological Diversity and the International Treaty on Plant Genetic Resources for Food and Agriculture encourage the design of benefit-sharing schemes to support conservation through rewarding farmers for their innovationsthough mechanisms for doing so are still under discussion.

National and local policy *instruments* that promote development and farmer management of crop biological diversity may also be feasible. A "win-win" policy solution occurs when crop biodiversity is maintained on farms for the benefit of future *generations,* while farmers themselves benefit today from a wider set of crop variety attributes for *consumption* or sale.

Under those circumstances, private and social benefits coincide. This chapter begins by reviewing how economic concepts can assist in identifying promising locations for on-farm management of crop biological diversity, supported by some findings from empirical analyses. Next, some of the policy *mechanisms* that have been invoked to support farmers in such locations are discussed and, where possible, evidence *regarding* their *effectiveness* is presented.

THE VALUE OF CROP BIODIVERSITY OF FARMS

As a *production* input and private good, seed is highly rival with low cost of exclusion. The genetic resources embodied in seed are nonrival, however, and the costs of controlling their use

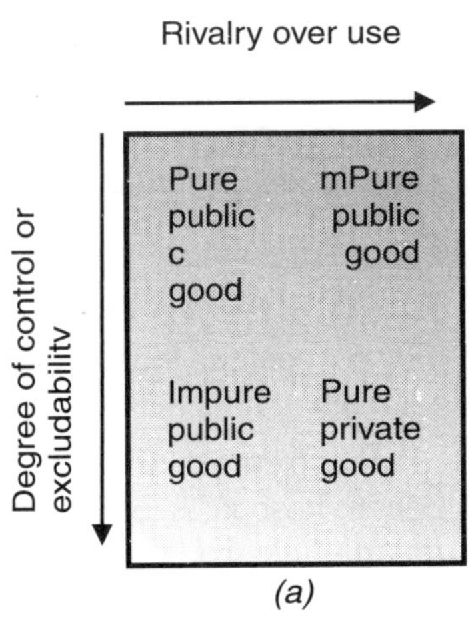

(a)

Characteristics of impure public goods that affect the form of institutional intervention required to manage them optimally

Characteristic	examples
Intragenerational regional	Waterways Satellite tranmissions
Global	regional fisheries
Intergenerational regional	ocean fisheries

(b)

Figure 4.1: Simplified taxonomies of goods.

on farms are relatively high. This means that two farmers cannot plant the same *handful* of seeds, but many farmers may grow the same variety *simultaneously.*

Controlling the flow of genes among fields is difficult, especially with *predominantly* cross-pollinating crops as they are managed by farmers in *lesscommercialized* agricultural systems. The combinations of seed types grown by farmers produce a harvest that they consume and/or sell and from which they derive private value, but the pattern of *genotypes* across the landscape contributes to the biological diversity of the crop genetic resources from which people residing elsewhere and in the future may benefit.

The public value of crop *biological* diversity includes insurance value for potential disasters and option value for any *unforeseen* events, such as changes in consumer tastes. The private value includes utility or *satisfaction* from the agronomic traits and *consumption* attributes that these *cultivars* provide to farmers as producers and/or as *consumers.*

In the special case of a commercial farmer producing for a well-defined market, that utility is related exclusively to profits derived from sales. Since the biological diversity of crop genetic resources is never fully apparent to the farmers who provide and use it and is *undervalued* in markets, farmers are unable to consider the *contributions* of all other farmers to genetic diversity in their *community* or elsewhere when they make their decisions.

Hence, biological diversity of crop plants has interregional and intergenerational *dimensions.* Economic theory predicts that, as long as crop diversity is a (desirable) "good", farmers as a group will generate less diversity than is socially optimal. *Institutional structures* are needed to compensate for the inability of markets to provide sufficient incentives for farmers to allocate their resources in ways that are consistent with the needs of society.

These structures will differ according to culture, as well as the temporal and spatial dimensions of the impure public good. Some societies have much stronger collective behaviour than others. All have norms related to *management* of genetic *resources.*

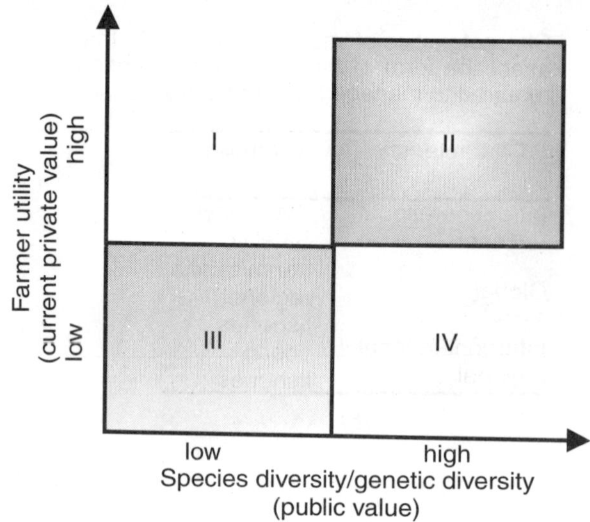

Figure 4.2: Sites with high benefit-cost ratios for on-farm conservation.

At the community level, depending on the social and economic conditions farmers face, community awareness *campaigns* may be sufficient to ensure that certain materials of genetic importance continue to be grown. By contrast, complex structures are necessary to mediate conservation interests at the *global* level because actors may not perceive that they share common interests.

The International Treaty on Plant Genetic Resources for Food and *Agriculture* and the Convention on Biological Diversity are elements of such structures, though these may not be *consistent* with local norms of use and access. Therefore, the extent of public *investment* and the policy *mechanism* needed to narrow the divergence between what *individuals* and *societies* perceive as optimal clearly depends on many factors.

IDENTIFYING PROMISING LOCATIONS FOR MANAGING CROP BIODIVERSITY ON FARMS

High Benefit-Cost Ratios

Not all global *locations* are equally promising candidates for managing crop biological diversity on farms. The highest benefit-cost ratios for onfarm *conservation* of diverse crop genetic resources occurs where the utility farmers derive from managing them as well as the public value *associated* with their biological diversity are high.

This occurs conceptually in area II of Figure elsewhere in this chapter. Since farmers are already bearing the costs of *maintaining* diversity de facto in those areas and they reveal a preference for doing so, the costs of public interventions to support conservation will also be least. Where genetic diversity is assessed as relatively low and farmers derive few benefits from it, there is no need to invest in any form of *conservation*.

Where the contribution to diversity is great but farmers derive little private value from it, *ex situ conservation* is the only option. Where there is little diversity but farmers care a lot about it, there is no need for public investment at all since no value is *associated* with conservation.

An example of an "I" location was found in the uplands of Nepal for rice. Farmers recognized many varieties but their genetic diversity when *characterized* was found to be relatively low. No modern varieties compete in that environment, so there were no opportunity costs associated with growing farmers' cultivars. An example of a "II" location was found in Mali.

There, despite 26 years of climatic change and drought, analysis of seed samples demonstrated that farmers maintained the same overall level of sorghum diversity, though its spatial distribution had shifted. This is strong empirical evidence that farmers depended on sorghum diversity to manage risk *ex ante* (as compared to those in more favoured, less isolated *environments* who manage risk through markets, *ex post).*

Geneticists also consider the range of traits found among those varieties to be important. Based on empirical findings, we know that in other places in the world, rural people depend on the diversity of their crops and varieties to cope with climatic risk, match them to specific soil and water regimes, and meet a range of consumption needs when markets are unreliable.

These locations are often characterized as "less favoured," or "marginalized"; the people who live in them are often considered to be poor on a global scale. With economic analysis, genetic analyses, and other scientific information, we can ascertain which locations are "promising candidates" for *managing* rare or diverse crop genetic resources on farms. Metrics can then be used to rank areas according to their expected social and private value.

To rank locations along the horizontal axis, diversity indices developed by *scientists* can serve as proxies for the public value of a set of crop varieties or *populations*. Indices are scalars constructed from any one of several types of data. For example, data may measure the physical *characteristics* of crop plants grown in controlled *experiments*.

Alternatively, data may *summarize* the patterns in DNA taken from plant tissue and observed under a microscope. Option values other than those subsumed in scientists' assessments of *information* value are not likely to be estimable but would *generally* be positive.

Adaptations of econometric models of variety choice, and nonmarket valuation methods combined with random utility models, enable us to rank candidate locations along the vertical axis using probabilistic statements. Some empirical evidence from related analyses is *summarized* next.

Predicting Locations where Landraces will Continue to be Grown

Broadly speaking, three generic factors are hypothesized to determine the likelihood that modern varieties are attractive to farmers and, hence, the opportunity cost of growing landraces: population density, agroecology, and development of *commercial* markets. The pattern of diffusion of modern varieties of wheat, rice, and maize *illustrate* this point.

Population density, or the ratio of the supply of labor to the supply of land, explains much about where the transition from low-yield, land-extensive cultivation to land-intensive, double, and triple crop systems has occurred. The genetic changes embodied in seed constitute one type of *intensification,* which refers more broadly to the increase in output per unit of land used in production (or yield).

Predictably, the adoption of modern rice varieties in the less industrialized world has been most complete in densely populated areas of their cultivation where *traditional* mechanisms for enhancing yields per unit area have been exhausted. Population densities interact with *agroecological* conditions in *explaining* the adoption of modern varieties.

Since the initial adoption and rapid diffusion of the first *semidwarf* varieties of wheat in the irrigated areas of the Asian *subcontinent* during the 1970s, more widely adapted descendants of these varieties spread gradually into environments less favoured for wheat production and rain-fed areas. Today, wheat landraces are cultivated extensively only in portions of the drier production zones of the West AsiaNorth Africa region and highlands of Ethiopia.

Similarly, the adoption of modern varieties of rice is virtually complete in irrigated areas and uneven in rainfed zones, while they are largely absent in the uplands and deepwater areas. In contrast with wheat and rice, maize is grown over a greater range of latitudes, altitudes, temperatures, and moisture regimes. Maize also has the greatest *proportion* of area in the less-industrialized world that is still planted to landraces.

For many of the *environments* in which maize is grown, suitable improved materials have not been developed by centralized breeding programs. The maize germplasm that performs well in temperate climates of industrialized countries cannot be introduced directly into the nontemperate regions of less-industrialized countries without *considerable* additional breeding for adaptation, nor are there always economic incentives for a commercial seed industry.

Thus, even if *adaptation* problems could be overcome through breeding, farmer demand for improved seed may be small. In maize, agroecological factors have interacted with the development of commercial seed systems in slowing the expansion of area in modern varieties. As the *orientation* of crop production shifts from subsistence toward *commercial* objectives, the locus of crop improvement and seed distribution also moves from individual farmers toward an organized seed industry composed of *specialized* private and public organizations. Maize has moved substantially faster than rice and wheat in terms of an increased reliance on commercially produced seed.

In a stylized depiction of the maize seed industries in developing countries, subsistence production is *characterized* by open-pollinated varieties improved through farmer selection and on-farm seed production with local seed markets governed by custom.

In a fully *commercial* system, the *predominant* seed type is a hybrid that is purchased annually. Seed is a globally traded product of specialized research that is both privately and publicly funded. The exchange of seed and the genetic resources used to improve it are enabled and protected by strict forms of intellectual property rights.

In rice and wheat, which are self-pollinating crops, the incentives for *privatization* of research have not been as strong as for maize, although this depends on the *institutional* and economic context. In the industrialized countries, profound changes in science and in intellectual property protection over the past 20-30 years have been associated with a higher rate of investment in agriculture by the private sector than the public sector and a shift in the *composition* of private investment from *agricultural* machinery and processing into chemical research and plant breeding.

Privatization is greatest in the maize seed industry in *industrialized* countries and is increasing in the *lessindustrialized* agricultural economies, but has only occurred to a limited extent for wheat-in Europe. Almost all of the seed research for rice has been and continues to be conducted by the public sector, and most research has occurred in Asia.

Historical experience in the diffusion of modern varieties of wheat, rice, and maize lead us to predict that, for highly bred, staple food crops, the opportunity costs of landrace cultivation will be lowest, and farmers' economic incentives to grow landraces higher, in less densely populated, *commercially* isolated areas for which either public or private breeding systems are unlikely to develop well-adapted materials. Where *conditions* are otherwise, the costs of designing *institutions* and *mechanisms* to encourage conservation are likely to be quite *high-unless* consumers demand specialized traits that are both unique to landraces and difficult to transfer genetically into modern varieties. potential.

On the other hand, the costs of designing *institutions* to support on-farm conservation is likely to be considerably lower for crops that are not so highly bred and have received less research attention but play significant roles in the strategies of smallscale farmers. These crops are sometimes known as "neglected" or "*underutilized*" species from the *perspective* of professional plant breeding and modern agricultural systems.

Predicting Locations where Crop Biodiversity will Continue to be Managed by Farmers

An extensive *microeconomics* literature on variety adoption and a growing set of case studies about on-farm management of crop biodiversity provide us with empirical evidence about the types of social and economic contexts that lead to a higher *probability* that farmers will continue to manage *biologically* diverse crop genetic resources.

The first factor, which is important on a local as well as a regional scale, is agroecological heterogeneity. *Environmental* heterogeneity has also been advanced as an *explanation* of farmers' continued use of landraces in Turkey. Across a series of villages with differing agroclimatic conditions, heterogeneity in agroecological conditions increased the number of different crops and the varieties of maize, beans, and squash varieties grown by farmers in the state of *Puebla, Mexico.*

In the highlands of Ethiopia, land *fragmentation*, soil erosions, and the numbers of plots on farms had effects on the variety diversity that varied among crops. The second factor that is important at both local and regional levels is the extent to which households trade on markets. The more removed a household is from a major market center, the higher the costs of buying and selling on the market and the more likely that the household relies *primarily* on its own production for subsistence.

Van Dusen (2000) found that the more distant the market, the greater the number of maize, beans, and squash varieties grown by farmers. In the highlands of Ethiopia, findings *regarding* market distances depended on the cereal crop, how the variable was measured (distance from farm to household, distance from household to all weather road, or distance from community to market), and whether the diversity among crops or the diversity within crops was the dependent variable.

In *environmentally* sensitive areas of Hungary, small-scale farmers of a more isolated site with poorer soils and fewer food markets valued ancestral varieties and higher levels of crop species richness more than farmers in an economically developed site with good road quality, fertile soils, and numerous food markets. Small-scale farmers' choice to grow more than one variety *simultaneously* is likely to reflect their need to address numerous concerns that no single variety can satisfy.

Case studies demonstrate that in many of the regions of the developing world where landraces are still grown, either markets for commercially produced seed, markets for the crop output, or markets for the multiple attributes farmers demand from their varieties are incomplete. This means that the traits demanded by farmers (grain quality, fodder, *suitability* for a certain soil type) cannot be obtained through the production of modern varieties or procured through impersonal market transactions, so that farmers must rely on their own or neighbors' production for their supply.

Knowing the crop characteristics that most matter to farmers, how these are *distributed* across varieties and populations, and to what extent farmers can meet their needs through market transactions is, therefore, important.

In Turkey, concern for bread quality in wheat, in addition to high *household* transaction costs such as transportation and uncertain prices, were associated with the choice to grow landraces rather than modern varieties. "*Promising candidates*" may also be locations where both modern varieties and landraces are grown, if growing both types in that location *represents* an economic *equilibrium*.

Though modern varieties have long been equated with a loss of diversity on farms, like any new or exotic type that is introduced, a modern variety can add to the portfolio of distinct agro-morphological types grown in a *community* precisely because it has been bred with the ideal type of other farmer-breeders or professional breeders in mind. Modern varieties may possess a trait not found in the local varieties grown in a community, suiting particular production niches but not others.

With *cross-pollinating* species, farmer seed *management* or deliberate *introgression* may mean that the introduction of modern varieties generates new types that exhibit traits from both. Farmers often choose to grow both landraces and modern varieties. Viewed in the conventional *microeconomic* literature as partial adoption, this observed pattern has been explained theoretically through attitudes toward risk and uncertainty, missing markets, and *differential* soil quality or nutrient response combined with fixity or rationing.

Though treated as a transitional period to full adoption, the coexistence of modern varieties and landraces may represent an economic *equilibrium* if one or several of these aspects persist despite economic change. Meng, Taylor, and Brush (1998) concluded that multiple factors, including missing markets, yield risk, grain quality, and agroclimatic constraints, influence the probability that a Turkish household will grow a wheat landrace; a change in any single economic factor is unlikely to cause farmers to cease growing it.

Zimmerer (1996) found that the capacity of farmers to grow diverse food plants (including maize) in Peru and Bolivia depends on whether they can cultivate them in *combination* with commercially developed, *high-yielding* varieties that were easier to sell because they had more uniform grain quality.

So far, *introduction* of modern varieties of wheat and maize has not meant that any single variety dominates or that modern varieties have replaced farmers' varieties in the Ethiopia highlands, most likely because they have limited adaptation and farmers face many economic constraints in this location. It is just as likely that small amounts of seed of modern varieties broadens the variety set of these farmers by meeting a *particular* purpose.

Neither the physical terrain nor the market infrastructure network is favourable for specialized, commercial agriculture. The relationship of household characteristics, such as wealth and income sources, to crop diversity depends on the *measurement* and empirical setting. Smale, Bellon, and Aguirre (2001) found that variety *attributes* such as *suitability* for food *preparation* (tortillas) far *outweighed* the importance of household characteristics in *explaining* the number of maize *landraces* grown by individual farmers and the average share of maize area planted to each.

In three sites in Nepal, based on a composite variable for wealth rank, Rana et al. (2000) found that poor households cultivate more coarse-grained, drought-tolerant varieties of rice, while wealthier households grew highquality varieties for premium market prices and special food *preparations*. In the state of Puebla, Mexico, Van Dusen (2000) found that the greater the wealth of the household, as measured by house *construction* and ownership of durable goods, the less likely the *household* is to plant a diversity set of maize, beans, and squash varieties.

Benin et al. (2003) found that while larger farms were associated with more varieties of any major cereal crop grown in the *highlands* of Ethiopia, livestock assets had conflicting effects among cereals. In many parts of the developing world, off-farm migration *generates* a growing proportion of the income of farm households.

Brush, Taylor, and Bellon (1992) found that off-farm employment was negatively associated with maintenance of potato diversity in the Andes, indicating that the *opportunity* cost of *cultivating* many varieties-which requires *laborintensive* seed selection and *procurement* tasks-is *significantly* higher where other employment *possibilities* exist.

Van Dusen found that overall diversity in the *milpa* system decreased as local labor markets intensify, or as more migration to the United States occurs, though these effects were not as *pronounced* when each crop was considered singly.

Yet off-farm income can also release the cash income constraint faced by some farmers, enabling them to shift their focus from growing varieties for sale to growing the varieties they may prefer to consume.

In Chiapas, Mexico, Bellon and Taylor (1993) found that off-farm employment was associated with higher levels of maize diversity. Meng (1997) found the *existence* of off-farm labor opportunities to have no statistically significant effect on the likelihood of growing wheat landraces in Turkey.

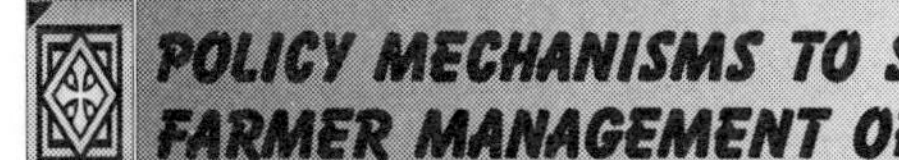

POLICY MECHANISMS TO SUPPORT FARMER MANAGEMENT OF CROP BIODIVERSITY

Policy mechanisms to support farmer management of crop biodiversity in locations that are "*promising candidates*" may be classified as either related to (1) the demand for or (2) the supply of genetically diverse or distinct crop varieties. Some *illustrative* examples are provided below, though many of these initiatives are new and their efficiency in meeting *conservation* goals given the level of *investment* required has not yet been assessed using cost-benefit analysis-at least in published *literature*.

Demand-Related Mechanisms

When markets are not well developed and transaction costs are high, the demand for distinct varieties reflects the tastes and preferences of local farmers who grow and consume them rather than those of distant urban consumers.

Farmers' demand for local landraces can be enhanced by improving the traits they identify as important, including disease resistance, abiotic tolerance, and *palatability* as food or fodder. In a project in the Philippines, molecular analysis revealed the genetic distinctiveness of the Wagwag group of traditional, *nonglutinous* varieties of rice that farmers also ranked highly according to *consumption* quality criteria. *Researchers* recommended a breeding intervention to address the long duration and low yield of the Wagwag group, reducing their *disadvantages* relative to modern rice varieties in the irrigated ecosystem.

Plant breeders can work with farmers to select superior local materials or transfer a preferred trait from exotic into local materials. In India, a *drought-tolerant*, locally adapted landrace was crossed with a modern variety with higher potential yield, and the offspring selected by farmers under their own conditions. Known as "*participatory*" plant breeding, such approaches can require *substantial* time investments by farmers.

Oaxacan farmers as a group obtained nearly a 4:1 benefit-cost ratio from participating in a project to enhance local maize landraces and their diversity, although from the perspective of a private investor, benefits did not justify the cost. In Oaxaca, as in other maize-growing areas of the less-industrialized world, the high rate of cross-pollination in maize is reinforced by the *simultaneous* flowering of contiguously planted, *fragmented* fields.

Under these- *circumstances*, yield advantages of improved *open-pollinated* materials and landraces are difficult to maintain unless a cost-effective, local system of seed multiplication, diffusion, and replacement is established. Smale et al. (2003) did not assess the public benefits of the project in Oaxaca, such as its contribution to *maintaining* the diversity of maize landraces, because of *measurement* difficulties.

Not all participatory projects are undertaken with the goal of enhancing crop biological diversity, however. The release of well-adapted varieties, whether they are developed with or without the active *participation* of farmers, may cause a decline in the diversity of varieties grown because they are so popular.

Research has indicated that farmer involvement in the later stages of variety selection is associated with better local adaptation of materials. Witcombe and Joshi (1995) argue that while participatory variety selection may not increase diversity, *participatory* plant breeding will necessarily enhance both intraand inter-variety diversity because of the methods employed.

In 1995, Loevinsohn and Sperling found that the linkage between participatory crop improvement and diversity was not well documented, though more published evidence is now being generated. As incomes rise and commercial markets develop, landraces may continue to be grown when there is consumer demand for some unique attribute that cannot be easily bred into or transferred to improved varieties and if seed *regulations* permit.

Advanced *agricultural* economies are *characterized* by growth in demand for an array of increasingly specialized goods and services. In general, though the income elasticity of demand for staple grains may be low or even negative, the income elasticity of demand for grain attributes is higher. For example, high income consumers spend more on rice by paying higher prices for varieties with preferred eating quality which they substitute for the lower-quality variety consumed when the income-level was lower.

Pingali, Hossain, and Gerpacio (1997) cite several examples of Asian landraces that are of higher quality and fetch premium prices in the market. In South Korea, the modern *tongil* variety was replaced by a relatively lowyielding, traditional *japonica* rice as consumers expressed a preference for *japonica* types by offering higher prices as their incomes rose. Niche markets or branding enable farmers to recoup the premiums consumers are willing to pay for unique attributes or qualities.

Farmers in niche markets do not control supply or prevent imitation. Branding enforces supply constraints through (1) some fixed and identifiable attribute such as a geographic origin, (2) *membership* in an *exclusive* producer group, (3) strict production standards or guarantee of process, or (4) control of an ingredient by an exclusive producer group through *intellectual* property rights. Restricted labeling systems have long been used to ensure consumer quality and authenticity for meat, cheese, and wine products in Europe.

Falcinelli (1997) describes how labeling systems may promote conservation of *farro* (einkorn, emmer, spelt) and lentils in Italy. In a study of the Label Rouge system for quality poultry production in France, Westgren (1999) cautions that the investment in consumer education required to support *significant*, sustainable price premia is costly and may not be fully *internalized* in any single market supply chain.

Though common in the European Union, farmer brands are relatively rare in the United States. Brush (2000) reports an example of a successful "green *marketing*" program for ancestral maize of Cherokee farmers in the United States.

Vidalia onions of Georgia are another example based on geographic origin related to superior quality. Labeling and marketing systems require public investments unless consumers are willing to pay price premiums large enough to cover the costs, and require fully commercialized, *well-articulated* markets for product attributes.

Hayes and Lence (2002) report several necessary conditions for the successful differentiation of farmer-owned brands, including good transmission of price signals, a scale of production large enough to justify the costs of creating a differentiated image among consumers, and capacity to prevent imitation. These conditions are not likely to be met easily in most contexts where diverse landraces are grown in *less-industrialized* economies, especially for staple foodcrops that are grown by many atomistic farmers.

For example, Gauchan, Smale, and Chaudhury (2003) found that except for traditional Basmati (aromatic high quality) rice, most rice landraces in upland Nepal are traded in small volumes through informal channels where price signals are weak. If desirable attributes can be clearly linked to specific landraces, the landraces themselves must be readily identifiable and their attributes *maintained* through careful seed *multiplication* and production.

Even if branding is successful, protecting one landrace may not have desirable implications for diversity conservation-especially if seed regulations require that landraces be uniform and stable like modern varieties. In either advanced or *less-industrialized* economies, public awareness initiatives can serve to increase farmer and consumer knowledge about the benefits generated by on-farm conservation, enhancing their demand for products and seed.

In Nepal, Vietnam, and the Andes, diversity fairs have been used to bring together farmers from one or more *communities* in order to exhibit the range of materials they use and raise awareness of the value of crop diversity. Rather than awarding prizes for the best individual variety (e.g., on the basis of yield or size), diversity fairs award farmers or cooperatives for the greatest crop diversity and related knowledge.

In some communities, *gatherings* similar to diversity fairs are already customary events, so the incremental costs are low. Though such initiatives are gaining *popularity* in seed projects, there appears to be little evidence concerning their cost effectiveness or impact.

Supply-Related Mechanisms

Possible *"supply"* solutions encompass a range of seed market and genetic *improvement* options, as well as, potentially, protection of local varieties through Farmers' Rights legislation. India is one of the first countries in the world to have passed a legislation granting rights to both breeders and farmers under the Protection of Plant Varieties and Farmers' Rights Act in 2001. As of yet, however, the requirements for registering farmers' varieties have not been elucidated.

Ramanna and Smale (2004) have argued that while this multiple rights system aims to equitably distribute rights, it could pose the threat of an "anticommons tragedy" where too many parties independently possess the right to exclude, resulting in the underutilization of crop genetic resources and discouraging the very innovation it was intended to promote.

In more isolated areas or more difficult growing *environments* of *lessindustrialized* agricultural economies, agroecological and environmental factors exert a more decisive influence on crop biological diversity than commercial markets. From one year to the next, farmers may lose their seed stock due to disastrous harvests, or *diminishing* seed quantities may threaten the genetic viability of the variety. Initiatives aimed at supporting the range and total supply of local seed types have been suggested in response to such situations.

Community genebanks provide a mechanism for storing valuable landrace germplasm in a local *ex situ* form so that farmers' have more direct access to seed when they need it. Typically small in size, community genebanks can only maintain a limited number of accessions and replicates. The economic feasibility of *community* banks is also *undermined* by the high covariance of local crop yields, which means that many farmers in a *community* face similar seed deficits and surpluses.

Cromwell further cautions that access to seed through the *community* system is not always egalitarian, since seed is often hoarded or may be of poor quality. If farmers prefer to store seed on an individual basis but desire access to knowledge about the location of other seed types in surrounding villages, biodiversity registers are one low-cost, modest alternative.

A *community* biodiversity register is a record of landraces cultivated by local farmers. In addition to the names of the farmers who grow them and place of origin, the register may include data about the *agro-morphological* and agronomic characteristics of landraces, agroecological adaptation, and special uses. Registers can serve as an information tool that reduces the *transactions* costs of locating and *exchanging* diverse materials, but do not solve the problem of a shortage in seed supply relative to demand.

National governments, nongovernmental organizations, and donors have invested in local-level seed projects as a means of delivering a better range and quantity of seed types where state seed enterprises have been ineffective and the commercial seed sector has been too slow to grow or too limited in focus. Reviewing these efforts to date, Tripp concluded that few of these projects have achieved the goal of establishing viable smallscale seed production enterprises.

He cites as a principal obstacle the failure to recognize that seed provision requires more than multiplication. Projects internalize the costs of managing contacts involved in obtaining source seed and establishing quality control procedures, arranging for seed conditioning, and, in particular, marketing the seed. When projects cease to be funded, the effectiveness of the seed enterprise falters.

He argues that more must be done to support institutional growth and strengthen farmers' links with markets and institutions that are already in place. One prerequisite for building such *institutions* is an *understanding* of existing seed exchange networks, whether formal or informal. Another is an understanding of how public systems might be adapted in order to promote the use by farmers of a broader range of materials.

In Nepal, informal research and development (IRD) has been used to test, select, and multiply seeds. A small quantity of seed of recently released and/or nearly finished varieties is distributed to a few farmers in a *community* to grow under their own conditions with their own practices. First practiced by Lumle Agricultural Research Centre (LARC), this approach has now been adopted by other organizations in Nepal and India for variety testing and *dissemination*.

Such approaches incur no substantive additional costs but speed the time to use of varieties since they shortcut release procedure. They are likely to enhance diversity since each farmer receiving seed adapts it through his or her own seed selection practices.

Recent work by *nongovernmental organizations* has sought to identify the ways in which seed markets can be stimulated as a means of introducing and supporting local diversity. Rather than impose seed *provisioning*, attention has been shifted to reinforcing existing farmers' systems and seed system recovery.

CONCLUSIONS

It does not make economic sense to trade productivity for biological diversity when it means thwarting the opportunities of poorer farmers in less-industrialized economies. "Promising candidates" for on-farm management of crop biodiversity are sites where the local crop genetic resources are ranked highly with respect to both farmer utility and their *biological* diversity, and where empirical analysis predicts that farmers are likely to continue cultivating them.

When genetic analyses confirm that relatively high levels of diversity are found in a location where farmers are choosing to specialize in fewer modern varieties for commercial sale, programs to conserve landraces may be *costly* in terms of private opportunities foregone and public *expenditures*.

Instead, careful introduction of larger numbers of modern types with distinct agronomic and consumption attributes, or participatory variety selection, may both benefit farmers and contribute to productivity enhancement over the longer term. *Remaining* landraces might be conserved *ex situ* following the criteria recommended for optimal sampling. In order to save misdirected public funds, it is equally important to know in which locations *remaining* landraces have little genetic interest. Not all landraces embody potentially valuable diversity. If farmers care about them, they will continue to grow them. If not, they will discard them, with few implications for society. On a global scale, for highly bred, staple food crops (e.g., rice, wheat, and maize), *historical* factors such as labor to land ratios, agroecological features, and *commercialization* explain to a large extent in which regions landraces are still grown and will continue to be grown.

At a more localized scale of analysis, studies suggest that, while the effects of incomplete local markets for crop products and market distance are fairly predictable, the effects of other economic variables, such as the extent of off-farm *employment*, income and wealth status, are not easy to predict a priori unless *researchers* already have *extensive* knowledge about farmer decision-making and local economies among candidate sites.

Typically, it will therefore be necessary to undertake more empirical research at the household and community level once candidate sites have been identified, while controlling *statistically* for the regional-level factors described above.

Only as more case studies accumulate can generalizations be drawn. Though some of the policy *instruments* mentioned here have been designed to support on-farm conservation through *enhancing* either the demand for or supply of diverse seed types, the balance sheet from the field has not yet been tallied.

5 Chapter

CONSERVING BIODIVERSITY

Chapter elsewhere in this book *focused* on indicators for funding *conservation* activities. This chapter discusses the economics of the distribution of funds from the International Treaty on Plant Genetic Resources for Food and *Agriculturehereafter* denoted by IT-discussed earlier in the book.

While economic benefits accruing from the distribution of these funds for *conservation* activities are almost impossible to ascertain, political *pressures* motivate this *distribution* to be made.

The question then becomes what is the most *economically* efficient method of *distributing* the funds among countries or among world regions, given the available data. The rest of this chapter is organized as follows.

Elsewhere in this book is an overview of the *economics* of investment in Plant Genetic Resources for Food and Agriculture (PGRFA) conservation. Elsewhere in this book, a proxy *indicator* for regional allocation of conservation funds is developed. Section 4 addresses mechanisms for *in situ* conservation. The last section is a discussion and conclusion.

PRINCIPLES OF VALUATION OF THE CONSERVATION INVESTMENT

A reduction in biodiversity matters for two reasons: (1) the *vulnerability* of a crop and its current *productivity*; and (2), the potential for farmers to cease *cultivating* a variety with untapped, potential use value. Any variety, whether bred with *conventional* methods or with *techniques* of genetic *transformation*, will be widely grown by farmers if they view it as superior to those they currently cultivate.

With respect to the first concern, if that variety, or a set of varieties, is uniform with respect to certain genes conferring biotic resistance, then cultivation of these varieties over a widespread area increases the probability of a mutation in the disease pathogen that overcomes the source of genetic resistance. Once that occurs, widespread cultivation of varieties with that same source of genetic resistance also contributes to more rapid spread of infection.

In other words, *uniformity* with respect to resistance genes can make a crop more vulnerable to economically meaningful crop losses-but this would be true for *conventional* and genetically modified seed, as well as for traditional varieties. Though *traditional* varieties or landraces are typically composed of more heterogeneous populations or mixtures, *historically* there are important cases of epidemics in *landraces* such as in the Indian *subcontinent* for the rusts of wheat.'

In fact, these epidemics were part of the motivation for early scientific plant breeding programs. *Furthermore*, diversity in genetic backgrounds and other resistance *mechanisms* (not confined a single gene) are often very important in explaining different disease reactions among varieties. Depending on how modern breeding techniques are used, they can maintain or decrease genetic diversity.

These techniques may contribute to *maintaining* diversity in systems dependent on traditional varieties if they enable the insertion of traits to overcome specific *disadvantages* into landraces that developing country farmers value for their consumption traits or *agronomic performance*. If so, the relative economic value of these landraces to local farmers may be enhanced.

On the other hand, if a company with breeding skills purchases a seed company and inserts the desired traits only in a subset of varieties that become dominant, then biotechnology may reduce the diversity of genetic materials used in plants. The economic value of PGRFA stems from its value as an input to the *agricultural* production process.

Genetic erosion is the loss of genetic diversity, including the loss of individual genes, and the loss of particular combinations of genes such as those *manifested* as locally adapted landraces. The term can be used in a narrow sense, i.e., the loss of genes or alleles, and in the wider sense, including the loss of varieties.

The main cause of genetic erosion in crops is the *replacement* of local varieties by improved or exotic varieties and species. Erosion can occur when a smaller number of varieties replace a larger number of older varieties and/or the newer variety has a different gene base from the old one. This chapter defines genetic erosion as the loss of genetically distinct varieties. Chapter elsewhere in this book provides the general *economic background* behind the economic valuation of the genetic resources for agriculture.

The appendix to this chapter presents a more detailed, and hence, more complex, *analytical* model of the economics of the conservation of PGRFA. In the model in the appendix, an increase in accessions to a gene bank is a function of *conservation investment*, and the change in *agricultural* supply is a function of the change in accessions.

According to the model, a reasonable criterion for choosing the optimal level of conservation investment is to choose the level of *investment* that *maximizes* producer plus *consumer* surplus in a *dynamic* context. The comparative statics of the model show that the marginal value of an additional dollar of conservation is a function of the marginal change in varieties conserved per *additional* dollar of investment as well as being a direct function of the change in welfare due to a change in *accessions.*

The economically efficient allocation of conservation funds among regions would be the one in which the marginal value of conservation investments is equated across regions. The marginal value of an additional dollar of conservation investment is not only a direct function of the marginal change in welfare due to a change in accessions, but it is also a function of the marginal change in *cultivars* conserved per additional dollar of *investment.*

The latter is *certainly* higher in areas of high diversity than in areas of low *diversity.* Hence, the *analysis* in the *appendix* can be seen as making a case for *concentrating* limited conservation resources in areas of high diversity, especially when little is known about the *quality* of the diversity with regards to *agricultural* uses.

Note that it is likely easier in principle to generate data on the marginal change in the number of cultivars conserved per additional dollar of investment than on the marginal change in welfare due to the change in the number of these *accessions.*

PROXY INDICATORS OF THE VALUE OF DIVERSITY

This section describes a proxy indicator for the importance of a region as a primary center of diversity. With the *exception* of a perhaps a case study or two, figures on crop *production* by variety are unavailable. The probability of variety loss is *unknown,* as is the marginal economic value of adding (*subtracting*) an additional variety (to) the set of varieties that make up a species.

In practice, the best one can do is to estimate a *third-best* proxy for the optimal allocation of the conservation funds. On a global scale, all we know is roughly where areas of high diversity are located for various species, and the best we can do is to rank regions by a proxy for the importance of their *agri-biodiversity* to the global *agricultural* economy.

Doing so can provide a rough index for targeting conservation funds. For the purpose of this chapter, and given the state of the available data, there is no reason to *strongly* promote any *particular* measure of diversity. For *policymaking* purposes, it is useful to divide the world into major regions of diversity, hoping to use some expert advice to assign as objectively as possible an area's relative *importance* to the world's major crops.

Whether one follows the approaches of Vavilov, Harlan, or Zeven is not a point for this economics paper to make. The basic goal of this discussion in this section on centers of diversity is that there appears to be enough research done on the subject of centers of diversity to make a plausible case that some botanical expert(s) could make diversity assignments to the world's major regions. Our indicator is based on the principle that plant genetic variability is not uniformly *distributed* throughout the world.

In the 1920s, the Russian geneticist Vavilov noted that the level of inter- and intraspecific genetic *variability* varies *geographically* across the world. He *identified* the geographic areas with the highest genetic variability in cultivated food crops. Vavilov thought that areas of maximum genetic diversity represented centers of origin and the origin of a crop could be identified by the simple process of *analyzing* variation patterns and plotting regions where diversity was concentrated.

Although his proposed centers of origin are widely accepted even today, his notion is *somewhat* simplistic and it turns out that even though many crops do exhibit centers of diversity, these centers of diversity have little to do with centers of origin. Domesticates can, and did, originate in one region and then develop much of their diversity in another.

However, while this distinction is of importance for *anthropological* and other reasons, it is of little practical import to the purposes of this chapter. For *conservation* of *landraces-as opposed* to wild *relatives-we* assume that centers of diversity are more important than centers of origin. However, centers of origin might be of *importance* in wild species conservation, at least in the case of species to which the concept is relevant.

Conservation of wild varieties has a separate set of policy mechanisms from that for domesticated varieties that, by definition, require the intervention of man. *Mechanisms* for wild species conservation include the set-asides, nature preserves, etc., discussed in the usual context of wild species *conservation*.

The term "region of diversity" is currently used to refer to the variability generated by crops during their dispersal from point of origin. Thus, a plant *population* can be described at any point in its *evolution* by frequency of genes and genotypes, which illustrates its historical evolution. The regions of crop diversity are areas with high *variability* in number of alleles and genotypes.

The genetic composition of these populations represents varying *adjustments* to *ecological* and social imperatives (Palacios). Given that these data on regions of diversity are the only relevant genetic data we have on a world scale, the best we can do is develop an indicator for the world (or OECD countries) value of agricultural production ascribed to primary regions, or centers, of diversity. The first columns of Table elsewhere in this chapter present geographic *distribution* of centers of *diversity* derived from Zeven and de Wet on the basis of the centers *identified* by Vavilov.

The fourth column presents the price per metric ton (in *international* dollars) for each commodity, and the fifth and sixth columns present the total world (and OECD) *production* data for each crop. The rational for the OECD figures is that if the OECD countries are paying for the lion's share of this conservation effort, they may be most interested in focusing on the crops of interest to them.

Given the price and quantity data, the total value of each crop is calculated. This value is then assigned to the *geographic* region *corresponding* to the center of diversity for each crop, with the total value ascribed to each center of diversity being the sum of the values of the crops for which the region is a primary center. This process produces the last column to the right in Tables elsewhere in this chapter. However, because a crop may have more than one geographic center of diversity, this measure has only an ordinal *interpretation*.

The alternative is to *normalize* the measure to sum to one by dividing each instance of a primary center for a crop by the number of primary centers for that crop (i.e., for a given crop, the weight assigned to a primary center falls as the number of primary centers increases).

Based on the *assumption* that each center provides an equal contribution as a center of diversity (which is all that can be done given the lack of data), imposing this *normalization* provides the measures in the last column in Tables elsewhere in this chapter with a cardinal interpretation, and the *normalized* indicator, when *expressed* as a fraction of total agricultural value, sums to

Table 5.1: Crops, their primary centers of diversity, and international price and production.

Crop	Primary centers of diversity	$/Mt	World production	OECD production
Rice	E./S.E./S. Asia/W. Africa	190	577,349,526	31,298,772
Wheat	W. & C. Asia	144	591,632,321	262,804,453
Sugar:				
Cane	S.E. & S. Asia/Pacific	34	1,253,253,810	121,072,256
Beet	Mediterranean/Europe	17	260,886,055	192,686,066
Maize	C. America	124	614,003,156	321,448,834
Soybeans	E. Asia	234	159,822,505	79,520,120
Potatoes	S. America	110	293,377,361	111,438,719
Cassava	S. America (Brazil-Paraguay) & C. America	68	162,072,352	11,096
Sorghum	Africa	124	61,044,434	21,358,847
Millet	Africa (excl. C. Africa)/S.E./S./E. Asia	158	28,606,826	260,298
Barley	W. & C. Asia/Mediterranean	114	138,454,668	96,457,199
Sweet potatoes	S./C. America	85	138,914,332	2,143,844
Oil palm	W. Africa	165	96,894,224	35,000
Rape/mustard	Mediterranean/Europe/E. Africa	328	36,423,345	21,907,635
Beans		445		
Phaseolus	S. & C. America	445	21,196,276	5,151,141
Vicia	C. Asia	445	5,581,953	1,294,213
Groundnut	S. America	491	33,750,854	2,098,819
Banana	S.E. & S. Asia/Indian Ocean	154	55,988,655	2,233,423
Plantain	S.E. & S. Asia/Indian	96	29,969,764	0 Ocean
Cotton	S. & E. Africa/C. Asia/S. & C. America	175	51,566,253	13,726,239
Coconuts/copra	Pacific/S.E. Asia	106	47,398,344	1,302,500
Yams	S.E. & S. Asia/Africa	85		
Oranges	E. Asia	202	35,753,627	181,000
Grapes	Mediterranean/W. & C. Asia	n.a.	63,140,901	22,582,779

(Table Contd.)

Apples	Europe/C. Asia	n.a.		
Sesame	S. & C. Asia/E. Africa	626	2,515,592	85,500
Olives	Mediterranean	n.a.		
Oats	Mediterranean/Europe	n.a.		
Rye	W. Asia	n.a.		
Tomato	S. America	n.a.		
Cocoa	S. America	663	2,914,830	43,968
Sunflower	N. America	292	24,833,755	7,703,127
Date	Mediterranean/W. Africa	n.a.		
Grapefruit	S.E. Asia	n.a.		
Pea	W. Asia/E. Africa	n.a.		
Onion	C. Asia	n.a.		
Paprika	Caribbean	n.a.		
Pineapple	S. America	n.a.		

one. Table elsewhere in this chapter ranks the regions in *descending* order according to their value as primary centers of diversity based on the values in Tables elsewhere in this book. Comparing the *normalized* indicator for the OECD with that for the world, we see that for the OECD, Central Asia and then West Asia rank highest as primary centers, while for the world, Southeast Asia and then South Asia are the highest.

Table 5.2:. Total number of primary centers per world region and world value of agricultural production ascribed to primary centers of diversity (expressed as a fraction of total world value of the agricultural commodities).

Region	*Number of primary centers*	*Fraction of world value ascribed to primary center (non-normalized)*	*Fraction of world value ascribed to primary center (normalized)*
Central Africa	2	0.019	0.004
East Africa	6	0.069	0.016
Southern Africa	4	0.044	0.008
West Africa	5	0.257	0.084
Indian Ocean	5	0.049	0.012
South Asia	7	0.316	0.086
Southeast Asia	7	0.322	0.090
East Asia	4	0.301	0.143
Pacific	2	0.087	0.031
Mediterranean	3	0.059	0.021

(Table Contd.)

West Asia	2	0.185	0.087
Central Asia	5	0.209	0.096
Europe	2	0.030	0.011
South America	7	0.168	0.126
Central America	5	0.215	0.172
Caribbean	0	0.000	0.000
North America	1	0.013	0.013
			Total: 1.000

The result is due to Central and West Asia being primary centers for wheat, which is of greater importance to OECD countries than rice, while Southeast and South Asia are primary centers for rice. Not surpris-ingly too, the *Mediterranean* and European centers rank higher in the OECD value than they do in the overall world value.

The low rankings associated with Central Africa, North America, and the *Caribbean* are the same for both the OECD and the world as whole. These results suggests that, with no other *information* being available on the general state of PRGFA, the OECD might be inclined to allocate a high share of funds to conservation efforts in higher ranked regions such as Central Asia, but if their view was more magnani-mous, they may give Southeast Asia higher *consideration.*

In other words, if the OECD is paying for the conservation activities, they may allocate funds differently than world ranking would dictate. However, if their goals include global food security, they may well want to follow world values. Central America ranks highest in the normalized ranking primarily because it does not have to share its title as center of diversity for maize with any other region.

Table 5.3: OECD value of agricultural production ascribed to primary centers of diversity (expressed as a fraction of total OECD value of agriculture at the farm gate).

Region	***Fraction of OECD value (non-normalized)***	***Fraction of OECD value (normalized)***
Central Africa	0.017	0.003
East Africa	0.079	0.022
Southern Africa	0.033	0.006
West Africa	0.055	0.013
Indian Ocean	0.019	0.004
South Asia	0.067	0.019
Southeast Asia	0.068	0.019
East Asia	0.186	0.157

(Table Contd.)

Pacific	0.027	0.009
Mediterranean	0.137	0.049
West Asia	0.312	0.144
Central Asia	0.331	0.151
Europe	0.067	0.026
South America	0.116	0.096
Central America	0.286	0.265
Caribbean	0.000	0.000
North America	0.014	0.014
		Total: 1.000

CONSERVATION MECHANISMS

In situ conservation may be achieved through several possible *mechanisms* that are discussed in the chapters in the first section of this book, as well as in the final two chapters. Hence, possible mechanisms are discussed only briefly here. One possibility is to directly pay farmers to *undertake* conservation activities (perhaps through *mechanisms* like the Conservation Reserve Program in the United States), but doing so will probably not work well, *particularly* in developing countries.

Table 5.4: Regions ranked in descending order according to their value as primary centers of diversity.

	Region: OECD value		***Region: World value***	
Rank	***(Non-normalized)***	***(Normalized)***	***(Non-normalized)***	***(Normalized)***
1	Central Asia	Central America	Southeast Asia	Central America
2	West Asia	East Asia	South Asia	East Asia
3	Central America	Central Asia	East Asia	South America
4	East Asia	West Asia	West Africa	Central Asia
5	Mediterranean	South America	Central America	Southeast Asia
6	South America	Mediterranean	Central Asia	West Asia
7	East Africa	Europe	West Asia	South Asia
8	Southeast Asia	East Africa	South America	West Africa
9	South Asia	Southeast Asia	Pacific	Pacific
10	Europe	South Asia	East Africa	Mediterranean

(Table Contd.)

11	West Africa	North America	Mediterranean	East Africa
12	Southern Africa	West Africa	Indian Ocean	North America
13	Pacific	Pacific	Southern Africa	Indian Ocean
14	Indian Ocean	Southern Africa	Europe	Europe
15	Central Africa	Indian Ocean	Central Africa	Southern Africa
16	North America	Central Africa	North America	Central Africa
17	Caribbean	Caribbean	Caribbean	Caribbean

Indirect methods appear more feasible. For example, programs that encourage farmers to adopt or continue *sustainable* farming practices may increase the competitiveness of landraces versus modern varieties. Decentralized *breeding* programs, niche market development, trademark or other schemes for tying quality or certain traits to source are some of the policy options.

The diversity of a production system dominated by modern varieties can be enhanced through policies designed to encourage the release of more *genetically* diverse *materials* at a higher rate, where higher rates of release of modern varieties contribute to "diversity in time".

Other policies might *enhance* the spatial diversity in modern systems by *favouring* greater numbers of more evenly distributed, *genetically* different varieties. Optimal *investment* strategy would require that precedence in the *distribution* of the funds be given to countries that do the best job at *cataloging* their varieties with information of use to breeders, given that the value of *germplasm* increases with the *information* associated with it.

Of course, for this approach to be undertaken, *less-developed* countries (LDCs) would require *technological* assistance with database development, etc. In terms of *mechanisms* for *distribution*, the fund can support specific *in situ* conservation projects that are chosen by competition from an open call for conservation proposals.

CONCLUSION

As discussed in other chapter of this book, the economically efficient allocation of *conservation* funds across regions is, not *surprisingly*, the one that *equates* the marginal value of an additional dollar of investment across the regions. When using centers of diversity to target *conservation* dollars, one question is whether the marginal value of an additional dollar of conservation investment is higher in an area with high diversity than in an area with low diversity.

Given the principle of *diminishing* marginal returns, it is possible that the marginal economic value of an additional variety conserved is higher in an area with low diversity. However, the marginal value of an additional dollar of *conservation* investment is also a function of the marginal change in varieties conserved per additional dollar of *investment*.

The latter is likely higher in areas of high diversity than in areas of low diversity. Hence, the case is made for concentrating limited *conservation* resources in areas of high diversity. From the idealized economic model presented anywhere else in this book steps down to *assessments* that can be made using available data. The basic genetic diversity data we have on a global scale are the association of *geographic* regions with primary centers of diversity for certain crops. This *information* allows us to generate a basic guide to the value of *agricultural* production associated with these centers of diversity. While this information does not incorporate the risks associated with genetic erosion, and the concept of primary centers may lead to imprecise measures, it does give us a measure of the economic importance of crops associated with these regions as centers of diversity.

As such, it may be useful as a rough guide for the IT in making allocations of the *conservation* funds to world regions. The current data on centers of diversity do not identify areas at the country level or lower. Hence, other *indicators* are needed to make the allocation from world region levels to country levels.

Given the lack of scientific information, the IT will mostly likely use the *subsidiarity* principle in *allocating* the funds, i.e., they will allocate the funds to the world regions, and each region will decide how to allocate the funds to individual countries. Instead of directly transferring funds down to the country level, given that there are no indicators and *mechanisms* available for transferring the funds in *anything* near a first-best fashion, perhaps the most efficient approach would be to ask countries to submit concrete proposals for use of the funds in *conservation* activities.

These proposals could then be *prioritized* for funding based upon their merits. Of course, such a mechanism could be used directly by the IT, and it could evaluate conservation project proposals directly. However, regions may argue over their shares of the *conservation* funds, and some regional indicator, such as the one developed in this chapter, will be necessary to the IT in making some *judgments* in the parceling out of the funds among regions. The indicators of centers of diversity discussed in the third section of this chapter are broad enough that they can only serve as a rough guide for conservation targeting.

Smaller areas, known as "*microcenters,*" might be especially useful to concentrate limited conservation dollars on. Harlan (1992) defines microcenters, for either crops or wild plants, as relatively small regions, 100 to 500 kilometers across, which may be packed with high variation of one to several crops in *comparison* to adjacent areas.

Some of these have been destroyed, and the rest are threatened with replacement with modern cultivars *(ibid)*. A global indicator based on microcenters is not feasible until they are studied on a more systematic basis. However, the analytical *framework* developed here could easily be *desegregated* by *microcenters* if such data became available.

In other chapter of this book provides more concrete suggestions for *mechanisms* for distributing funds at the country level, as well as for policy mechanisms for conservation programs. The final allocation of the conservation funds will be affected to some extent by the relative bargaining power of actors, both at the regional level and at the country level. Furthermore, as implied in The Global Plan of Action's call for the sharing of benefits to be "fair and equitable," there may be significant demand to distribute at least a portion of the funds on equity grounds to countries most in need of development aid. The next chapter addresses in detail the *qualitative* and policy-relevant aspects of *bargaining* power and equity in the allocation of these funds.

6

Chapter

IMPACT OF CROP DIVERSITY

Agricultural *productivity* enhancements have often been based upon development and dissemination of a small number of highly competitive plant varieties or animal species and, thus, have been associated with a decrease in crop genetic diversity (CGD). Mechanical *innovations* such as tractors and combines, supported by breakthroughs in chemical *fertilizers* and pesticides, have facilitated the mono-cropping of vast areas of land, so it is not surprising there is concern that the advent of *agricultural biotechnology* may exacerbate these trends.'

In this chapter, we argue that agricultural biotechnology may instead offer unique opportunities to preserve CGD, but the speed and *extent* to which this potential is realized depend upon institutional factors, including the distribution and level of protection afforded to intellectual property rights (IPRs), transaction costs associated with licensing, *technology* transfer, and biosafety regulations. As is common in the literature, CGD is used to describe the genetic diversity of *agricultural* crops.

The number of different varieties or landraces being used by farmers is an important indicator of the *in situ* CGD of a particular crop species. *Biotechnology* introduces a *fundamental* change in the way that seeds and other genetic materials can be produced. With *traditional* breeding techniques, existing varieties are selectively combined to develop new varieties or hybrids.

This is a lengthy process and involves a *significant* degree of *randomness*. The outcome is usually a novel variety that has a number of new traits and characteristics, not all of which are desirable. Biotechnology, however, allows the targeted introduction of selected genetic materials into existing crop varieties. Once the genetic sequence coding for a desirable trait such as insect resistance has been identified, a "*transformation* event" is created by transferring this genetic sequence to a particular receptor variety.

Additional genetically modified varieties (GMVs) are then developed by crossing existing *conventional* varieties with this transgenic receptor variety. Although several *backcrossing generations* are necessary to eliminate unwanted characteristics, this process is far quicker, easier, and cheaper than developing a new *conventional* variety through cross-breeding.

It usually results in GMVs that are virtually identical to their conventional counterparts except for the new desirable trait. In other words, *biotechnology* permits a separation between the act of developing a specific agronomic trait and the breeding of a particular, locally adjusted variety.

As such, it has important *institutional* and economic implications that will affect the diversity of crop plants in *agricultural* production. In this chapter, we evaluate the potential for adoption of seeds and genetic materials developed using *biotechnology* and assess the impact on the diversity of crop varieties produced under alternative industrial and policy structures

We examine the roles of IPRs and the research capacity and efforts of private and public sectors in *determining* the utilization of agricultural biotechnology to obtain policy implications for agricultural research efforts. The next section provides an overview of our framework and main findings. It is followed by a detailed *mathematical* derivation of the main results. The final section presents a summary of our results, and it *demonstrates* their validity using available information on a number of GMVs of various crops grown in different countries.

SUMMARY OF MODEL AND MAJOR FINDINGS

The legal framework in the United States, Europe, and other developed countries has gradually evolved so that those who decipher genetic structures, discover the functions of genes, or identify mechanisms to alter genes can register patents and own the IPRs for the *utilization* of these discoveries. Private parties have an incentive to conduct research leading to new discoveries because they expect to gain *financially* from selling the rights to utilize the IPRs, or to utilize them directly in their own commercialization efforts.

However, the extent to which IPRs are protected and traded varies among nations, and these variations may affect the way the products and processes of biotechnology are managed and utilized. In our conceptual analysis, we consider the case of an agricultural industry sector that produces a single crop where producers are *characterized* by high degrees of heterogeneity.

Variation in land quality, *topography,* and climatic conditions, even within a region, may result in growers adopting different varieties of the same crop. We assume that, *technically* speaking, all the existing varieties can be modified using biotechnology. This modification, for example, may reduce susceptibility to pests and diseases or increase the *efficiency* of nutrient uptake.

For simplicity, we assume that the effect at the farm level is an increase in yields. It is assumed that *utilization* of a biotechnology-based innovation, such as *a Bacillus thuringiensis* (Bt) gene that codes for the expression of an *insecticidal* protein in plant tissue, requires a large fixed cost in *infrastructure* for technology development, a modest fixed cost to obtain the capacity to *incorporate* the technology into each specific variety, and a relatively small variable cost of seed production.

We argue that *biodiversity* and the impact of a new *biotechnology-based* innovation on grower and consumer welfare depend on the manner in which its introduction and pricing *strategies* relate to the existing cropping system. In particular, we *distinguish* between *situations* where private sector companies introduce the technology and *situations* where it is introduced by the public sector.

We also *distinguish* between situations where traditional local varieties are replaced by "generic" GMVs and situations where specific genes are introduced to local varieties, which continue to be planted in the new, modified form. The generic GMV may be an imported variety that, on average, performs well, but since it may not be highly suited to each location's conditions, will likely not perform as well as GMV versions of local varieties.

These features will determine the outcomes of farmer and consumer welfare and biodiversity preservation. When a private company introduces GMVs, we assume that it charges a *monopoly* price. This price is set at a level where marginal revenues are equal to marginal costs of producing the modified seeds. When farmers are heterogeneous in their conditions, the impacts of the technology may vary. GMVs will be adopted only where the gain from adoption is sufficient to cover the technology fee.

The technology fee will increase with the variable cost of *modification* (and when the cost of modification is assumed to be fixed, only the cost of seed production is variable). Therefore, low variable costs increase the adoption of GMVs. When a local variety is genetically modified (GM), the effect of *environmental heterogeneity* is likely to be an increase in the yield effect and cost savings compared to the case in which the local variety is replaced by a generic GMV.However, the *modification* of each variety entails a fixed cost.

In deciding whether to genetically modify a specific variety or to offer growers a generic GMV, the private company will compare the extra cost of *modification* with the extra revenues earned from a modified variety relative to a generic variety. Two factors that will affect this decision are the cost of modification and the size of the market. When modification costs are high or when the market is small, it will be more profitable to sell a generic GMV.

Thus, regions where the *crop-breeding* industry has low capacity and the cost of modification is relatively high are more likely to adopt generic GMVs, and their introduction may lead to a reduction of crop biodiversity. Thus far, private sector companies have developed and introduced most of the GMVs globally. The history of the Green Revolution suggests that eventually public sector *institutions* will also develop and introduce GMVs, and these varieties will be distributed through small seed companies.

We analyze outcomes where the public sector makes the choice of whether or not to introduce a GMV to a region, and whether there will be a genetic modification of the local variety or a generic GMV imported from elsewhere. The public sector organization is assumed to *maximize* the domestic economic surplus, including seed producers, farmers, and consumers.

Under these scenarios, the seed companies will charge *competitive* prices that are less than the price charged by a monopoly seller. As a result, adoption levels are likely to be higher than under the monopolistic scenario if all the other parameters are the same. Furthermore, since the total domestic surplus is larger than the profit of monopolistic seed *companies,* the public sector is more likely to invest in development of GMVs than a private sector monopolist.

The public sector is more likely to develop local GMVs rather than import generic GMVs, since the development of local GMVs is likely to have a larger yield effect that benefits growers and consumers rather than seed producers. The decision whether to introduce a local or an imported generic GMV, in any context, depends on the *difference* between the gain and the cost of development.

Countries with more advanced crop-breeding capabilities and *sufficient* public sector resources for developing GMVs are more likely to modify existing varieties, which will lead to preservation of biodiversity. On the other hand, in *situations* where public sector resources are limited and where investment in GMVs is not profitable for the private sector, the public sector may develop or import a small number of generic GMVs and, in spite of limited adoption, it may lead to a loss of biodiversity.

A MODEL

An agricultural sector is producing a single crop in M locations, with varying climatic and *agroecological* conditions. Let *j* be a location indicator. Then *j* takes values from 1 to *M*. Before GMVs are *introduced*, farmers at each location use the best *traditional* variety given their conditions, and we assume that each location has its own distinct *traditional* variety.

For simplicity, we assume that the gross benefits of production are *measured* in *monetary* terms. Output price denoted by P is assumed to be constant. In the analysis, we consider several scenarios with respect to market and properties of genetic materials. Let s be a *scenario* indicator; $s = 0$ is the initial situation where only *traditional* varieties are used; $s = m$ is the case in which locally optimized varieties are modified *(GMV)*, and $s = g$ is the *situation* in which only one generic modified variety *(GMV) is* adopted.

Let X^s_j denote output *produced* at location *j* under scenario *s*, which is assumed to be a function of land and genetic materials. The production function f^s_j (A)denotes the output *produced* on *A* acres of land in location *j* and scenario *s*. The marginal *productivity* of land for scenario *s* is $Mp^s_j(A)=\partial f^s_j/\partial A$. It is assumed that $Mp^s_j(A)$ and $\partial Mp^s_j/\partial A < 0$ (decreasing marginal *productivity* of land in a location), which may reflect *heterogeneity* of land quality within a location. Prior to the introduction of GMVs, when *s* is = 0, the demand for land of variety *j* at location *j* is denoted by $PMp^0_i(A)$.

The marginal cost of land grown with variety *j* is assumed to be constant and is denoted by Mc_j, and the seed cost prior to *biotechnology* is assumed to be 0. Before the *introduction* of biotechnology, the acreage allocated to variety *j*, A^0_j, was at the point when marginal benefits equal marginal cost of acreage, or $(PMp^0_j(A^0_j) = Mc^0_j)$.

Assumptions about the GM Technology and Seed Industry Structure

Before *analyzing* several outcomes from GMVs, let us specify *assumptions* about their benefits and cost structure. We assume that each traditional variety can be modified, but it may also be replaced by a "generic" GMV. The exact outcome depends on costs, constraints, and decision making about seed supply. For *simplicity*, we assume that GMVs improve seed productivity. This corresponds to *yield-increasing* GMVs (e.g.).

Bt cotton in India or South Africa). Let $Mp^m_j(A)$ denote the marginal productivity *(Mp)* of land planted with *GMVj*. We will assume that (a) GMVs have higher marginal productivity than the *traditional* variety, $Mp^m_i(A) > Mp^0_i(A)$ and (b) the marginal *productivity* gain declines with *A*,

$$\partial(Mp_i^m(A) - Mp_i^0(A))/\partial A < 0$$

This assumption that better-quality lands gain more from genetic modification is done for convenience, and most results hold for broader circumstances.

It corresponds to situations, say, where a certain percentage of the crop is lost to pests. GMVs reduce pest damage and thus provide higher gains to locations with higher potential output. If traditional varieties are replaced by a generic GMV, then $Mp^s_i(A) = \partial f^s_i(A)/\partial A$ is the marginal *productivity* of land in location *A*. We assume that the *Mp* of the *unmodified* version of the generic variety is less than that of variety *j*. Because modification increases the marginal *productivity* of the target variety, it is assumed that the marginal benefit of the generic GMV, while less than that of the GMV_i is greater than the traditional variety. Hence,

$$Mp_i^m(A) > Mp_i^s(A) > Mp_i^0(A)$$

and also that

$$\partial(Mp_i^s(A) - Mp_i^0(A))/\partial A < 0.$$

The introduction of the GM technology is associated with several cost categories. The first is the fixed cost to introduce the GMVs at the crop level. It consists of research, testing, registration, and *regulatory* compliance costs to introduce, say, Bt cotton in South Africa or Bt sweet potato in Kenya. We assume that this cost has been incurred and consider two other costs.

(a) F_i = fixed cost to modify variety *j*. This cost includes both the technical cost to insert genetic materials into a specific variety and the cost of IPR transactions and biosafety regulation. Countries with a more advanced breeding sector are likely to have lower modification costs.

The IPR component depends on specific circumstances. If a private company already controls a local variety, it will not face any IPR cost. On the other hand, it may have significant cost if a competing company controls a specific variety.

(b) V_i = per acre costs of GMVs of variety *j*. We assume that these variable costs are constant. They include the physical variable cost of producing GMV seeds and the IPR and marketing costs that a *manufacturer* has to pay in order to be able to sell the seeds. Let V_g be the per acre cost of the generic seed variety when it is used. It is assumed to be smaller than V_i. It is imported from a low-cost *production* center and does not *require* payment to owners of the rights for local seeds.

The difference decreases as the seed sector producing GMV_J becomes more advanced and when the seed producer does not face IPR costs for the use of the local variety. The *outcomes* with GMVs depend on the cost to introduce them as well as the structure of the seed industry and the constraints it faces. We consider two patterns.

The first pattern applies to countries where public research *institutions* develop GMVs, which are sold to farmers by a competitive seed sector. In the second pattern, GMVs are introduced and marketed by a monopolist. This scenario is appropriate for the developed world, where major *companies* such as Monsanto control the GMVs *available* in the market. In the case of a monopoly, the seed industry may face IPRs and other constraints that impede its capacity to modify *traditional* local varieties. Recognizing these *constraints,* we derive *outcomes* for four stylized scenarios presented below.

Competitive Markets for Seeds of GM Local Varieties

Assuming that the fixed cost of introducing the technologies are covered by the public sector, competitive seed sellers will charge farmers V_i per land unit of GM seeds.

The GMVs may be either fully or partially adopted. In the case of partial adoption, some land will continue to be grown with traditional variety j. Let A^1_i be total acreage of variety *j* (traditional and GM), and let A^{m1}_i be the GM acreage.

The acreage of the traditional variety *j will* be equal to $A^1_i - A^{m1}_i$. There are three possible outcomes under competition:

- $C^m_i 1$: No adoption when $PMp_i^m(0) - PMp_i^0(0) < V_i$. In this case, $A_i^{m1} = 0, A_i^1 = A_i^0$.
- $C^m_i 2$: Partial adoption when $PMp_i^m(0) - PMp_i^1(0) > V_i > PMp_i^m(A_i^0) - PMp_i^m(A_i^0)$.

  Here $A_i^{m1} < A_i^0$ where $PMp_i^m(A_i^{m1}) = PMp_i^0(A_i^{m1}) + V_i$ and $A_i^1 = A_i^0$.
- $C^m_i 3$: Full adoption when $PMp_i^m(A_i^0) - PMp_i^0(A_i^0) > V_i$.

  $A_i^1 = A_i^{m1}$ when $PMp_i^m(A_i^{m1}) = Mc_i + V_i$.

The *assumption* that the *Mp* gap between the GMV and traditional variety declines with acreage is the key to the above results. If gains from adoption cannot cover the variable cost even for the first acre, adoption will not occur.

If marginal gains from adoption are greater than the variable cost, but only for a subset of the acreage, there will be partial adoption; and if the marginal gains from adoption are greater than the variable cost for all acres, there will be full adoption and the acreage under the crop might even *increase* (recall that we have assumed the total output market is sufficiently large that this change has no impact on output prices).

Figure elsewhrere in this chapter depicts case $C^m_i 1$ with no adoption when $A_i^1 = A_i^{m1}$ when $PMp_i^m(A_i^{m1}) = Mc_i + V_i$. Figure elsewhrere in this chapter depicts case $C^m_i 2$ of partial adoption, and Figure elsewhere in this chapter depicts case $C^m_i 2$ with full adoption.

These figures demonstrate that high variable costs may discourage adoption entirely, while reduction in these costs may result in partial or, eventually, full adoption. In the figures, the same notation should be used as in the text, that is, replace *Mb* by *PMp*.

The introduction of GMVs affects output and net benefits of land utilized with variety *j*. The output when GMV_i is available becomes

$$X_i^{m1} = \begin{cases} f(A_i^0) & \text{if } C_i^m 1 \\ f_i^m(A_i^{m1}) + f_i^0(A_i^0 - A_i^{m1}) & \text{if } C_i^m 2 \\ f_i^m(A_i^{m1}) & \text{if } C_i^m 3 \end{cases}$$

The net change in social benefit is

$$Nsb_i^{cm} = P(X_i^{m1} - X_i^0) - V_i A_i^{m1} - F_i \qquad (1)$$

and it represents the difference between the production gain and the variable and fixed cost of the new variety. Since up until now almost all crop biotechnologies have been developed and commercialized by private companies in monopoly situations, *introduction* of GMVs by the public sector has rarely occurred in reality.

However, the case of Bt cotton in China closely *corresponds* to this scenario. Although Monsanto and Delta and Pine Land (D&PL) have introduced U.S. Bt cotton varieties in China, the Chinese Academy of *Agricultural Sciences* has developed and commercialized its own Bt cotton technology, which can be freely used by public and private sector breeders.

Due to weak IPR *protection*, the *Monsanto technology* also has been incorporated into Chinese cotton varieties by local organizations, without payment to the company. At present, there are 22 officially registered local Bt varieties and five imported ones available on the market.

Adoption of Bt varieties has occurred in *approximately* 35% of the Chinese cotton sector and is increasing rapidly. There is no indication that GM technology has a negative effect on cotton biodiversity. On the contrary, the Bt varieties imported from the United States appear to have broadened the local germplasm base.

A similar situation could occur in other countries that do not protect IPRs but have a strong breeding capacity. If foreign GMVs are introduced in these countries, breeders can freely use these varieties to cross-breed the transgenic traits into their own germplasm.

The trade-off, however, is that without IPR enforcement, private seed industry development is hampered and technology transfer from abroad is discouraged.

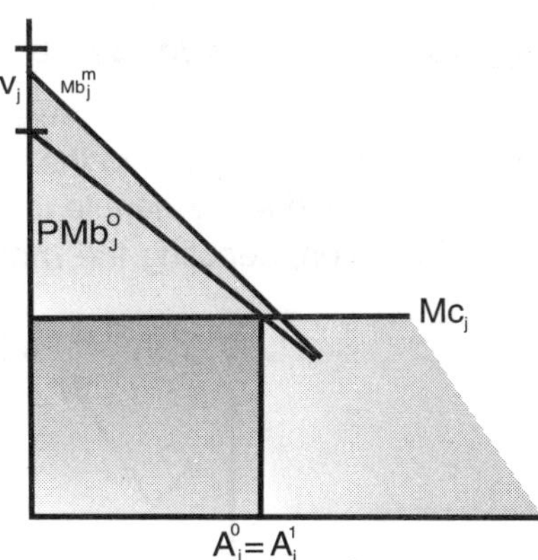

Figure 6.1a: No adoption

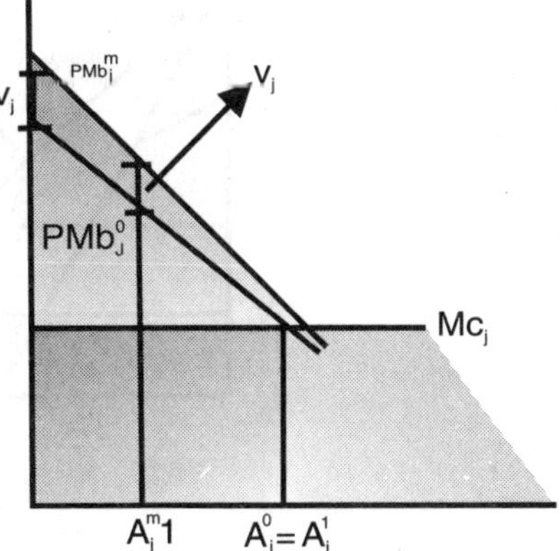

Figure 6.1b: Partial adoption

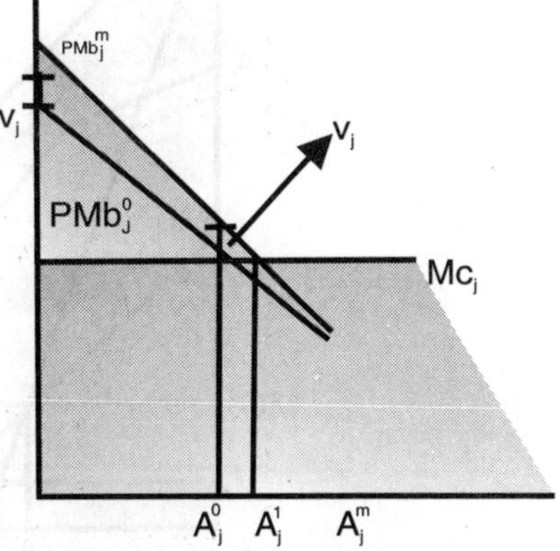

Figure 6.1c: Full Adoption

Monopolistic Markets for Seeds of GM Local Varieties

Consider the case when GMVs are produced and marketed by a monopolist. The monopolist is assumed to have access to the traditional local varieties and to modify them. Let A_j^{m2} denote the area of GMV_j and let A_j^2 denote total area (*traditional* and GM) of variety *j*. In this case the inverse demand function, denoting the *maximum* price (W_j^m) farmers are willing to pay per acre for GMV

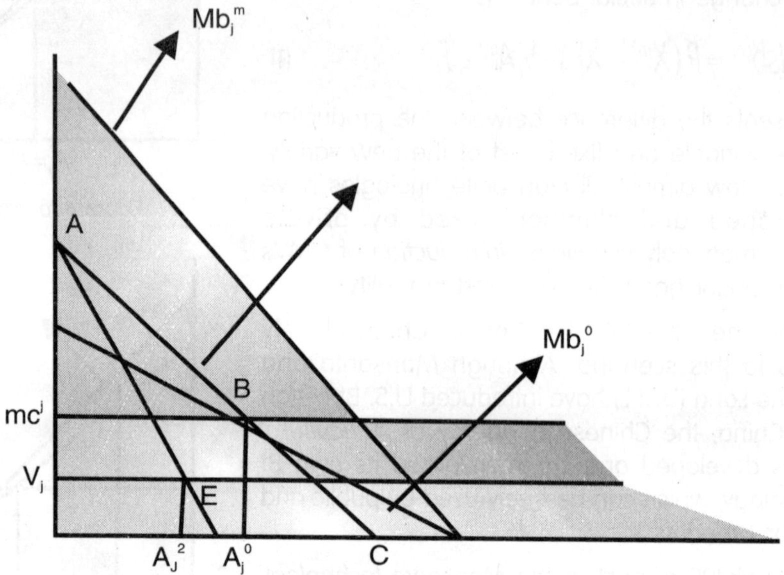

Figure 6.2a: Partial adoption.

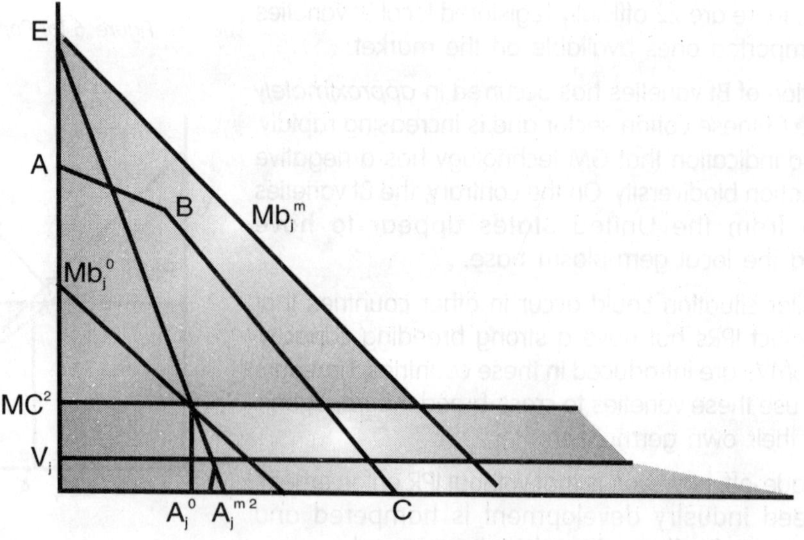

Figure 6.3b: Full adoption.

seeds, as a function of acreage, is

$$W_i^m(A) = D^{-1}(A) = \begin{matrix} P[Mp_i^m(A) - Mp_i^0(A)] & \text{if } A < A_i^0 \\ PMp_i^m(A) - Mc_i & \text{if } A > A_i^0 \end{matrix} \quad (2)$$

The marginal revenue from the sale of seeds for A acres is

$$MR_i^m(A) = \begin{matrix} P\left[Mp_i^m(A) - MP_i^0(A) + A\frac{\partial}{\partial A}[Mp_i^m(A) - Mp_i^0(A)]\right] & \text{if } A < A_i^0 \\ P\left[Mp_i^m(A) + A\frac{\partial}{\partial A}\right][Mp_i^m(A)] - Mc_i & \text{if } A \geq A_i^0 \end{matrix} \quad (3)$$

This inverse demand curve indicates that buyers will not be willing to pay more than (a) the difference between the marginal benefits per acre of GMVs and *traditional* varieties when both are viable $(A < A_i^0)$ and (b) the difference between marginal benefits of land with GMV and marginal cost of land when only GMVs are economical. The possible adoption patterns of GMV_i under monopoly includes

- $M_i^m 1$: <u>No adoption</u> if $PMp_i^m(0) - PMp_i^0(0) < V_i^m$.
- $M_i^m 2$: <u>Partial adoption</u> if $PMp_i^m(0) - PMp_i^0(0) > V_i > MR_i^m(A_i^0) - MR_i^0(A_i^0)$.

  In this case, $A_i^{m2} < A_i^0$, at A_i^{m2}; $MR_i^m(A_i^{m2}) - MR_i^0(A_i^{m2}) = V_i, A_i^2 = A_i^0$.
- $M_i^m 3$: <u>Full adoption</u> if $Mc_i > V_i$. At $A_i^{m2}, MR_i^m(A_i^{m2}) = Mc_i + V_i$.

The monopolist will sell the amount of seeds where its marginal revenue is equal to the variable per unit cost. When the marginal revenues intersect V_i at a quantity smaller than A_i^0 (case $M_i^m 2$), there will be partial adoption and when the marginal revenues intersect V_i at a quantity greater than A_i^0, there will be full adoption.

It can be verified that (a) higher gains in marginal productivity (high $PMp_i^m(A) - PMp_i^0(A)$) result in an increase in adoption of the GMV, and (b) adoption rates under monopoly are smaller than under competition. This is so because the monopoly price $PMp_i^m(A_i^{m2})$ for GMV_i will be greater than the competitive price, V_i. The profit of the monopolist, presented in Eq. (4), is smaller than the net social benefits considered by public sector decision makers when *determining* whether or not to assume the fixed cost of introducing a new variety.

Thus, a monopoly outcome will provide a *less-than-optimal introduction* and adoption of GMV_i. There may be cases when profit does not cover the fixed cost of modification. Then, a monopolist will not introduce GMV_i, even though the net social benefits might be positive. Figure elsewhere in this chapter denotes the monopoly outcome for the case of partial adoption. Curve ABC denotes demand for GMV seeds and has two segments—AB is $PMp_i^m - PMp_i^0$ and BC is $PMp_i^m - Mc_i$.

The marginal revenues AE, associated with AB, intersect V_i to establish $A_i^{m2} < A_i^0$. Figure elsewhere in this chapter denotes the monopoly outcome for the case of full adoption. With the marginal benefits of using GMV, PMp_i^m's are much higher than those of the traditional variety. Demand for GMV seeds is represented by ABC, and the relevant marginal revenues are EF, which intersects with V_i at $A_i^{m2} > A_i^0$.

The output of the industry in a monopoly situation is

$$X_i^{m2} = \begin{cases} f_i^0 (A_i^0) & \text{if } M_i^m 1 \\ f_i^0 (A_i^0 - A_i^{m2}) + f_i^m (A_i^{m2}) & \text{if } M_i^m 2 \\ f_i^m (A_i^{m2}) & \text{if } M_i^m 3 \end{cases} \tag{4}$$

The price of seeds is equal to $P\left[MP_i^m (A_i^{m2}) - MP_i^0 (A_i^{m2})\right]$ and $PM_i^m (A_i^{m2}) - MC_i$ if $M_i^m 3$. Thus, taking into account the fixed cost *i*, net profit of the monopolist is

$$\pi_i^m = \begin{cases} 0 \text{ if } M_i^m 1 \\ \left[P\left[Mp_i^m (A_i^{m2}) - Mp_i^0 (A_i^{m2})\right] - V_i\right] A_i^{m2} - F_i \text{ if } M_i^m 2 \\ \left[PMp_i^m (A_i^{m2}) - Mc_i - V_i\right] A_i^{m2} - F_i \text{ if } M_i^m 3 \end{cases} \tag{5}$$

Whether the monopolist has access to all local varieties and markets GMVs itself or the technology is licensed to other seed producers does not matter for the scenario outcome. Technology licensing under strong IPRs is rather typical for GMVs in the *United States*. For Roundup Ready (RR) soybean and RR and *Bt corn, biotechnology* firms have issued nonexclusive licenses to all breeders and seed companies interested in endowing their own breeding lines with the transgenic traits.

Due to high demand and a relatively low fixed cost to modify local varieties, numerous GMVs are available on the U. S. market. The number of *RR soybean* varieties increased along with increasing technology adoption rates. In *2002*, around *200* different seed companies marketed over 1,100 RR soybean varieties, which were adapted to diverse local conditions.

On average, this implies an area of less than 20,000 hectares per variety. Likewise, several hundred RR and Bt corn hybrids are available from seed companies of all sizes, with an average area of less than 10,000 hectares per hybrid.

Although the patent owners capture a significant share of the rent through monopoly pricing, U. S. farmers and consumers also benefit *significantly*, as the licensors have not set prices to take advantage of *variations* in marginal product across regions or varieties. For RR soybean and Bt corn in Argentina, the scenario is similar.

Although IPR protection is weaker than in the United States, the technologies have been licensed to various seed companies that incorporated them into their own breeding lines. Today, there are seven different companies providing *56* RR soybean varieties and four *companies* providing over 20 different Bt corn hybrids in Argentina. Most of this *germplasm* has been locally bred or adjusted, and the total number of soybean varieties and corn hybrids in Argentina did not change *significantly* since the introduction of GM technology.

Especially in the case of RR soybeans, farmers are the main *beneficiaries* because Argentine *legislation* allows the on-farm reproduction of seeds. Thus, average royalty payments are relatively low, so that the situation has some elements of the social optimum scenario.

Competitive Markets for Seeds of a GM "Generic" Variety

In some countries, a limited capacity to modify GMVs or cost *considerations* may lead even the public sector to introduce a generic GMV, imported from abroad, instead of genetically modifying local varieties. The results for the *competitive* and *monopolistic* markets for *GMy'* can be modified accordingly. If the seed industry is competitive, the possible outcome includes

- $C_i^g 1$: <u>No adoption</u> when $PMp_i^g(0) - PMp_i^0 < V_i$.
- $C_i^g 2$: <u>Partial adoption</u> if $PMp_i^g(0) > V_i > PMp_i^g(A_i^0) - PMp_i^0(A_i^0)$.

  Here $A_i^{g3} > A_i^0$ when at A_i^{g3} $PMp_i^g(A_i^{g3}) = PMp_i^0(A_i^{g3}) - V_i, A_i^0 - A_i^{g3}$ in a traditional variety.
- $C_i^g 3$: <u>Full adoption</u> when $PMp_i^g(A_i^0) - PMp_i^0(A_i^0) > V_i$. At $A_i^{g3}, Mp_i^g(A_i^{g3}) = Mc_i + V_i$.

  Output in this case is

$$X_i^{g3} = \begin{cases} f_i^0(A_i^0) & \text{if } C_i^g 1 \\ f_i^g(A_i^{g3}) + f_i^0(A_i^0 - A_i^{g3}) & \text{if } C_i^g 2 \\ f_i^g(A_i^{g3}) & \text{if } C_i^g 3 \end{cases}$$

and net social benefit is

$$NSb_i^{Cg} = p[X_i^{g3} - X_i^0] - V_i A_i^{g3} - F_g \tag{6}$$

The generic GMV has a lower marginal productivity than GNV_r, but its variable cost is lower. Thus, comparing $C_i^m 1 - C_i^m 2$ with $C_i^g 1 - C_i^g 2$ suggests that more (less) acreage will be utilized with the GMV_i than the generic GMV if the marginal benefit gain $P\left[Mp_i^m(A_i^{m1}) - Mp_i^g(A_i^{m1})\right]$ is greater (smaller) than $Mc_i - (V_i - V_g)$. When the variable cost of GMT_i *is* very high due to an undeveloped crop-breeding sector (and the yield disadvantage of generic variety is not overwhelming), adoption rates of a generic GMV will be higher.

In these situations, it may be that, despite lower yield per acre, the actual output of the generic GMV will be greater than those of the GMV_r. In most situations, however, we do not expect the extra cost of GMV_i to be dominant and expect both adoption and output to be higher with a GMV_r. At this stage, there are no empirical examples of a generic GMV being introduced by the public sector. As indicated before, the introduction of most GMVs worldwide has been done by private monopolists, and outcomes with generic GMVs in such situations are discussed below.

Monopolistic Markets for Seeds of a GM "Generic" Variety

If a monopolistic firm that controls the GM technology is precluded from access to local varieties, or if the fixed cost of modifying local varieties is too high relative to expected profits, a generic GMV will be introduced to the area grown with variety *i*. In this case total acreage is A_i^4. and acreage of the generic GMV is A_i^{g4}. The inverse demand $W_i^g(A)$ and marginal revenue functions $MR_i^g(A)$ for this case are defined similarly to the ones in Eqs. *(2)* and (3), and only the indicator *g* replaces *m*. The possible outcome includes

- M_i^g1 : <u>No adoption</u> when $PMp_i^g(0) - PMp_i^0(0) > V_i$.
- M_i^g2 : <u>Partial adoption</u> occurs when $PMp_i^g(0) - PMp_i^0(0) > V_i > MR_i^g(A_i^0) - MR_i^0(A_i^0)$.

  At $A_i^{g4} < A_i^0, PMp_i^g(A_i^4) - PMp_i^g(A_i^4) = V_i$ and $A_i^0 - A_i^{g4}$ is the non-GMV area.
- M_i^g3 : <u>Full adoption</u> when $MR_i^g(A_i^0) - MC_i > V_i$. In this case adoption of A_i^{g4} is when $MR_i^g(A_i^{g4}) - MC_i - V_i = 0.$

With this notation, the output and the profits of the monopolist are given by

$$X_i^{g4} = \begin{cases} f_i^0(A_i^0) & \text{if } M_i^g 1 \\ f_i^g(A_i^{g4} + f_1^0) + (A_i^0 - A_i^{g4}) & \text{if } M_i^g 2 \\ f_i^g(A_i^{g4}) & \text{if } M_i^g 3 \end{cases} \tag{7}$$

$$X_i^{g4} = \begin{cases} 0 & \text{if } M_i^g 1 \\ \left[Mp_i^g(A_i^{g4}) - Mp_i^0(A_i^{g4}) - V_i\right]A_i^{g4} - F_i & \text{if } M_i^g 2 \\ \left[Mp_i^g(A_i^{g4}) - Mc_i - V_i^g\right]A_i^{g4} - F_i & \text{if } M_i^g 3 \end{cases}$$

Comparison of M_i^g1 to M_i^g3 with M_i^m1 to M_i^m3 suggests that, under a monopoly, adoption of GMV_i is greater (smaller) than adoption of the generic GMV if the gain in marginal revenues of the GMV_i $\left[PMR_i^m(A_i^{m2}) - MR_i^g(A_i^{mg})\right]$ is greater than the extra variable cost. Thus, there may be a situation when adoption of GMV_i will be less than that of the generic variety. There can be different reasons why rights cannot be freely traded.

Lack of IPR protection may prevent the innovator from trading because it is difficult to enforce compliance of licensing agreements. A similar situation might occur when IPRs are strong but high transaction costs hinder a smooth transfer. Due to diverging objective functions, licensing agreements can be difficult to negotiate for the private innovator, especially when germplasm is owned by the public sector.

The Choice of GMV or a Generic GMV

Thus far, we have analyzed the area to be planted with GMVs in the four scenarios considered. However, a more fundamental question is whether to introduce a generic GMV at a given location or to modify the local varieties. The decision rule for a monopolist is different from that of a public sector entity, which introduces seeds to be distributed by competitive seed companies.

We will solve the public sector case first and then consider the monopolist problem. Let ∂_j^m be an indicator variable that assumes the value 1 if option GMV_i is selected and 0 otherwise. The indicator variable ∂_j^g assumes the value 1 if option *GMV* is selected and 0 otherwise.

The public sector can either (a) modify variety *j*, (b) introduce a generic GMV, or (c) neither. The public sector aims to maximize net social welfare so its decision problem for region *j* will be

$$NSb = \max_{\delta_j^m \delta_j^g} NSb_j^m \delta_j^m + NSb_j^g \delta_j^g$$

subject to $0 \leq \delta_j^m + \delta_j^g \leq 1, \delta_j^m, \delta_j^g$.

The public sector will select GMV_i if net social benefits with this technology are positive and greater than Nsb_i^g. The generic GMV is chosen when it generates positive social surplus greater than Nsb_i^m. From conditions (1) and (6) the public sector decision whether to introduce GMV_r *or a* generic GMV hinges on three factors—revenue differential $P[X_i^{m1} - X_i^{g3}]$, variable cost differential $[V_i A_i^{m1} - V_g A_i^{g3}]$, and fixed cost differential, F_i. GMV_i is selected when its extra revenues are greater than the extra variable and fixed cost that its introduction entails.

When the production advantage of the local variety is not substantial or the breeding sector is not well developed (so F_i is high), the public sector will prefer to introduce a generic GMV to location *j*. In cases where the monopolist controls the introduction of GMVs, let δ_i^g and δ_i^m be defined similarly. The optimization problem becomes $\max_{\delta_i^m, \delta_i^g} \delta_i^m \pi_i^m + \delta_i^g \pi_i^g$

subject to the constraints $0 \leq \delta_i^m + \delta_i^g \leq 1$.

The monopolist will elect to introduce GMV_i if $\pi_i^m > \pi_i^g$ and $\pi_i^m > 0$. The generic variety will be introduced if $\pi_i^g > \pi_i^m$ and $\pi_i^g > 0$.

A comparison of Eqs. (5) and (7) suggests that in determining whether to genetically modify the local variety or to introduce a generic GMV, the monopolist compares the likely extra revenues of GMV_i with the extra fixed and variable cost its introduction entails.

The generic GMV is likely to be introduced

(a) in locations with small acreage, where volume of seed sales will not cover the extra cost of GMV_i

(b) in cases when the yield differences between the generic and local varieties are not substantial, and

(c) when the variable and fixed costs of modification are substantial. The high fixed cost may reflect cost of access to local varieties or undeveloped local breeding sectors that make it worthwhile to import a generic GMV.

Case of Inelastic Demand for Agricultural Commodities

The price of internationally traded agricultural commodities, such as corn and cotton, is determined according to international demand and supply. Neglecting transportation costs and quality differences, the assumption of price-taking behaviour is appropriate for most regions in the world. Yet, there are likely to be situations where commodities, especially food staples, are primarily produced for local consumption within a region.

This may occur in regions with high transportation cost to major markets or low degrees of integration to the global economy for other reasons. These regions are not likely to be served by monopolistic seed companies but, rather, by small companies or direct provision of seeds by public extension programs. When GMVs are introduced in these situations, then the output price effect may be significant. Suppose the demand for output at location j is $X_j = D_j(P_j)$, where P_j is output price in region j. The initial equilibrium with traditional variety j consists of acreage A_i^0, output X_i^0, and price P_i^0.

These values are determined solving simultaneously the equilibrium condition in the output market, $P^0 = D^{-1}(X_i^0)$ ($D^{-1}(X)$ in inverse demand) ; the equilibrium condition in the land market $P^0 Mp_i^0(A_i^0) = Mc_i$; and the production function $X_i^0 = f_i^0(A_i^0)$. Suppose GMVs are sold by small seed companies (or distributed by extension programs) at price V_r Let P_i^{m1} be output price under competition when GMV_i is introduced.

The equilibrium conditions in this case determine output price P_i^{m1}, total output X_i^{m1}, total acreage A_i^1 and acreage with GMV_r A_i^{m1}. A procedure that can yield the equilibrium value consists of using the formulas we developed to obtain A_i^{m1}, A_i^1, and X_i^{m1} for cases of fixed output prices for a range of plausible values of P_i^{m1}. P_i^{m1} will clear the output market, so that $P_i^{m1} = D^{-1}(X_i^{m1})$ is the equilibrium output price.

The equilibrium price P_i^{m1} also establishes acreage level $\hat{A}_i^0$ solved from $P_i^{m1} MP_i^0(\hat{A}_i^0) = Mc_i$, which is the amount of land that would have been utilized if the initial price was P_i^{m1}. Since the GMVs are assumed to increase yield, it can be shown that their adoption (even partially) will increase supply and, with negatively sloped demand, $P_i^{m1} < P_i^0$, which implies that $\hat{A}_i^1 = \hat{A}_i^0$.

Thus, in the case of partial adoption, when the total acreage $\hat{A}_i^1 = \hat{A}_i^0$. adoption of *GMV* will *reduce acreage*. There may be cases of reduced acreage even in cases of full adoption. The reduction in output price associated with the introduction of GMV_i will lead to increases in consumer surplus, denoted by ΔCS_i^{cm}

$$\Delta CS_i^{cm} = \int_{X_i^0}^{X_i^{m1}} D^{-1}(X)dx \tag{8}$$

The change in net social benefit becomes

$$NCb_i^{CM} = \Delta CS_i^{CM} + P_i^{m1} X_i^{m1} - P_i^0 X_i^0 - A_i^{m1} V_i - [A_i^1 - A_i^0] MC_i - F_i$$

While consumers will gain from the *introduction* of *GMV,* the impact on farmers is mixed. They produce more, yet receive a lower price. They may plant fewer acres but have to pay a technology fee. The case with inelastic demand suggests that the introduction of GMVs to a location by the public sector may raise social welfare, reduce farmed acreage (and thus may improve *environmental conditions*), and improve consumer well-being but not *necessarily* help farmers. The analysis of the impact of introducing a generic GMV is similar to that of *introducing* a modified local varicty to location *j*.

The generic variety, with its lower yield effect, will have less of an impact on output prices and may lead to a smaller reduction in acreage. Its *introduction* may benefit farmers more than the introduction of GMV_j and, thus, the introduction of a *GMV* will benefit consumers more.The *likelihood* of introducing *GMV* increases, the smaller the *fixed* costs to modify variety *j* and the variable cost to produce *GMV* seeds. The analysis of the public sector-led scenario is useful to provide some *intuition* about the private monopoly case, when demand for output is *negatively* sloped.

Private companies that introduce new GMVs globally are aware of the output price effect of the new innovation and its impact on their sales and revenue. We do not analyze the choices formally here but, rather, compare the outcomes when *technology* is provided by an idealized public sector that maximizes global welfare, versus when it is provided by a monopolist. Under the monopoly, we expect low rates of adoption and, thus, lower *aggregate* output, and a higher output price.

The profit of the monopolist is likely to be less than the aggregate social welfare that also includes consumer and farmer surplus. Thus, the monopolist will not introduce GMVs in some cases where they would have been introduced by the public sector. *Furthermore,* since generic *varieties* have a lower yield effect and lower fixed costs, the *monopolist* is more likely to introduce the generic varieties than modify local ones. The decision of whether or not to introduce a GMV at a location depends both on expected *productivity* gains as well as the fixed costs of introduction.

Reduction in the costs of introduction, due to improved efficiency of the breeding sector or a decrease in the regulatory costs, is likely to increase *introduction* of GMVs in general and modified local varieties in particular. In some cases, there may be substantial costs to the public sector to introduce the *technologies* because of domestic capacity *limitations.*

Then, the private sector may introduce a GMV, even though the private benefits are smaller than the public benefits. In other situations, low private benefit and high costs of introduction by the public sector may prevent the introduction of a GMV to a region. In these situations, if the public benefits of the GMV are sufficiently high, it may be *introduced eventually* as a result of policies that will enhance the capacity of the public sector and reduce its technology *introduction* costs.

IMPACT ON CROP BIODIVERSITY

The effect of the introduction of GMVs on crop biodiversity depends on the extent that traditional local varieties are replaced by a small number of GMVs. *Continued planting* of local varieties, even in GM form, can be a mechanism for preserving crop *biodiversity*. Our analysis has identified a wide array of circumstances in which local *varieties* may be preserved after the introduction of agricultural biotechnology.

In situations where the revenue gains from genetic modification of local varieties relative to a generic variety are substantial, and fixed and variable costs of modification and production are low, local varieties will be modified and biodiversity will be preserved. Even in situations where the *introduction* of agricultural biotechnology will lead to *replacement* of areas of local varieties with a generic GMV, adoption of the modified varieties need not be complete.

In particular, when a *monopolistic* private company controls the technology, it will charge a technology fee that may not warrant adoption of the technology on much of the land. Hence, a significant portion of the acreage will continue to be planted with traditional varieties. Full adoption of GMVs rarely occurs, and with partial adoption local varieties may be preserved.

DISCUSSION AND CONCLUSIONS

Biotechnology may preserve CGD more than conventional breeding. The reason is that biotechnology allows for separation between the act of *developing* novel crop traits and the process of breeding plant varieties. As a result, a given *biotechnology innovation* may be incorporated into a large number of plant varieties.

This chapter has shown some of the conditions under which this might happen. Modern biotechnology is still a fairly recent phenomenon, so that *empirical* evidence about the actual impact on biodiversity is limited. Table elsewhere in this chapter shows adoption levels and the number of GMVs available in different countries for selected innovations.

So far, widespread adoption occurred only for RR soybeans and Bt corn in the United States and Argentina, and Bt cotton in the United States and China. In all these cases, the *technology* has been *incorporated* into a large number of varieties, which supports our general hypothesis that biotechnology can preserve CGD. In other empirical cases, technology diffusion is still at an early stage so that conclusive statements are difficult to make.

Biosafety regulations for GM crops can play an important role in this respect. The cost of regulatory compliance has become a major component in the overall budget to develop new biotechnologies. In most countries, only the transformation event is regulated, so that the regulatory cost for each technology occurs only once, regardless of the number of varieties into which it is incorporated later on. However, in countries such as India, each GMV is *regulated* separately. Such *varietyspecific* approval *procedures* may foster loss of CGD and can be challenged on this basis.

Table 6.1: Number of available varieties fordifferent GM technologies in selected countries (2001/2002).

Country	Technology	Area under technology	Number of local varieties/ (ha.)	Number of imported hybrids varieties/hybrids
USA	RR soybean	22 million	> 1,100	0
	Bt corn	7 million	> 00	0
	Bt cotton	2 million	19	0
Argentina	RR soybean	10 million	45	11
	Bt corn	0.7 million	15	6
	Bt cotton	22,000	0	2
China	Bt cotton	1.5 million	22	5
India	Bt cotton	40,000	3	0
Mexico	Bt cotton	28,000	0	2
South Africa	Bt cotton	20,000	1	2

Including locally adjusted ones.Based on the conceptual analysis and empirical observations, we suggest a fourfold classification of situations according to the expected biodiversity outcome.

1. Strong IPRs, a strong breeding sector, and low transaction costs. Most situations within the private sector in developed countries, and some advanced developing countries such as Argentina, Brazil, and South Africa, belong in this category. The private technology owner will license the innovation to different seed companies that incorporate it into many or all local varieties, so that CGD is preserved. Adoption will be fairly widespread, and the innovator captures a rent through royalty payments. This outcome is equivalent to case II of the above analysis.
2. Strong IPRs and a strong breeding sector, but high transaction cost to trade rights. High transaction costs can occur, particularly when licensing contracts between a private technology owner and a public breeding *organization* have to be negotiated. Examples are biotech companies trying to reach agreements with international agricultural research centers, or public breeding stations that serve certain regional niche markets. If an agreement cannot be reached, the most likely outcome is that the biotech company will directly introduce GMVs that are not locally adapted. A widespread adoption of these varieties would lead to a loss of CGD, which is the outcome described under case IV of our analysis.
3. *Weak IPRs and a strong breeding sector.* Countries such as China and India belong into this category. Because IPRs are weak, every breeder or seed company can use commercialized GMVs, in order to cross-breed the technology into their own germplasm.

Thus, many different GMVs will be available on the market. Due to the competition, the innovator's ability to capture rents are limited, so that farmers and consumers are the main beneficiaries. This outcome almost *corresponds* to case I, the social optimum.

4. *Weak IPRs and a weak breeding sector.* This situation is typical for most of the least-developed countries in Africa, Asia, and Latin America. In none of these countries have GM crops been commercialized so far. If biotechnology developed by the private sector abroad should reach these countries, the most likely outcome is that foreign GMVs are directly introduced without adaptation. A widespread adoption of these varieties would lead to a loss of CGD, which is the outcome described in case IV of the analysis.

There are different policy implications for each category. In category (1), strong IPRs and smooth transactions ensure that biodiversity is preserved. In category (2), widespread realization of the benefits of biotechnology will likely require *international* efforts to create an effective mechanism that reduces the *transaction* costs of trading IPRs.

An intellectual property clearinghouse is one model that may be effective, if it is designed in a manner that addresses the needs of the poor as well as environmental concerns, while recognizing that much of the *technology* will be developed by profit-driven firms in the developed countries. In category (3), the outcome is socially optimal in the short run, but the situation might look differently from a dynamic perspective. Lack of IPRs deters international technology transfer and innovation in the private sector.

If biotechnology is entirely acquired and provided by the public sector, this might be less problematic. However, this may not be feasible or even advisable in more advanced developing countries. It may require *significant* amounts of funds and retard the evolution of private seed companies.

Our analysis suggests that, in such situations, introduction of IPR protection and enabling sale of rights will be desirable. The alternative will lead to takeovers and concentration in the seed industry, which would be associated with a loss of CGD and underutilization of the economic potential of biotechnology. For situations in category, the implications are somewhat different.

Realizing the biotechnology opportunities in least-developed countries will require that there is a minimum absorptive R&D capacity. Since a local private seed industry is hardly existent, funding for the incorporation of biotechnology into local varieties will have to come from *noncommercial* sources that recognize the total gain in social welfare rather than strictly private financial returns.

In practice, this means that the existing system of *agricultural experiment stations* may increasingly become centers of biotechnology adaptation and application. Also, the international agricultural research centers could play a bigger role in this respect. *Lumpsum* royalties to private innovators will have to be paid if such technologies are being used. In some cases such royalties might be waived for *humanitarian* purposes.

An *international* clearinghouse mechanism would also be very *beneficial* in these situations. In summary, biotechnology-based innovations in agriculture have the potential to preserve CGD, yet the actual impact will depend on the specific *institutional* conditions, R&D capacities, IPR policies, and biosafety regulation schemes in the individual countries.

PROMOTING BIODIVERSITY

An increase in awareness of the importance of the environment and the threats it is facing, as well as an appreciation for the value of ecological, economic, and social services it provides, has led to rising concerns about *biodiversity conservation*.

Biodiversity is an environmental good, as well as an indicator of the presence of other environmental goods, and thus its conservation has assumed great *importance* in the effort to improve *environmental* management and ecosystem health.

Since much of the most valuable and threatened biodiversity resources are located in developing countries, policies to promote conservation and sustainable use frequently have to be studied within the context of economic development. This chapter *discusses* some of the major issues related to biodiversity conservation and sustainable use.

We will especially focus on agricultural biodiversity, and specifically crop genetic diversity, picking up and *amplifying* themes raised in other chapters in this volume. Else anywhere else in this book we discuss various categories of *biodiversity* conservation and their implications for *conservation* priorities. Section 3 looks at the different objectives that conservation programs may have and the types and recipients of conservation values.

In section 4 we describe various types of program and policy *mechanisms* through which conservation may be obtained, with section 5 providing a discussion of the most likely and effective payment mechanisms associated the varying means of conservation. In section 6 we focus on issues of efficient targeting and *management* of conservation funds. We *conclude* the analysis in section seven.

THE VALUE OF BIODIVERSITY CONSERVATION AND PRIORITIES FOR CONSERVATION

The 1994 *Convention on Biological Diversity* states that biological diversity means variability among living organisms and includes diversity within species, between species, and of ecosystems. In this chapter we are especially interested in agricultural biodiversity, a vital subgroup of general biodiversity.

Agricultural *biodiversity*: "*encompasses* the variety and variability of animals, plants, and micro-organisms on earth that are important to food and agriculture which result from the interaction between the *environment*, genetic resources and the *management* systems and practices used by people.". In contrast to wild biodiversity, agricultural *biodiversity* contains a large human capital component, where genetic diversity depends on a *combination* of human and natural selection pressures.

Therefore, we adopt a broad interpretation of agricultural biodiversity *conservation* and sustainable use to include species and ecosystems, as well as the management practices which sustain them. Protection of the human capital required to identify and utilize genetic resources is as important as protecting the resources themselves in designing agricultural *biodiversity conservation* strategies, as opposed to the *conservation* of wild biodiversity, where human knowledge is a much less important component of conservation. The associated benefits of the natural and human resources comprising *agricultural biodiversity* is a natural means of *prioritizing* conservation programs.

These benefits may be divided into use and nonuse categories. The *assessment* of usefulness is from a human perspective, which has often been criticized, particularly in the context of wild biodiversity. With *agricultural* biodiversity, since the resource itself is the result of human selection applied in *conjunction* with natural selection with the *intention* of providing *something* useful to humans, assessing the value of the resource from the human perspective is quite appropriate. Agricultural *biodiversity* conservation yields several types of use benefits, manifested as both public and private goods.

Several studies have shown that higher levels of agricultural diversity provide important services to farmers in the form of *insurance* against production risks, the ability to spread labor *requirements* over a production season, adaptation to heterogeneous production conditions, and the possibility of producing for differing final consumption outlets, including market or *selfconsumption*.

Higher levels of biodiversity may generate reduced pest incidence, improved soil nutritional levels, crop *pollination*, and hydrological functions. All of these *characteristics* fall into the category of private goods-the farmer's *maintenance* of biodiversity impacts their own production and *consumption* outcomes. However, agricultural biodiversity also provides important services to local and global populations through the *maintenance* of the gene pool, which is the basis for the development of new crop varieties.

This capacity allows farmers and plant breeders to develop varieties to adapt to changing production and *consumption* conditions over time. Some of these use benefits are known and to some extent *quantifiable*, but much of the use benefits from *agricultural* biodiversity are in the form of option values, which have not yet been realized. Option values are associated with the possible future uses of biodiversity resources that may be captured with future knowledge and conditions.

The preservation of biodiversity also *generates* nonuse benefits. Some individuals have a strong bequest motive in their willingness to pay for preserving biodiversity, but this motive may also imply a *preference* for a future use in which case it is *essentially* the same as an option value of *biodiversity*.

In other cases individuals may hold an existence value for biodiversity that is derived from the knowledge that valued species and ecosystems exist. The use and nonuse values of *conservation* are expressed in various forms of *agricultural* biodiversity, which also have implications for targeting criteria under conservation programs. A simple categorization of these forms follows below:

Species and Varieties

Species (including crop varieties, animal breeds) provide both use and nonuse *biodiversity* values. Species can be grouped into those that are known and utilized, those that are known but not *utilized*, and those that are unknown.

Species that are utilized

By definition, these have a use value, although frequently this value is nonmonetary. This category includes species of plants and animals that are utilized in the production of food, fiber, oils, etc. It includes nonharvested species essential for *agricultural production* such as soil microbiota, pollinators, etc., as well as harvested species such as crops and livestock. Wild relatives of domesticated varieties may also fall into this category, as they are frequently an important source of value to rural populations.

These species and varieties provide the basis for biological production systems-e.g., the basis of food and *agricultural* production. They also constitute a storehouse of genes, which enables the development of technologies that allow for increased yields, *overcoming* disease, adjustment to adverse conditions, etc. The continued collection of species and varieties and the *documentation* of their properties have become even more valuable with the development of biotechnology, which allows for transgenic species and variety development. *Secondly, biotechnology* can identify desirable properties of *organisms* that may lead to innovations that will benefit a wide array of species.

Known but not utilized species

Many of the species that are *documented* or cataloged are not a source of economic benefit, and most are not likely to be commercially utilized. Besides their important intrinsic value, some species have the potential to be sources of *significant* economic benefit in the future, and others may have genetic *structures* that will be *beneficial*. Thus, species in this category may have *significant* option values.

However, since conservation is a costly activity, and the number of known species is substantial, not all species will be preserved, especially when the cost of preservation *significantly* outweighs the benefit. *Weitzman* (1998) proposed a *framework* for assessing these trade-offs, in which priorities for species conservation are derived from a formula that includes the distinctness of the species, the utility of the species in terms of value to humans, the degree to which the species' potential for survival is enhanced by *conservation* activities, and the costs *associated* with the conservation.

Unknown species

Most species are not known, and they may hold many surprises in years to come. The uncertainty regarding unknown species is such that their current market value cannot be estimated. However, from a social perspective, it is worthwhile to invest resources both in their preservation and in discovering and documenting their properties. One of the biggest challenges is how to target conservation activities to *bioresources* with the highest potential benefits given the degree of uncertainty about their value.

Diamond (1997) argues that only a minute fraction of all the species in the world have been domesticated and, while domesticated species are crucial to our civilization, their close relatives may have genetic content that provides protection against disease and which can improve the *performance* of agricultural crops.

Ecosystems

A functional understanding of biological systems or genetic properties cannot be obtained without understanding how species evolve and interact within ecosystems. Regev, Shalit, and Gutierrez (1983) have shown that the population *dynamics* and evolution of each individual species is dependent on the well-being of other species that are either consumed or preyed upon by that species. Science is relatively young, and until now much of the effort in the *biological* sciences has been directed towards *obtaining* an *understanding* of *microlevel* processes.

As we document the genetic structure of many species and gain a better idea of how organisms perform individually, understanding the interactions among species will become the main challenge of science and a key for achieving new technological developments. Therefore, the preservation of ecosystems is a targeting concept that should be distinguished from preserving individual species.

Knowledge and practices

Knowing that species exist and even documenting their genetic structure is not very valuable unless their function and benefits are known. *Indigenous people, farmers, scientists,* and others throughout the world have accumulated knowledge and systems to manage species and crop systems in a *beneficial* manner, and some of this knowledge is *disappearing* with modernization. Preservation of this knowledge is *sometimes* even more urgent than the preservation of species.

Adoption of modem technologies and practices may lead to the loss of knowledge of traditional technologies and practices. Features of these practices are very valuable and may provide clues to the future capacity to manage resources sustainably.

THE OBJECTIVES OF CONSERVATION AND SUSTAINABLE-USE ACTIVITIES

Programs to promote the conservation and sustainable use of agricultural biodiversity may be intended to meet one or more environmental or social objectives, focused on preserving one or more of its associated values and components.

The objective of conservation programs determines both the design of the activities and in the establishment of mechanisms for the financing of such efforts. We summarize some of the main types of program foci below.

The Promotion of Sustainable Production Systems and Support of Local Populations

Programs designed with this objective focus on *preserving* and *enhancing* the private benefits associated with agricultural biodiversity. The impetus here is on *maintaining* biodiversity, and the knowledge associated with it, for the purposes of enhancing farmers' capacity to respond to *varying* and complex production and market *conditions*, as well as *preserving* ecosystem functions which directly impact farm *productivity*.

Such programs are based on the notion that the conservation of *agricultural biodiversity* is the most effective means to enhance the *sustainable* production capacity of farm populations, *particularly* among low-income producers operating under marginal production conditions and facing frequent failures in both input and output markets.

It is argued that the degree of heterogeneity and risk present in such environments requires high levels of genetic diversity in crops and animals for successful and sustainable production systems. An added *advantage* of such programs is that they also may generate *significant* option values from the on-farm conservation of *agricultural* biodiversity, by *preserving* a dynamic system of interaction between natural and human selection factors.

Thus, the benefits of such programs would be realized not only by the farm *communities* involved in the *implementation*, but also the global *community* and future generations. The design of programs falling under this criterion may focus at the species level, such as participatory plant breeding or seed system enhancement programs directed at major *subsistence* crops, or at the ecosystem level, such as programs designed to enhance crop variety availability in highland or drought-prone *environments*. They also *frequently* involve a component of local knowledge preservation.

Farmers have developed production systems, including rotations and pest *management* strategies, that have enabled them to utilize biological resources effectively, and preserving the knowledge of these systems is part of agricultural biodiversity conservation. The *documentation* of landrace varieties, their *characteristics*, and use is one means by which local knowledge is conserved.

Maintaining the Option Value of Biodiversity Conservation

This is a primary objective of many conservation efforts. Under this objective, the focus is usually on *preserving* genetic diversity. As Weitzman (1998) argues, species may be perceived as carriers of genes; thus, preservation activities should emphasize *maintaining* the broadest base of genetic *combinations* possible. One way to evaluate a species or variety under this criterion is to evaluate their genetic uniqueness and relative distance from others.

From this perspective, species that have close substitutes may be less valuable than those that are genetically unique. However, from another perspective, the value of genetic material is derived from the products that they generate. In the case of agricultural biodiversity, preserving closely *substitutable* varieties may be valuable because some of the genes that distinguish varieties may have a unique value in controlling diseases or *improving* food quality.

Knowledge is also an important aspect of conservation programs focused on the option value of biodiversity. Human knowledge is necessary for the identification of useful *phenotypic* and genetic characteristics of species, as well as for their development into new and useful varieties and breeds. The knowledge required is both science-based knowledge on species *characterization* and breeding, as well as the knowledge of local communities on the identification and use of species under varying types of *environmental* interactions.

Modern crop varieties developed for monocultural agricultural systems rely on a subset of genetic material that is especially valuable under current technological conditions and under good production conditions. However, future *improvements* in cultivation practices may reduce the cost of adopting more diverse systems of production, and changes in climatic conditions and/or preferences may require *modification* of crops and cropping systems.

However, future improvements in cultivation practices may reduce the cost of multicropping, and changes in climatic conditions and/or preferences may require modification of crops and cropping systems. The capacity to modify production systems will depend on the availability of the genetic material, as well as knowledge regarding interaction among crops, nondomesticated species, and ecosystems.

Benefits derived through the preservation of knowledge for the development of future varieties and breeds will be *realized* through product *improvement* cost reduction to both *consumers* and producers.

Preserving the Existence Value of Biodiversity

Programs with this objective are more common for wild biodiversity rather than agricultural biodiversity conservation, due to higher ratio of existence to use values with the former as *compared* with the latter. However, it is likely to be important in agricultural *biodiversity conservation* as well, particularly in the *preservation* of *unknown* species that have the potential to be useful in future *applications*.

Programs, which involve the conservation of ecosystems and evolutionary processes, have the advantage of allowing for the conservation of unknown as well as known species. By protecting a diverse set of ecosystems and their functions, *presumably* a wide range of diversity of unknown species will also be maintained.

MECHANISMS FOR BIODIVERSITY CONSERVATION

The multiple objectives of resource conservation, and the uncertainty associated with their *performance* and outcome, has led to the *development* of a wide range of practices for promoting the conservation and sustainable use of agricultural biodiversity. They differ according to the degree of human intervention in the natural system, ranging from the highly managed *ex situ* gene and seed banks to *undisturbed wilderness* areas.

In this section we describe some major forms of conservation, while in the following section *mechanisms* for their financing are discussed. Any one system of *conservation* and sustainable use may adopt combinations of methods with other elements that will allow learning or produce other benefits associated with the conservation efforts.

Seed Banks and Gene Banks

As discussed in Chapter 8 in this book, gene banks are a relatively *inexpensive* means of conserving genetic resources with the potential to be an effective means of conserving option values *associated* with genetic resources of known species. Costs of *conservation* vary by crop and consist of a large fixed cost component, indicating a need for greater *coordination* and in some cases consolidation for more effective *management*.

An important *coordinating mechanism* for *ex situ* sites is the *International* Network of Ex *Situ* Collections managed under the auspices of FAO. This network involved 12 centers of the Consultative Group on International *Agricultural Research* (CGIAR), which placed most of their collections (some 500,000 accessions) into the International Network. The participants agreed to hold the designated germplasm "in trust for the benefit of the international community," and "not to claim ownership, or seek intellectual property rights, over the designated *germplasm* and related information."

Ex situ collections range in the degree to which they are accessible to local populations, from small community seed banks, highly dependent on frequent flows of seeds in and out of the community, to government and *international* collections tending to be more remote but with a much wider scope of coverage.

Experience from the field has indicated that seed banks need to be more closely aligned with farming communities, as well as integrated into ongoing research activities carried out by research institutions. A framework for cooperative relationships between public and private gene banks and breeders collections should be established, but the details of such a framework are subject to further research.

Botanical Gardens and Experimental Stations

Botanical gardens provide for the protection of genetic materials, allowing scientists to monitor progress and control inventory while at the same time enabling some *evolutionary* processes to occur. At present, there are *approximately* 1,500 botanical gardens worldwide and the vast majority maintain *ex situ* collection. We use the term "*experiment station*" for research units that have plots and collections of plants (or animals), which they preserve and experiment with.

In some cases, experiment stations are affiliated with botanical gardens, while in other cases they may collaborate with gene banks by displaying and experimenting with different types of species and varieties. Experiment stations can play a major role in *analyzing* the functions of genetic materials and in renewing and *expanding* the use of resources.

In situ Conservation Projects

Chapters elsewhere in this chapter have all discussed the central role farmers play in preserving crop genetic diversity through their selection and planting of crop varieties. In addition, farmers are often important agents of other forms of *agricultural* and wild biodiversity conservation. In recent years several programs have been established to provide incentives to farmers to maintain diverse *production* systems.

In a few cases this has involved direct payments to farmers for maintaining diverse crop varieties, one example being the *Global Environment Facility* funded project: *A Dynamic Farmer-Based Approach to the Conservation of African Plant Genetic Resources* implemented in Ethiopia from 1992 to 2000. Frequently such programs seek either to increase the availability and productivity of diversity to farmers, or to increase the returns to diverse production systems through the development of markets where some sort of premium would be paid for diversity.

Adding value through the development of markets for the products of local varieties is a means by which the returns to farmers of growing diverse varieties can be increased Programs and policies to increase diversity availability are discussed in point 6 below.*In situ* conservation programs may also be focused on *preventing* or slowing processes that lead to the loss of on-farm diversity, which in some situations is likely to be the most effective means of promoting *in situ* conservation.

However, a dilemma is raised when these same processes lead to economic development. The adoption of modern crop varieties and integration into markets have been identified as potentially important drivers of the loss of crop genetic diversity on farm; yet, this same process also yields tremendous benefits to the farming populations.

One proposed solution to achieving dynamic efficiency and equity for *in situ* programs is to enhance the private values of genetic diversity to farmers such as developing markets for diversityrelated traits, payments to farmers for *maintaining* diverse systems, or *enhancing* the *productivity* of local varieties. Equally important is reducing the costs of access. In the following section, we look more closely at programs that are intended to reduce the costs of diversity by increasing its availability at various points in the seed system.

Programs and Policies that Increase the Availability of Crop Genetic Diversity

Improving the performance and availability of genetic resources to farmers, *particularly* the poorest, is a critical means of achieving food security and reducing poverty. In this section we focus on policies which affect the *availability* of crop genetic diversity, defined as all the genetic materials of plant origin of actual or potential value for any particular crop. Since these resources are embodied in seeds and planting materials, factors affecting seed *distribution* become relevant as well.

At the farm level, the availability of such resources is driven by the type of material developed and released from formal sector plant-breeding systems, the *distribution* patterns of seeds embodying diverse genetic resources, as well as *interactions* in the informal system of seed exchange and use among farmers.

Increasing the Genetic Base of Crop Breeding

Most of the modern varieties and genetic populations with which plant breeders in the formal sector work consist of elite germplasm, which has been carefully built up over periods of perhaps 10-50 years. These lines can be destroyed by crosses with *unimproved* germplasm, providing a disincentive to breeders to introduce new materials, particularly when *improvements* can be made within the existing populations. In some cases the result is dependent on an increasingly narrow germplasm base for crop improvement.

One possibility for broadening the genetic base of formal sector breeding is public investment into "pre-breeding" or genetic enhancement activities, involving the *introduction* of new characteristics from crop wild relatives or development of specific selection where desirable inbreds can be obtained. Such lines could then be made available to plant breeders, resulting in an increase in the genetic base of modern variety development. However, at present *insufficient* resources are allocated to the *diversity-enhancement* research, and the overall trend in investments has been one of decline.

Better Incorporation of Local Materials and Knowledge into Formal Breeding Systems

Modern plant breeding methods have been highly beneficial to farmers operating in favourable environments, or those who can profitably modify their environments to suit new varieties. However, the results of such breeding programs do not result in superior *performance* under the unfavourable *environmental* conditions that most low-income farmers operate under.

They also result in the adoption of uniform plant varieties over large areas, and thus the erosion of genetic diversity in planted crop varieties. Limitations imposed by large genotype x environment (GxE) interactions are considered to be among the main factors *contributing* to the poor performance of modern varieties in marginal areas.

Decentralizing variety selection and testing to target environments is an important means of promoting the specific adaptation of crop varieties to *varying* production conditions. One aspect of such decentralization would involve the *international* agricultural research centers assigning more crop selection work to national programs.

Jana (1999) proposes "*biodiversity friendly*" breeding for gene-rich areas, including mass reservoirs and bulk populations, and Cleveland and Soleri (2002) have proposed a *reorientation* of the formal breeding process along the lines of GxE interaction, in order to utilize more CGR and provide outcomes more appropriate to farmers.

Greater participation of farmers in breeding programs is required, which also allows for better incorporation of local knowledge into breeding strategies. In addition, national programs then need to *decentralize* research further by *extending* into farmers' fields, particularly in unfavourable environments. This latter step also generally requires the *participation* of farmers in breeding programs, which also allows for better incorporation of local knowledge into breeding strategies.

Such decentralized breeding strategies result in greater *maintenance* of genetic diversity both within and among the varieties produced. A high degree of variation between selection *environments* and users results in a high degree of variation in the selection of breeding material in different selection sites and, *ultimately*, a high degree of crop genetic diversity made available to farmers.

Harmonization and Reform of Variety and Seed Regulations

Crop variety and seed regulatory frameworks are generally designed to promote the development and delivery of high quality and reliable commercial varieties. *Regulations* usually cover variety testing and release, as well as seed certification and quality control. A major objective of such regulations is the development and *distribution* of varieties that are distinct, uniform, and stable, as well as seeds that are viable and healthy.

At present, most regulatory frameworks are set up at a national level, with little regional integration. Consequently, the flow of varieties between countries in similar agroecological zones

Table 7.1: Selected benefits from crop genetic diversity, likely suppliers andconsumers at various points in the seed system, and implications forpossible incentive mechanisms to stimulate conservation .

Benefit of diversity	*Supplier*	*Consumer*	*Payment mechanism*	*Note*
Increase in agricultural productivity and sustain-ability of productions systems; particularly in marginal environments	Farm communities	Farmers, particularly in marginal environments	Seed prices in cash transactions	Diversity is a source of new varieties through both in formal and formal systems of breedinl In both cases breeding generates value-added to the germplasm which farmers may pay through seed prices of exchange value.
	International and national research systems		Facilitation of seed and information flows	Facilitation of seed and information exchanges may include seed banks, participatory plant breeding and community seed registers, regulator; measures which facilitate seed flow
Input to breeding new seed varieties for comer-cialization	Farmers with *in situ* collections	Public/private plant breeders; Agribusiness and pharma-ceutical firms	Access fees In-kind technology transfers	Requires information on probability of benefits to set costs cst uarima paid by final consumers of agricultural or pharmaceutical products. Values may be quite low due to high substitutability among genetic resources.

(Table Contd.)

Benefit of diversity	Supplier	Consumer	Payment mechanism	Note
	Seed banks-*ex situ* collections	Public sector breeders	User fee	Requires infor mation on probability of benefits set fees; no fees set at present. Values may be quite low due to high substtutabfty among genetic resources.
	Seed banks-*ex situ* collections	Private sector breeders: commercial seed com-panies; agri-business and pharmaceutical firms	Royalties	Currently these resources are free; for a viable system need to set up a system to track use of *ex situ* materials in commercial seed production
Availability of differentiated products to final consumers	Farmers	Consumers of agricultural products	Price premium for diverse varieties/crops	Need market development for niche markets and consumer education.
Option value maintaining genetic resources for possible future use and for existence value	Mostly developing countries	International organizations on behalf of global public Developed countries	Support to *ex situ* gene banks and *in situ* conservation programs transfers	Conservation funds to support *ex situ* and *in situ* conservation; technology

is limited by the need for *time-consuming* trials and evaluations in each country. The integration of regulatory systems based on *environmental* conditions rather than political boundaries offers the potential for substantial technological spillovers, which are sorely needed by the resource scarce national agricultural research systems of most developing countries.

Such *regionalization* of varieties offers the possibility of developing a wider range of varieties suited to specific environments (e.g., better capturing the GE effects), resulting in better *performing* varieties under farmers' *conditions,* as well as a broader genetic base among released varieties.

Seed certification regulations are also made at a national level to ensure the stability and *uniformity* of a given variety. The content of such regulations varies by country but, in general, results in limits on the number and nature of varieties that can be multiplied. Such regulations that are strictly enforced can also block the adoption of alternative breeding and seed production strategies, such as the *multiplication* of landrace varieties or *participatory* plant-breeding programs.

Greater *flexibility* in seed certification to allow for *nonuniform* varieties, or exemption of certain types of seed production from certification *requirements* may allow seed systems to better meet farmers' needs, by increasing the levels of diversity and, thus, choices made available to them.

Emergency Seed Provision

In disaster-prone areas of the developing world, the provision of seeds has become an *increasingly* frequent response to emergency situations. Emergency seed supplies are thus an increasingly important source of *germplasm* to low-income and vulnerable farm populations.

The most common form of emergency seed aid is the direct seed provision program, which involves the importation of certified seeds to the country *experiencing* the disaster. Under these programs, the seeds provided tend to be limited to a narrow range of crops and varieties.

More recently, there has been a move towards local procurement of emergency seed supplies, including the use of local seed markets and merchants under voucher programs. There has been little evaluation of the impact of emergency seed *distribution* programs on local seed systems and crop genetic diversity, although concerns about negative impacts have been raised, particularly for areas that have *experienced* repeated inflows of emergency seed supplies.

Programs which build on greater reliance of local materials and seed systems and utilize both formal and informal parts of the seed system are likely to reduce the possibility of eroding local genetic variability, and are also more likely to *complement* long-run development efforts in local seed systems.

Complementary Resource Control

Biodiversity conservation may be threatened by a lack of complementary or supporting resources, in particular, water and *environmental* quality. Thus, an essential component in the design of conservation systems is the provision of sufficient complementary resources to attain the desired conservation goal. For example, assuring a *continuous* supply of good quality water or providing access to water reserves in periods of drought can be a critical form of *agricultural* biodiversity conservation.

Similarly, policy mechanisms (zoning, taxation, direct control) or incentives (purchasing of development rights) may be needed to divert or prevent pollution damages in areas designated for conservation and sustainable use.

Knowledge Banks

Knowledge about the functioning of ecological, agricultural, and biological systems are also major objectives of biodiversity conservation programs, and conservation systems need to be designed with mechanisms to obtain, preserve, and distribute such knowledge. As mentioned above, some knowledge conservation programs are integrated with other forms of *conservation*, such as with community seed bank programs.

In other cases knowledge banks are being set up to allow for wider distribution networks. The

Center for Indigenous Knowledge for Agriculture and Rural Development (CIKARD) at Iowa State University is one such example. CIKARD focuses its activities on *preserving* and using the local knowledge of farmers and other rural people around the globe.

The goal is to collect indigenous knowledge and make it available to development *professionals* and scientists. Nineteen other centers for the preservation and *documentation* of indigenous knowledge have been set up at regional and national levels.

FINANCING BIODIVERSITY CONSERVATION AND SUSTAINABLE USE

The expansion of biodiversity and, in particular, crop genetic diversity *conservation* efforts, requires well-designed mechanisms for financial support. One feature of many conservation programs is that they require farmers and land-users in developing countries to forego certain *production* activities of benefit to them, in order to generate benefits to individuals outside their region.

The beneficiaries in many cases are corporations, *environmental* groups, and citizens of developed nations, who generally come from much higher income groups, and are more likely to be willing to pay for conservation benefits, *particularly* the option and existence value aspects. *Conservation* programs may have a negative impact on equity, without proper attention to the distribution of the costs and benefits of programs and the appropriate design of financing mechanisms.

In many cases conservation efforts require the establishment of financial schemes where the gainers from conservation pay those who bear the costs. A second important issue to consider in the design of financing *mechanisms* is that many of the benefits of biodiversity *conservation* have the properties of a public good.

For example, genetic or biological knowledge can be utilized *simultaneously* by many and, until recently, there were few barriers to access to some aspects or manifestations of this knowledge. Without some kind of intervention, public goods will be underprovided, as no incentives exist to provide a good where no profits can be captured. In the past, the solution was to mobilize the public sector to generate such knowledge, through publicly funded research and development programs.

More recently, *technological* and institutional changes have resulted in the ability to assign property rights to biological and genetic knowledge in the form of intellectual property rights. This has created more incentives for the private sector to generate such forms of knowledge, as they stand to reap significant benefits. However, concerns have been raised about the impacts of assigning *intellectual* property rights on the accessibility of knowledge, particularly as an input to the development of new varieties and breeds. Several mechanisms for *overcoming* these types of barriers are being designed or set up, and are discussed in other chapters of this book.

Another concern about the *privatization* of biological and genetic knowledge is the impact on agricultural research and development programs and new variety development aimed at poor populations. Such groups do not represent lucrative markets, and thus their needs will not be targeted under private research programs. As discussed in other chapter of this book, this implies a greater need for the public sector to focus on such issues. Even when biodiversity conservation results in outcomes that *exclusively* benefit a specific and identifiable agent, the magnitude and timing of these benefits may be uncertain.

In many of these cases, there may be a significant lag between conservation efforts and the realization of benefits. For example, the decision not to cultivate a land parcel may preserve species that only years later will become essential for the development of a desired and valuable medical product. When outcomes of conservation activities are highly uncertain, it may be easier to raise funds for their support, if at least part of the payment is dependent on the actual outcomes.

For example, a payment scheme for providing a company access for a reserve for *bioprospecting* may include both a fixed fee as well as a royalty tied to actual benefits derived.

In Table elsewhere in this chapter below, we present a *categorization* of selected benefits from crop genetic diversity *conservation*, the likely suppliers and consumers of such benefits, and the implications for payment mechanisms.

The analysis is quite general, with only four broad categories of benefits included. These include genetic diversity as a base for improving agricultural *productivity* and sustainable production systems, particularly in marginal areas, genetic diversity as a source of inputs to *commercialized* breeding systems, genetic diversity as the basis for producing differentiated consumer products, and genetic diversity in the provision of options and existence values.

Although the first two categories of benefits both involve genetic diversity as an input to current breeding systems, we have *differentiated* them because of likely differences in the ability (and willingness) of the consuming population to meet the costs of research and development associated with breeding.

In the first category, we focus on farmers as the consumers of the benefits from diversity, while in the second we look at another point in the seed system where the consumers are breeders and commercial seed enterprises that use diverse genetic resources in developing products, although eventually these products would be sold to farmers as well.

This analysis indicates a wide range of potential payment mechanisms between supplier and consumers of genetic resources, and some indication of their implications for stimulating conservation.

THE OPERATION OF CONSERVATION FUNDS

As the analysis in the previous section indicates, conservation funds are likely to be an important source of finance, particularly where option and existence values of *conservation* are the key objective. Even where royalties or access fees are assessed, they may be placed into some type of conservation fund, as is the case with royalties from commercialized products under the International Treaty for Plant Genetic Resources.

Such funds may be managed by countries, private businesses, nonprofit agencies, and *international* agencies. In this section we look at some of the key issues which arise in managing such funds.

Forms of Biodiversity Conservation Payments

Here we will focus primarily on various forms of *in situ* conservation and distinguish between the outright purchase of resources versus periodical leasing or payments for *environmental* services. In general, outright purchases are appropriate for *stimulating* efforts that require a long-term investment or commitment, while leasing is more appropriate for measures that require a continuous incentive to maintain.

Of course, economic, social, and political conditions will also be key determinants of the most appropriate form of conservation purchases. Outright purchases are more effective if the new buyer has the ability to enforce rules and control of the resources that they purchase. Several examples of this can be seen in wild *biodiversity conservation*, where *environmental* groups purchase the rights to a primary forest from a government.

If these groups lack the means to control intrusions into it, and enforce the desired management regime, the program may be ineffective. If, instead they leased the forest for a certain time period, then the local government would have greater incentive to insist on proper *conservation* because of the future earnings at stake. In the context of *agricultural* biodiversity *in situ* conservation, outright purchases are not likely to be a widely used option.

Conservation groups could purchase farmlands and cultivate diverse varieties; however, since *in situ* conservation involves conserving the interaction between human and natural pressures on genetic populations, it would be difficult to conserve the human side of the equation in what is essentially an artificial socioeconomic *environment*. Leasing or periodical purchases of environmental services is more appropriate when a purchaser is interested in behavioural *modifications* of a given environment which require frequent (say, annual) activities on the part of the seller, or which are easily reversible.

For example, if the objective is to have farmers maintain a diverse set of crop varieties in an evolutionary setting, a one-time fixed payment to farmers is not likely to be sufficient to ensure continuing participation of the farmers.

Establishing a system where producers are paid according to their actual activities as they occur over time is likely to lead to better follow-up and a more effective result. In the case of *in situ* conservation of crop genetic diversity however, this type of payment system is difficult to implement, due to difficulties in *establishing* the value of *maintaining* any one variety in production. In addition, monitoring costs associated with such programs can be quite high.

In many cases it may be more effective to fund *complementary* activities that support the preservation of crop varieties in the field, such as niche market development, participatory breeding programs, and so on.

Targeting-Based Quantitative Analysis

A primary challenge of conservation funding is how to target purchases to maximize the impact of a given budget. Some of the principles for analysis and data collection required to answer this question have been addressed in the emerging economic literature on the *management* of bioresource purchasing funds, which has been used to analyze the Conservation Reserve Program (CRP) and water quality programs in the United States.

The basic premise of this approach is that an agency has a certain amount of money that it must to use to purchase, rent, or modify the use of environmental resources. These resources can be land or water rights. The question is how to target the resources. To solve this question, one needs to take into account the *quantification* of *environmental* benefits and the costs associated with changes in behaviour.

Quantification of the Environmental Benefits Associated with Modification of Behaviour at Various Locations

This information can be represented by indices of *environmental* quality. In the case of the CRP in the United States, indices of environmental quality improvement included such items as the reduction in soil erosion, increases in quality and diversity of native plants, increases in populations of migrating birds, etc. OECD[3] has also developed indicators of environmental quality based upon the pressure-state-response (PSR) mode!. Work has already begun at FAO on developing indicators of agricultural biodiversity, as part of the *Global Plan of Action for Conserving Plant Genetic Resources for Food and Agriculture*.

The types of indicators being developed include measures of crop genetic erosion and vulnerability, number and kind of threatened species relevant to food and agricultural production, number and kind of wild relatives of species relevant to food and agriculture under *conservation* programs, areas under *in situ* conservation, and degree of genetic integrity of ex *situ* accessions.

In light of the discussion in this paper, it may be useful to distinguish between areas which yield high biodiversity conservation benefits in the form of option values versus those which yield high values in terms of improvements of current production systems, as the payment mechanisms and costs associated with conservation in the two areas will be quite different.

A key question is the extent to which the value of genetic diversity in improving cropping systems coincides spatially with the value of genetic diversity as an option value for future development.

Quantification of Costs Associated with Inducing the Desirable Changes in Various Locations

Moving towards changes in behaviour requires payment of some kind. As discussed in other chapter of this book, the most significant cost to farmers in providing *in situ* conservation services is foregoing agricultural productivity gains that may be obtainable with modern variety adoption. Areas with the highest *conservation* costs, therefore, may be expected to be those where adoption is possible, but has not yet occurred.

Areas with the lowest costs of conservation are those where no option of modern variety adoption exists. These tend to be marginal production areas and cropping systems for which no modem varieties have been developed, or where modern varieties do not perform as well as local varieties. As noted in our analysis above, these also tend to be the areas with the highest potential benefits from genetic diversity as an input to the development of sustainable production systems.

In many cases, we may end up with a situation where the private incentives to preserve diversity coincide with the public values. If the most effective way of increasing the productivity and profitability of farmers operating under such conditions is to enhance the availability and performance of genetic diversity, then farmers and the breeding systems serving them will have strong incentives to conserve diversity. With the development of information technologies, the costs of *establishing* databases on the costs and benefits of conservation funds is declining over time.

Studies on the costs and distribution of *environmental* benefits of resource *conservation* efforts suggest that there is significant heterogeneity in the distribution of benefits. For example, 10% to 15% of the land base considered for preservation of native plants in the United States provided up to 90% of the potential benefits.

Similarly, the costs of *purchasing* resources vary greatly and, again, a relatively small percentage of the resources may possess most of the economic value and may absorb most of the costs. Efficient conservation fund *management* will target funds to locations that provide the highest rate of conservation per dollar spent.

Thus, locations that have the highest ratio of per acre benefits to per acre costs would be selected. Sometimes, for convenience or political/economic reasons, fund *managers* may target the cheapest resources. This support will maximize acreage that may be enrolled in a land-based conservation program with a given budget, but the effectiveness of this strategy depends on the correlation between environmental benefits and the cost of land.

If land provides a high level of biodiversity conservation and there is a high positive correlation between cost of purchasing/leasing/payment per land area and environmental quality, this strategy may be inefficient because some of the included land provides very little additional conservation value.

On the other hand, if there is a strong negative correlation between the cost of bringing the land into a conservation program and its associated biodiversity level, then acreage maximization and targeting of the lowest cost lands may also maximize *environmental benefits* purchased with a given budget. Another targeted approach aims to conserve locations that provide higher conservation benefits per acre, regardless of cost.

This approach may be suboptimal if *environmental* benefits are *negatively* correlated with economic costs, but it may result in the most efficient outcome when there is a strong *correlation* between *environmental* benefit and cost.

For example, if land potentially highly productive under modern varieties provides high *in situ* conservation benefits, such that the relative advantage in conservation provision is higher than in production (e.g., the value of the conservation benefits are greater than the opportunity costs of foregone production), then funding farmers to preserve cropping and variety patterns which generate *in situ* conservation will be optimal.

Pitfalls in Managing Conservation Funds

Wu, Zilberman, and Babcock (2001) argue that in some cases conservation funds may affect the prices of food and other commodities sufficiently so that resources previously not used for production will start being utilized-or "leakage" occurs. Thus, we may have a paradoxical situation where farmers are paid to reduce *utilization* or intensity of use on certain lands, creating pressures to bring other lands into intensive production.

Under this type of scenario, increased levels of agricultural biodiversity *conservation* could lead to reduced levels of wild biodiversity conservation. The designer of a *conservation* fund has to *recognize* this possibility and also provide incentives against extension of production into wilderness areas.

Recent empirical research has indicated the *complexity* of putting such incentives into place, and the need for measures on both macro- and microlevels. The design of agricultural biodiversity conservation activities has to recognize and address potential impacts on food availability.

Particularly in remote areas in developing countries, which are largely self-sufficient, any reduction of *agricultural* production due to conservation activities may have a negative affect on food consumption, at least for some part of the population.

Thus, *mechanisms* may be needed to increase the productivity or value of the land that stays in production and to enable increased conservation without affecting the food security and economic well-being of the local population. However, two strategies which have been adopted for addressing this *concern-agricultural intensification* and integrated conservation and development projects (ICDPs)-have proven to have major problems in achieving the intended goals of both increasing food security and *biodiversity* conservation.

Experience with these programs has shown the critical necessity of assessing the driving forces of land-use management decisions by local populations and their potential responses to changes in technology, institutions, and policies. Wu, *Zilberman*, and Babcock (2001) also argue that conservation funds may be the dominant resource buyer in the region, and *minimizing* their cost in acquiring resources could result in *monopsonistic* pricing strategies.

In such cases, resource prices will be lower than if there was competition among buyers' resources for environmental and conservation purposes, and the net affect is that the owner of the resources, who may be small farmers, may be *compromised*. This indicates the necessity for careful assessment of the potential impacts of purchasing funds, especially in regions where such funds play a dominant role in the local economy. While market power *considerations* suggest that it is preferable for *resource-purchasing conservation* programs to be restricted by size, there may be biological considerations which would require a minimum size of land parcels in order to take advantage of increasing returns to scale in the generation of *environmental* amenities

As Wu and Boggess (1999) have shown, when the scale of conservation projects is sufficiently small, then an increase in the marginal productivity of conservation is associated with an increase in size. Only when size is beyond a certain threshold will marginal benefits from expansion of the project decline. That suggests a lower bound on scale of *conservation* projects and indicates that small-scale conservation funds may be most effective by *specializing* in a small number of *sufficiently* large projects, rather than spread resources among a large number of small projects.

CONCLUSIONS

In this chapter we have argued that agricultural biodiversity conservation generates several types of benefits, which are realized by different groups in society over time, and this is an important basis for prioritizing, designing, and financing *conservation* programs. We have noted that agricultural biodiversity conservation has potentially high use values to farm populations in highly heterogeneous and marginal production areas in terms of *generating* increased *productivity* and sustainable production systems, and these areas will also likely be *significant* providers of option and existence values from *in situ* conservation.

An important means of achieving efficient and equitable agricultural biodiversity conservation is identification of areas where there are high potential *productivity* gains to be made from *increasing* and enhancing the diversity available to farmers, as well as those which are likely to provide the highest option values of *conservation* and targeting these for priority under conservation funding.

We have also discussed the effectiveness of various types of payment mechanisms for conservation, depending on the supplier and consumer of the good, as well as its nature. We emphasize the wide range of actors who are and *potentially* could become involved in conservation through the use of a wide range of *mechanisms* that go well beyond the traditional concepts of *conservation* activities.

A key theme throughout our discussion has been the importance of *recognizing* human knowledge as a key component of *agricultural* biodiversity and, thus, the necessity of *incorporating* means for knowledge preservation as much as the physical conservation of agricultural *biodiversity*.

8

Chapter

MANAGING GENETIC DIVERSITY

In this book we have brought together a unique set of *analyses* on managing agricultural biodiversity and biotechnology in the context of development. One of the key features of the book is using the common thread of plant genetic resource management as a point of departure in *analyzing* the potential for, and barriers to, jointly managing *agrobiodiversity* and *biotechnology* to achieve a range of development-related goals. These include increasing agricultural productivity and *sustainability*, reducing poverty, and *improving* the *conservation* of genetic diversity.

Policies governing the management of both genetic diversity and biotechnologies have the potential to affect the ability of countries to achieve any of these objectives, depending on prevailing socioeconomic and environmental circumstances. One of the strengths of the approach taken in this volume is that it allows for an *assessment* of where there are overlaps, synergies, and *contradictions* in policy approaches to managing biodiversity and *bio-technology*.

Just as importantly, the approach helps to identify where there is not likely to be any interaction between biodiversity and biotechnology *management*, and if not, the types of policy intervention needed in each separate arena to achieve desired outcomes. Increasing the *productivity* of *agricultural production* systems through the development and *dissemination* of improved genetic resources is a primary means of accelerating economic growth and addressing the problems of food insecurity and poverty in developing countries. Biotechnologies provide one important vehicle to achieve this improvement, albeit with several caveats.

Both institutional and technological obstacles need to be overcome to realize its potential. Since molecular biotechnology is a major innovation, the full ramifications of its impacts are still unknown, and safeguards for preventing *undesirable consequences* are *necessary*, but difficult to formulate in the presence of uncertainty. Advances in biotechnology have also generated radical *institutional* changes in plant breeding with a major shift of funds and control from the public to private sector.

Harnessing the benefits of biotechnology for developing *countries* and poor farmers thus requires new *institutions* and new ways of managing existing *institutions*. There are other means of improving genetic resource *productivity* in agriculture besides *biotechnology*, which may be more effective, *particularly* in marginal production areas. These include *approaches* such as production ecology, participatory plant breeding, or reducing the costs of accessing a diverse set of genetic resources by increasing the supply of diversity.

However, these strategies face their own set of constraints, including cost *effectiveness*. Biotechnology-based approaches to improving genetic resource productivity are not mutually exclusive with *alternative* approaches: In fact, *enhancing* the *productivity* of conventional approaches is likely to be one of the most valuable contributions of the technology to development. Regardless of the approach taken to improve the access to and performance of genetic resources, greater focus on *improvements* in the development and delivery of genetic resources that meet the specific *production* and *consumption* constraints of the poor is necessary in order to achieve poverty-reduction goals.

The conservation and sustainable *utilization* of *agricultural* biodiversity is another important policy objective in developing countries, which is separate but linked to that of increasing the productivity of genetic resources for food and agriculture. Agricultural, biodiversity is a broader concept than genetic diversity, *encompassing* human knowledge and *ecosystem* functions as well as the genetic variability of plant, animal and *microorganisms*.

Its conservation generates benefits that are realized globally as well as nationally and locally. As signatories to the International *Treaty on Plant Genetic Resources* and the Convention on Biological Diversity, many developing countries have assumed *obligations* to promote the conservation and sustainable utilization of agricultural biodiversity. The relevant decisions that face developing country policymakers are how to optimize the benefits from agricultural biodiversity *management*, including the potential for increasing productivity and sustainability of agricultural production, as well as receiving *compensation* for providing public *environmental* goods and services.

Much of the world's valuable plant genetic diversity is located in developing countries and the conservation of this resource generates both private and public goods. To the extent that *maintaining* genetic diversity results in private benefits to the farmers who provide it through their planting decisions, incentives to conserve exist, although *oftentimes* rapidly eroding under processes of social and economic change.

Maintenance of the public good aspects of diversity conservation (e.g., reduced vulnerability to pest and diseases, and options for future genetic inputs to plant-breeding efforts) requires some type of policy intervention. One way to generate a socially desirable level of conservation is by setting up *mechanisms* to allow for flows of payments from the *beneficiaries* to the providers.

However, two major problems arise with such mechanisms. First, there are concerns that establishing property rights and rights to compensation for diversity conservation will lead to a reduction in the free exchange in genetic materials among crop breeders and farmers that has prevailed thus far, thus reducing their capacity to generate new varieties, and ultimately reducing farmer access to genetic resources in developing countries. Second, even if payments are desirable, payment *mechanisms* are difficult to design due to difficulties in valuing the benefits associated with conservation.

This debate over compensation and benefit sharing is part of a bigger *discussion* about the ownership of genetic materials and the benefits that farmers, breeders, and other groups obtain from conserving agrobiodiversity, which have been the focus of the *International* Treaty on Plant Genetic *Resources* and for Food and Agriculture whose implementation *mechanisms* have yet to be designed.

The challenges outlined in the paragraphs above are being faced by developing country *policymakers* in a rapidly changing and high-stakes environment, where current policy choices may have significant consequences for current and future *generations.* This book has been designed to provide insight into the key problems of managing agrobiodiversity and biotechnology efficiently and *equitably* in the context of economic development.

It is structured in an *incremental* fashion, first looking at the key forces and factors which are shaping the overall "rules of the game" under which biotechnology and biodiversity can be managed, and the impact of these on the ability to achieve efficient and *equitable management* regimes. Next, specific considerations of *biodiversity* conservation, biotechnology development and *dissemination*, and sharing benefits from genetic resource *management* are addressed.

The last part includes a series of policy-oriented chapters drawing upon the analyses in earlier sections. In the first part of the book, the chapters describe a series of major changes in global economic and *environmental* settings. On the environmental side, there is a decline in the global natural capital asset base of plant genetic diversity, together with a rising *appreciation* of their value and *institutions* designed to promote them. On the economic side, increased integration of global agricultural markets gives rise to changes in the structure of production and marketing, resulting in the *expansion* of markets and the economic *opportunities* associated with them.

However these same changes may also result in increased barriers to market participation, particularly among small-scale and low-income producers. Three key lessons can be *summarized* from the chapters in the first part. The first is that markets are *increasingly* important as *mechanisms* for *transmitting* incentives for production and consumption decisions, as they are expanding in terms of participants on both the supply and demand side for a wide range of agricultural input and output products as well as for *environmental* goods and services.

Secondly, demand and supply are increasingly determined at supra-national levels-e.g., consumers and suppliers beyond national borders have increasing impact on market signals at a national and subnational level.

One example is the rise of environmental concerns in developed countries leading to increased willingness to pay for *agricultural* products grown under specific *environmental* conditions, e.g., organic, no genetically modified organisms (GMOs), etc., affecting the production decisions of farmers in developing countries who supply *international* markets. Another is the potential impact of the privatization and *commercialization* of genetic resources in developed countries on the cost of accessing these resources in developing countries.

The third, and perhaps most key point, is that we cannot rely on market forces alone to *generate* socially desirable levels of poverty alleviation, agricultural *biodiversity* conservation and biotechnology development. In some cases this is because markets are non-existent for the socially desirable goods and services, as is the case with agricultural biodiversity conservation.

The value and sources of diversity are difficult to identify and quantify and, thus, so is the establishment of marketbased payment *mechanisms*. In other cases the problem arises from distortions and poorly *functioning* markets. Examples here include the inability of poor farmers to express their demand for improved genetic resources in commercial seed markets due to their limited *purchasing* power and poorly functioning seed, credit, and other markets.

Increased reliance on a market-based research and *development* system in this era of increased privatization would bypass the needs of the poor farmers in technology development. Another example is concentration and vertical integration in *agricultural* input and output markets that lead to noncompetitive and inefficient markets. Finally, markets are a means of achieving efficient, but not necessarily equitable, allocations of resources and thus interventions may be required to achieve socially desirable levels of equity.

The tension between the increasing importance of markets as a means of improving genetic resource *management* for both conservation and *development* on the one hand, contrasted with the increasing recognition of a need for policy interventions to either improve, supplement, or *substitute* for markets on the other hand, is a theme that recurs throughout the analyses presented in this volume.

TOWARDS EFFICIENT AND EQUITABLE STRATEGIES FOR CONSERVING AGRICULTURAL BIODIVERSITY

Agricultural biodiversity is a major and valuable form of natural and human capital, comprised of several *components*, including plant genetic diversity, which has been the focus in this book. The gains from the conservation and *enhancement* of *agricultural* biodiversity spread far beyond the location such activities take place. Agricultural biodiversity has strong public good properties, but its benefits are *uncertain* and vary across locations and over time.

Thus, market forces by themselves will lead to a socially undesirable rate of loss, and global collective action and cooperation are required to efficiently manage agricultural *biodiversity*. Several means of attaining conservation have been *identified* in this volume, associated with varying costs and benefits.

The socially desirable levels of activities in conservation and enhancement of crop biodiversities may be most efficiently achieved through *compensation* in exchange for these activities. Ideally, *conservation* funds should be allocated across the portfolio of potential activities, in order to maximize the expected benefits of *agricultural biodiversity conservation*, and *coordination* between the various forms of conservation promoted in order to enhance cost effectiveness. In reality this is difficult to achieve, due to several issues raised in the chapters of this volume.

These include the following: the valuation of conservation benefits, the *identification* of criteria for *establishing* efficient conservation programs, the design of mechanisms to provide incentives to developing countries for conservation and developing means of incorporating diversity conservation into overall agricultural and economic development concerns and strategies. A key question which arises in this design of effective policies for *conservation*, is just how exactly should diversity be defined-what is it that we are trying to conserve? The answer is complex, depending on the type of value focused upon, as well as *assumptions* about how best to generate or maintain it. Chapter elsewhere in this book discusses the *controversies* over defining genetic narrowing in crop genetic diversity, noting several relevant dimensions, including spatial vs. temporal diversity, *variation* within vs. among varieties, and variation within landrace vs. modern varieties.

Chapter elsewhere in this book states that the biological diversity of crops *encompasses phenotypic* as well as genotypic variations, resulting in differences in the perception of crop genetic diversity between farmers and plant breeders. The author also notes the importance of conserving rare alleles in centers of crop origin, which requires a different type of *conservation* strategy than one targeted at *maintaining* high levels of varietal *heterogeneity*.

Chapter elsewhere in this book describes several forms of agricultural diversity, including species and varieties, ecosystems and human knowledge, all of which, the authors argue, are important to consider in conservation programs. Clearly *agricultural biodiversity* conservation generates several types of goods and services and *conservation* programs will vary *depending* on which are of key concern.

However, there is considerable uncertainty about the most effective means of *generating* goods and services from conservation, as well as uncertainty about the relative values of these services. Thus one of the biggest problems in designing effective *conservation* programs is defining what should be conserved and where. Several chapters in the book provide insight into where and how the conservation of *agricultural biodiversity* in general, and plant genetic diversity *specifically*, can be most effective.

A variety of conservation methods exist, ranging from *ex situ* gene collections to *in situ* farm-based diversity *management*. However the high degree of *uncertainty* associated with both the private and public values of diversity, as well as a lack of information about the actual and *opportunity* costs involved, means that conservation efforts are often not efficient. The private benefits associated with plant genetic diversity conservation are realized by farmers whose maintenance of diverse cropping *systems* can be thought of as the outcome of a constrained utility *maximization* problem.

These values are described and analyzed in some detail elsewhere in this book. In other chapter of this book *summarizes* the results of several studies where risk *management*, responsiveness to highly *heterogeneous* production conditions, labor management, and preferred *consumption* characteristics have all been found to be important determinants of on-farm diversity.

These private values of diversity are determined by agroecology, population density, and the level of commercial market development. Chapter elsewhere in this chapter adds another important *determinant* of the private values of crop genetic *diversity*: the seed system, which affects the availability and *accessibility* of genetic resources and information at the farm level. The incidence of natural disaster and political strife that can disrupt supply systems can also be important determinants the private value of *maintaining* crop genetic diversity.

The market failure in diversity conservation arises from the fact that conservation generates several types of public goods. One is in the form of reduced vulnerability to pests and disease incidence, which occurs mostly as a local public good, but also with potentially wider benefits.

Much of the use benefits associated with diversity conservation have not yet been realized and, as such, remain as potential. In such cases, the benefits of biodiversity conservation are primarily in the form of an option value as is discussed in other chapters. Chapter elsewhere in this book presents one approach to the *measurement* of this value, using an *empirical* example from teak breeding.

This chapter *concludes* that the value of increasing the number of potential parents for *breeding* is actually quite low at the margin, measured in terms of changes in the consumer and producer surplus. This chapter raises the important question of how much effort and cost should be made in *maintaining* crop genetic diversity as a source of input to future breeding efforts.

The question is still open to considerable debate, and is likely to vary *considerably* among crops. Rausser and Small (2000) argue that the option values of *agricultural* genetic resources are likely to be sufficiently high to support market-based bioprospecting activities, since researchers have prior knowledge about where the most promising leads are likely to be found.

Chapter elsewhere in this book discusses various criteria for assessing option values, and discuss the disincentives to plant breeders in broadening the genetic base of their breeding lines. Public sector interventions to promote genetically diverse "pre-breeding" activities could lead to higher option values for crop genetic resources. Moving towards *consideration* of criteria for designing conservation programs, one approach identified is minimizing associated costs.

Chapter elsewhere in this book examines this issue in detail in the context of *ex situ conservation*, which is the term applied to all conservation methods in which the species or varieties are taken out of their traditional ecosystems and are kept in an environment managed by humans. An estimated 6.2 million accessions of 80 different crops are stored in 1,320 gene banks and related facilities in 131 countries at local, national, and *international* levels.

Chapter elsewhere in this book also discusses the inefficient *management* of these facilities, finding significant differences in the degree of national commitment and expenditures on PGR conservation, which are not necessarily tied to the level and value of domestic genetic diversity. The chapter concludes that better collaborative relationships are the primary vehicle for reducing costs and improving the *management* of *ex situ* sites at a regional level, between public and private entities, and within the multilateral system. The primary costs *associated* with *in situ* conservation are opportunity costs, which are addressed in other chapters.

Chapter elsewhere in this chapter provides a conceptual framework for assessing public/private tradeoffs in maintaining *in situ conservation*, *differentiating* between situations where the private and public values of diversity maintenance coincide, versus come into conflict. Ostensibly, situations where they coincide require no intervention to maintain desired levels of conservation.

This implies that the least-cost means of *in situ* conservation is to focus on areas where private values of diversity are high and, thus, opportunity costs of conservation are low. However, in a dynamic setting, problems arise as high private values of diversity conservation are often negatively associated with processes of economic development, *particularly* increasing integration of farmers into markets.

Assessing the future opportunity costs farmers may face in *maintaining* diversity given efforts to promote economic development thus becomes a critical issue. Reducing the future opportunity costs farmers may face in maintaining on-farm diversity, and thus providing incentives for its maintenance, can be achieved by either addressing the change in conditioning factors that reduce the value of diversity or through *compensation* programs. A key strategy for reducing future opportunity costs of conservation is to increase the supply of diversity and reduce access costs.

This strategy includes increasing the supply of a diverse range of improved crop varieties (that is, more diversity in modern, i.e., *genetically uniform, varieties*) as well as *enhancements* to existing varieties and populations that encompass a high range of diversity (e.g., enhance the performance from varieties that encompass genetically diverse populations, landraces, and seed lots). Increasing diversity supply is an issue which is addressed throughout the book, with various pathways identified.

The chapters in Part II analyze the potential for changes to traditional and conventional breeding systems that may lead to higher levels of diversity supply. Participatory plant breeding, broadening the genetic base of *conventional* breeding *programs*, and the *establishment* of community seed banks and registers are all examples of programs that fit here. A major problem identified with these programs is their cost effectiveness.

The inability of such programs to cover costs does not mean they are undesirable. However, some level of public support will be required to achieve the desired objective of *increasing* the supply of genetic diversity and thus increase the provision of both private and public goods *associated* with conservation. Chapter elsewhere in this book argues that plant genetic diversity conservation requires *consideration* of the costs and benefits of all available options.

The chapter analyzes a variety of mechanisms which may be *appropriate* for promoting efficient conservation in both *in situ* and *ex situ* situations, ranging from direct approaches such as payments to farmers for growing diverse crop varieties and royalty payments on genetic resource inputs to *commercialized* products, to more indirect methods such as provision of access to biotechnologies and other forms of technology and *institutional* support.

Agricultural research and development and plant-breeding *management*, seed regulation, input and output market *development*, information transmission, and seed provision under disaster conditions all have implications for the costs and values of *in situ* conservation, and these have not been well researched to date.

LINKAGES BETWEEN BIOTECHNOLOGY AND PLANT GENETIC DIVERSITY

Advances in biotechnology will have a *significant* impact on both the demand for, and supply of, plant genetic diversity *conservation*, and several chapters in the book address this issue. In this discussion clear definitions are critical: Within both biodiversity and biotechnology, there is a range of meanings, and the relationship between the two depends on which specific aspect is being considered. *Improving* information about the nature, source, and value of biodiversity is one critical function *biotechnologies* offer to the *improvement* of genetic diversity conservation.

As raised in several points throughout the book, lack of *information* is a serious problem hampering effective *agricultural* biodiversity conservation efforts. In other chapter, Virchow suggests that the value of genetic *collections* is reduced by the uncertainty regarding the properties and impacts of genetic materials stored in specific seed varieties. The existing and emerging tools of *biotechnology* will expand the capacity to utilize the information stored within *in situ* and *ex situ* collections, allowing analyses of the genetic content and potential of stored seeds.

Emerging techniques of molecular and cell biology, and in particular the tools of *computational* genomics, allow for rigorous *classification* and *documentation* of genetic materials and, thus, a reduction in the cost of accessing and utilizing genetic materials stored in various collections.

This improves the ability of researchers to identify promising genetic materials for incorporation into breeding products, which is likely to increase their marginal value and demand for *conservation*. The *production* and dissemination of GMOs are another aspect of *biotechnology* development likely to have significant impacts on agricultural *biodiversity* in general, and plant genetic diversity specifically. The introduction of GMOs may affect the number as well as genetic content of new varieties available for adoption in developing countries.

Adoption patterns will affect both spatial and temporal patterns of diversity through two processes: The replacement of one type of *germplasm* for another, and the integration of new genetic materials into existing gene pools through gene flows. The first is a *human-driven* process, dependent on the supply of and demand for GMOs.

The second is governed by the natural process of gene flow and integration. Ultimately, the impact on genetic diversity depends on (1) a series of forces which drive supply and demand patterns, (2) the baseline situation with regard to crop genetic diversity, (3) vulnerability of the crop to geneflow (reproductive characteristics, presence of weedy relatives), and (4) the way in which diversity is defined. Chapter elsewhere in this book looks at factors that *determine* the supply of GMOs in *developing* countries.

They argue that the capacity to adapt biotechnologies to local materials is a critical determinant of the potential benefits of GMOs in *agricultural* development as well as impacts on crop genetic diversity. The strength of intellectual property rights (IPRs) over plant genetic resources and their *enforcement* within a country, together with the level of *competence* in the plant-breeding sector and the level of transactions costs associated with accessing *biotechnologies*, are identified as the key *determinants* of the numbers and genetic content of GMO varieties likely to be supplied in developing countries.

Countries with strong IPRs, advanced breeding capacity, and relatively low *transactions* costs are most likely to develop a wider range of GMOs for any one crop, as the marginal costs of adding a *transgenic* trait to an increasing number of varieties of a sexually propagated species is smaller than the marginal benefits. In addition, the degree of local materials incorporated and thus conserved into GMO varieties is likely to be higher under these conditions.

The authors argue that GMO *development* under these conditions can lead to an increase in crop genetic diversity, as incentives exist to modify local *materials* with improved traits and generate several varieties, resulting in a higher number of improved varieties, with a higher content of local materials preserved.

The impacts on diversity also depend on what the GMO varieties are replacing; the implications are quite *different* if they are replacing a few conventionally bred modern varieties versus landrace populations. The introduction of GMO varieties may also affect diversity through gene flows from transgenics to other planted varieties. Managing undesired gene flows is an important aspect of biosafety regulation, but the degree to which gene flows pose a risk to *biodiversity* conservation and the degree to which regulations will be effective in *managing* such risks are still unknown. Apart from the technology and products of biotechnology per se, several authors raised concerns about the impacts of the institutional changes *accompanying* biotechnology on diversity conservation.

Chapter elsewhere in this book argues that biotechnology-induced changes in IPR regimes increase the privatization of knowledge and could increase the costs of accessing breeding materials. Therefore, stringent IPR regimes may well reduce the capacity of breeders in developing countries and the CGIAR centers to access new materials and technologies.

They also note that the absence of transparent and *well-functioning* biosafety *regulations* are likely to restrict access, as suppliers of the technology may be unwilling to enter such markets. Public sector access to genetic materials is a critical concern since it is this sector that will be focused on crops of most importance to the poor, which in many cases are not commercially attractive. IPRs have also been associated with increases in the number of new varieties developed.

In other chapter of this book argues that IPRs were a crucial stimulant to the development of private sector research and development in canola, leading to an explosion in the number of new varieties developed. However, Graff and Zilberman describe the current situation with IPRs in *agricultural* biotechnology as an *anticommons climate,* restricting both public and private sector access to technologies and thus development of new varieties.

In other chapter of this book discusses the implications of changing IPRs under impetus from the TRIPS agreement of the World Trade *Organization* on agricultural *biodiversity*, finding the potential for both positive and negative impacts.

These chapters indicate that the numbers, genetic content, and accessibility of improved varieties are *changing* in response to *institutional* changes associated with biotechnology; however, assessing the impacts on plant genetic diversity conservation is again a function of how diversity is defined. Overall, the analyses in this book indicate that agricultural *biodiversity* and *biotechnology* are co-evolving, with a number of different points of intersection.

The adoption of transgenic products may harm or enhance crop biodiversity. The new tools of biotechnology improve our capacity to interpret and utilize *agricultural* biodiversity. *Improvements* in the conservation of plant genetic diversity are likely to increase the productivity and value of *agricultural biotechnology*.

The analyses in this book suggest that recognition of the interdependency between biotechnology and biodiversity is critical to the achievement of sound policy design for the *management* of agricultural biotechnology and *biodiversity* in the context of economic development.

EQUITY ISSUES IN THE MANAGMENT OF PLANT GENETIC RESOURCES

Sharing the benefits (and costs) of plant genetic diversity conservation and *maintaining* access to genetic resources and biotechnologies for lowincome groups are critical concerns addressed throughout this volume. Equity (and efficiency) criteria would suggest that since much of the natural capital embodied in *agricultural* biodiversity is in developing countries, companies and nations in the North, which are potential beneficiaries of this conservation, should contribute to crop biodiversity conservation funds. Developed countries tend to be in regions whose original genetic endowment in the major *agricultural* crops was lower than in biodiversity hotspot areas.

As Tables elsewhere in this book demonstrate, primary centers of agricultural genetic diversity are mostly in developing countries. Thus, private breeders largely from developed countries develop and market varieties that rely on genetic materials that originated at some point (perhaps many generations ago) from the developing world.

Many less-developed countries (LDCs) or associated interest groups claim that these breeders are benefiting from utilization of their native landraces without *compensating* the farmers responsible for their *maintenance*. Furthermore, they assert that developed countries are benefiting more from the *utilization* of plant genetic resources for food and agriculture (PGRFA) from developing countries than do the LDCs themselves and that these LDCs are not being compensated in return for using these resources.

The issue has become particularly acute with the development of biotechnology and privatization of *agricultural* research and development. This perspective leads to active demand for compensation of farmers and others in LDCs for past conservation efforts. However, at least from the economics standpoint, there is some difference between biodiversity funds that aim to compensate for past conservation and funds that aim to encourage future conservation.

Paying LDC farmers for past conservation efforts is largely an equity issue given that insufficient data are available to establish compensation payments based on the economic value of conservation efforts in the past and, as such, one must appeal to equity, even though it is a weak mechanism for allocating funds. Paying for current and future conservation activities can have more potential to be made using notions of economic efficiency (i.e., making *conservation payments* such that the marginal benefit of conservation effort equals its marginal cost).

Chapter elsewhere in this book, for example, *demonstrates* a proxy measure for *economic* value that can at least be used as a rough mechanism for distributing conservation funds to world regions with an eye on *increasing* the economic benefits to society of conservation efforts. The analyses suggest that on efficiency as well as equity grounds, direct beneficiaries from *agricultural biodiversity conservation* would be made to reward the *providers* of the benefits, based both on actual and expected gains.

However, there are also significant benefits to *maintaining* a free flow of genetic resources among breeders and other researchers, and this is a difficult issue to address in the design of compensation and incentive *mechanisms*. On the one hand, improved property rights over genetic resources and their embodied values would facilitate the *establishment* of *exchange* and *compensation mechanisms*.

However, at the same time, economic efficiency and equity criteria suggest that the continued sharing of the benefits associated with these goods be promoted. While this book suggests some possibilities for *cost-sharing mechanisms*, their exact design still needs further research. Chapter elsewhere in this book describes in detail how issues of equity and benefit sharing have been incorporated into the design of the *International Treaty* on *Plant Genetic Resources* for *Food* and *Agriculture*.

The chapter discusses the economic, technical, and legal reasons for the establishment of a multilateral system to facilitate access to and sharing of benefits from the *utilization* of plant genetic resources. The chapter also discusses the role of various forms of *property* rights, including intellectual property rights and farmers' rights and how the two systems can *complement* each other to ensure that incentives to innovate are *maintained*, while at the same time ensuring the capacity of rural *communities* to benefit from their conservation of plant genetic resources.

BIOTECHNOLOGY: MAXIMIZING THE BENEFITS AND MINIMIZING THE COSTS

One of the major points made about biotechnology in this volume is that it is much more than just a tool for genetically modifying crop varieties. Aside from the crop sector, livestock, fisheries, and forestry *biotechnology* products of relevance to the poor are currently under development. As noted above, one of the key benefits of biotechnology is through increasing information on genomics and, thus, values of *biodiversity*, which are necessary for developing effective *conservation* and compensation strategies and programs.

The main focus of the potential benefits of biotechnology in economic development has been on the increased potential to generate breeding materials that are specifically relevant to the production and *consumption* conditions in developing countries, and in a much more targeted fashion and shorter time frame than is possible with *conventional breeding* methods. Several chapters in the book describe the experience that has already been seen with biotechnology adoption in developing country agriculture. Note the dominance of tissue culture technologies in developing countries, and their importance in generating disease-free plants.

Other chapters focus on the experience with GMOs in both developed and developing countries. Transgenics are in the early stages of their development, yet GMOs that control pests have high adoption rates for major crops in *Latin America* and *China. Nevertheless*, the adoption of transgenics in the majority of developing countries has been minimal, and no GMOs have been *introduced* for several major staples consumed by the poor (rice, wheat, cassava) in developing countries.

At this point it is not possible to draw firm conclusions about the potential impacts of the adoption of transgenics in developing countries. However, the current evidence provides valuable insights, including:

- Adoption patterns and impacts of GMOs vary over different economic and agronomic circumstances. Chapter 14 cites evidence on how differences in pest incidence, land quality, and credit availability affect adoption rates. The availability, *effectiveness*, and prior use of pesticides determine the extent to which GMOs reduce chemical use and affect output levels, and GMOs may increase agricultural production where other approaches have not been effective in controlling pest damage at lower *environmental* costs.
- By reducing the variability of crop yields, GMOs can serve as an insurance strategy allowing the farmer to cope with the randomness of pest infestation within and between seasons. The benefits of GMOs consist both of their average yield effect and yield riskreducing effects.

- The yield gains from the adoption of GMOs are likely to be smaller if the modified varieties are generic, as opposed to those based on local *materials* adapted to the local conditions. Chapter elsewhere in this book suggests that using GMOs not adapted to local conditions is likely to introduce new sources of yield risks.
- Whether transgenics increase yields or reduce pest-control costs, they tend to increase the overall supply of the crops. Chapter elsewhere in this book *suggests* that this may lead to reduction in the prices of the modified commodity, which will benefit consumers including urban population, the rural landless poor, and net-consuming farm households. However, lower prices may harm the nonadopting farmers.
- Transgenics are a highly divisible technology with low fixed costs and low management requirements-e.g., they have limited requirements for human capital inputs. These *characteristics* make GMOs accessible and attractive to small- and low-income producers. Nonetheless, the traditional constraints to technology adoption among the poor-such as lack of credit, poorly developed input and output markets, and the presence of risk-are likely to impede adoption among smallholders.
- The adoption of GMOs may generate *environmental* and human health benefits through the reduction of pesticide use, and yield effects may lead to reduction of land conversion to agricultural use and thus reduce deforestation and land degradation. These benefits have to be weighed against the risks that may be introduced with GMOs, such as irreversible changes in genetic populations through geneflow.

The substantial rates of adoption of agricultural biotechnologies in some developed and developing countries and their realized net benefits suggest that these *technologies* are likely to play a significant role in global *agriculture* as they evolve. Anywhere else in this book *highlight* the applications of *agricultural biotechnology* currently available and in the development pipeline, which could be highly beneficial to lowincome farmers in particular and to developing countries in general.

However, the degree to which these potential benefits of biotechnology are realized by poor farmers and developing countries is likely to be determined more at a macro than at a microlevel. The benefits of biotechnology to farmers and the poor will depend on the degree to which biotechnology innovations address *production* and *consumption* constraints, and are affordable and accessible to farmers. Farmer access to biotechnology is determined by the type, amount, and cost of technologies produced by plant breeders-either nationally or internationally.

As argued in other chapter of this book, these factors are, in turn, driven by the combination of intellectual property (IP) regimes, local breeding capacity, commercial seed industry development, and biosafety regulation regimes. Transactions costs (*affected* by IPRs and *biosafety regulations*) associated with obtaining breeding materials will determine the degree to which private sector materials would be available to local breeders, while local breeding capacity will determine the costs of adapting them to local conditions, and the development of the commercial seed sector drives the degree to which such innovations could be disseminated to farmers.

High transactions costs in obtaining breeding materials and local breeding capacity are the two most critical *determinants* of potential beneficial effects of biotechnology in developing countries, which can be addressed by institutional reforms at both the national and international levels.

Chapter elsewhere in this book gives one example of such an institutional reform, arguing that the transaction costs can be *significantly* reduced by *establishing clearinghouses* for IP, which will provide crop breeders with *information* on the status of IP over various crops and *technologies* and assist them to obtain access to it.

The recently established *Public Intellectual Property Resources* for *Agriculture* (PIPRA) is one example of a clearinghouse that aims to reduce the transaction cost constraints of agricultural biotechnology. Chapters elsewhere in this book also argue that reducing registration requirements for agricultural biotechnology will reduce transaction costs and lead to a more diversified portfolio of modified varieties.

A clear example of a policy that reduces transactions costs is the requirement of registration and safety testing only for new biotechnology events (such as development of a parent GMV, which through back crossing can lead to insertion of the modification from the parent into all the varieties of the crop) rather than for every modified variety.

Chapters in this book argue that the CGIAR centers have an important role to play in filling the gap created by a lack of local breeding capacity in many developing countries, as well as greater integration of NARS research work over agroecological regions. In many developing countries, *agricultural* biotechnology may not be the least cost or most efficient means of improving agricultural productivity.

The national breeding capacity, type of farming systems present, and constraints to increases in agricultural productivity are key *determinants* of the degree to which developing countries will benefit from agricultural biotechnology. The future of *agricultural* biotechnology and its impacts on economic development will be affected by the management of the human health and *environmental* risks associated with it.

Continuous research and monitoring of the potential risks involved are clearly critically important. The efficiency of the regulation of biotechnology applications would increase significantly with greater quantification and definition of the risks associated with these applications. However, the high degree of uncertainty and the lack of information on the risks prevent precise estimation.

Building this uncertainty into biosafety regulatory structures would provide more meaningful information than that associated with simply providing mean measures of risks. One way to overcome the lack of information at the initial stage may be to quantify the potential risks under plausible pessimistic scenarios, and assess their costs relative to the expected economic and environmental benefits of the technology.

It is important to recognize that, beyond a certain stage, estimates of outcomes and the technology itself will not improve significantly without field experience, which implies that the efficiency of assessment and regulation of technologies will be increased if they can incorporate adaptive learning and through taking advantage of findings in the laboratory as well as outcomes in the field.

Poorly designed biosafety regulations that lead to excessive delay in the introduction of biotechnologies may generate significant economic costs in terms of foregone opportunities for technological development, including learning by doing, and improvements in agricultural productivity. An important reference point for the development of biosafety regulations in the context of agricultural biotechnology is the Draft Code of Conduct on Biotechnology as it relates to Genetic Resources for Food and Agriculture.

The objective of the code is to *maximize* the positive effects and minimize the possible negative effects, of biotechnology. The draft Code is based on the results of two major surveys of stakeholders in 1993 and 2001, which identified the key issues of concern.

Issues currently being considered as possible components of the Code include access to and transfer of biotechnology, capacity-building, biosafety and *environmental* concerns, public awareness, development of appropriate biotechnologies for poor farmers and developing countries, ethical questions *regarding* new *biotechnologies*, genetic use restriction technologies ("terminator" technology), GMOs, gene flow and the question of liability, *voluntary* certification schemes, and possible FAO universal declarations on plant and animal genomes. A clear message that emerges from the analyses in this volume is that appropriately designed policies and institutions are essential for enabling agricultural biotechnology to fulfill its promise for developing countries.

One policy implication arising from the analyses presented is the potential benefits to be reaped from strengthening of the capacity of developing country agricultural research and development and seed sectors to introduce desired traits into local varieties, rather than relying upon imports of generic *transgenic* varieties.

A second policy implication that emerges is the need for regulations to manage the risks associated with the new technology as well as the importance of including cost considerations-particularly the costs of foregoing opportunities to improve productivity-when *designing* such regulations.

Thirdly, barriers to access the intellectual property needed for the development of transgenic crops for developing countries should be reduced through institutional arrangements for technology transfer and sharing of knowledge about IPR and technology management. A clear message that emerges from the analyses in this volume is that designing *appropriate* policies and *institutions* is essential for enabling agricultural biotechnology to fulfill its promise for developing countries.

First, benefits to developing countries will be greater if the capacity of the seed sector in these countries is enhanced to allow introduction of desired traits into local varieties than with simply *importing* generic transgenic varieties.

Second, economic efficiency suggests that the level of regulation of new varieties to allow control against risks has to be balanced against cost *considerations-particularly* the costs of foregoing opportunities to improve productivity-when *designing* such regulations.

Third, the most efficient way to reduce barriers to access the IP needed for the development of transgenic crops for developing countries would be likely be through institutional arrangements for technology transfer and sharing of knowledge about IPR and *technology management.*

DIRECTIONS FOR FUTURE RESEARCH

Several areas where new research is needed on issues related to managing plant genetic diversity and agricultural biotechnology for economic development have been identified throughout this volume. With regards to biotechnology, it is *important* to continue to assess the economic impacts of adoption of various types of *agricultural* biotechnologies as they evolve. To best assess these impacts, we need quantitative understanding on how the features of various technologies, the economic and environmental conditions in various *locations*, the institutional setup in general, and the policies associated with the new technologies affect their impacts in terms of pricing and welfare of various groups.

This research will allow identification of the countries and situations where investments in agricultural *biotechnology* are likely to generate significant returns in terms of agricultural productivity increases and poverty alleviation, relative to other potential strategies. The potential environmental side effects of *agricultural* biotechnology are a *continuous* source of controversy that will affect the future of this technology.

Identifying and assessing the risks *associated* with biotechnology adoption relative to potential benefits is a major priority and, more *importantly*, designing institutions for monitoring the environmental impacts of agricultural biotechnology and effectively regulating to control potential risks is a major policy challenge. It has also emerged as an area where more research is urgently needed. We also need to identify features of biotechnology products that are especially desirable from the perspective of the developing world and identify mechanisms that will help developing *countries* gain access to them, especially if they will not be pursued as part of the agenda of the private sector.

For example, it is important to understand to what extent can biotechnology enhance the micronutrient content of food consumed in developing countries and to what extent the biotechnology innovations that serve this purpose will be pursued by the private sector and, if they are not pursued privately, whether and how to provide the incentives for their introduction. We need a better understanding of the role and effects of regulatory regimes, including *environmental*, IPR, and market structure *regulations* on the evolution and adoption of new biotechnology products and their impact on the environment.

As new institutions for the *management* and regulation of biotechnology are introduced, we need research that assesses their performance and suggests design modification and reform. Specifically, more work on policy and *institutional* reforms necessary to facilitate the potential benefits of biotechnology to the poor is necessary-particularly in reducing the transactions costs associated with access under increasingly restrictive property rights for genetic materials and associated technologies. On the topic of genetic diversity conservation, first we need to have a better handle on the *contribution* and value of various forms of genetic resources and the costs associated with their loss.

One could use emerging information technologies to collect data on use of various genetic collections and analyze it statistically. It is crucial to understand how improved capabilities affect the usage and productivity of biodiversity in order to better their storage and distribution, an understanding which requires *interdisciplinary* research cooperation. Determining how to optimize the value of both *in situ* and *ex situ* conservation to developing countries requires better information on what these values are, as well as the costs associated with obtaining them, considered in the dynamic context of economic development. Some of this valuation work must be inferred indirectly from greater understanding of the improved economic value derived by bioresources.

Valuation work on plant genetic diversity has focused at the farm level in looking at household decision-making over a portfolio of crops and varieties. More work is needed on the local and global public good values of diversity in terms of reducing vulnerabilities to pests and diseases. In addition, further work on the value of maintaining diversity as an input to agricultural breeding programs is needed, following up and expanding on the work of Simpson and others.

Combining research on valuation and costs could be a highly useful guide to developing countries on targeting strategies for conservation. Together with an assessment of the most efficient conservation opportunities, there is a need for, analysis of the most effective and equitable mechanisms for providing incentives for conservation. Markets, due to their increasing importance as a mechanism for the allocation of resources, need to be analyzed in terms of their role in providing incentives and *disincentives* for conservation.

Here markets are taken in the widest sense, ranging from local commodity *exchanges* up to global markets for environmental goods. The efficiency of markets in allocating plant genetic resources and the implications this has for diversity at the farm and local level are areas where more research is needed. The efficiency and optimal design of market-based *mechanisms* for *maximizing* global public good values associated with diversity conservation are other areas where gaps in the economic literature exist.

However, market forces are not the only drivers of interest in assessing conservation incentives: The impact of nonmarket forces, *particularly* government *regulations* in the agricultural and seed sectors, is also a critical area for further research.

Regulations of interest range from biosafety, to seed certification and release *procedures*, to agricultural pricing interventions. Finally, an important area for further research is the equity implications of alternative *management* schemes for plant genetic diversity conservation and agricultural biotechnology.

Designing mechanisms to compensate farm communities for their past services in conserving and providing genetic resources to the formal breeding sector, which do not create new barriers to exchanges and thus reduces access, is a challenging area where more work is needed.

Designing incentives for *in situ* conservation, which address not only current but also future opportunity costs associated with conservation in the presence of economic development, is another *important* equity issue where the analysis in the book indicates the need for more economic research.

Finally, further analyses of the distribution of benefits and costs to agricultural biotechnology *investment* and adoption and the impact, *particularly* on low *productivity* agricultural *populations* relative to other means of productivity increases, is a highly important area of research both from an equity and efficiency standpoint.

9

Chapter

MECHANISM OF CONSERVATION

The seed is the structure in which a usually fully developed plant embryo is dispersed and that enables it to survive the period between seed maturation and establishment of the next *generation* as a seedling after it has germinated. The dry, *quiescent* seed is well equipped to sustain extended periods of unfavourable conditions.

Dormancy, defined as the failure of an intact viable seed to complete germination under conditions favourable for germination, is an adaptive trait optimizing *germination* to the best suitable time that enables the species to complete its life cycle.

The environmental conditions required for germination are not defined specifically, but in practice refer to conditions that allow *germination* of a nondormant seed batch of the species under *investigation*.

Dormancy varies in a quantitative way often described by deep and non-deep or strong and weak dormancy. Dormancy is the property of the seed, so the degree of dormancy defines which conditions should be met to make the seeds germinate.

Therefore, a more precise method of defining the dormancy status of a seed batch is to describe the environmental requirements for germination (*temperature* range or time of *after-ripening required* to *overcome dormancy*).

A complication in seed dormancy research is that the germination assays used are a measure of the integration of many events that happened in the history of the seed (dormancy) and the various environmental factors acting during germination. For a better *understanding* of seed dormancy and germination, it is important to distinguish between these two processes.

Germination (or *germination sensu stricto* or *visible germination*) is defined as embryo protrusion, which depends on embryo expansion (which is a growth, mainly cell expansion, process) driven by water uptake. After radicle protrusion, seedling *establishment* takes place, which requires *mobilization* of reserves and growth of the seedling. Seedling establishment is often included in the seed germination process. Different types of dormancy, including primary, secondary, seasonal, and coatimposed dormancy, have been defined. Primary dormancy is induced during seed development.

Dormancy is most likely induced or initiated during the later phases of seed development, as can be concluded by the absence of seed dormancy in mutants that have a strongly disturbed seed maturation, such as *ABA-insensitive 3 (abi3), fusca 3 (fus3),* and *leafy cotyledon (lec1* and *lec2).* Because virtually all of the cellular and metabolic events known to occur before the completion of *germination* in nondormant seeds also occur in imbibed dormant seeds, failure of the embryo axis to elongate seems to represent the mechanism of dormancy.

Secondary dormancy can be induced when imbibed seeds cannot germinate because of an unfavourable environmental factor; this indicates that dormancy induction mechanisms continue to operate even after loss of primary dormancy. Dormancy and *germination* are determined by balance of the growth potential of the embryo and the constraints imposed by the tissues *surrounding* it.

The balance between these forces and their relative contribution as well as differences in the response to environmental conditions results in the fact that dormancy can be very different between species. In many species, the seed envelope imposes a strong physical constraint to radicle protrusion.

This explains why envelope characters affect the dormancy status of the seed and also why weakening these envelopes (which can be the testa, the *endosperm* layer, or both) leads to germination. Dormancy and germination are strongly influenced by environmental factors, which are mainly light and temperature, as well as soil factors, among which nitrate is best known. These *environmental* factors can act during formation of the seed (*maternal factors*) and during the imbibed stage of the mature seed.

According to Vleeshouwers et al., changes in dormancy levels of imbibed seeds depend only on temperature. In addition, some factors may act during conditions of low metabolic activity due to low water content and explain the after-ripening effect, a strong factor *influencing* dormancy release.

How endogenous and environmental factors interact is largely unknown, with the exception of the induction of gibberellin (GA) synthesis during imbibition. The ecological significance and large variation in dormancy characteristics between species are described in reviews by Baskin and Baskin and references there. Seasonal dormancy can delay germination until conditions are appropriate for growth and *consequently* may influence fitness.

In temperate climates, delaying *germination* until spring (when the cold winter changes the dormancy status) can prevent winter mortality. However, autumn *germination* may have a selective advantage when the risk of winter mortality is low by enabling the plants to flower earlier or at a larger size.

DORMANCY AS AN AGRICULTURAL PROBLEM

Dormancy has been an *agricultural* problem since early farmers first started to domesticate wild plant species. Those features of dormancy that provide ecological advantages present agronomic *disadvantages* within a farmed system. At one extreme, many weed species show very high levels of dormancy when shed from the plant, thereby infesting land that subsequently requires long-term treatment to remove *succeeding generations*.

At the other extreme, lack of dormancy in crops is considered part of the *"domestication syndrome"* that provides the benefit of early seedling *establishment*, but the disadvantage of possible germination before harvest and reduced quality of the seed. Within crop species, competence for dormancy is associated with several agronomic problems.

The capacity of oil seed rape *(Brassica napus)* to display secondary dormancy means that in successive harvests from fields originally sown with rape, "volunteer" plants can appear as germination is triggered following ploughing and light-activated germination. Wild rice types (including red rice) can also present problems as weeds, in part because of dormancy characteristics.

Within the Triticeae tribe of the grass family, the important *agricultural* species barley, rye, and wheat (bread and pasta) can show extreme reductions in dormancy during grain development, leading to a complex set of traits together known as preharvest sprouting (PHS).

Alternatively, seeds can display high levels of primary dormancy, thus requiring heating and storage treatments before use, for example, in malting (where *uniform germination* is *essential*). Storage and heating *requirements* to remove moisture (and the *potential* for *sprouting*) from wheat seeds harvested under damp conditions are major environmental costs because of the energy used during storage.

PHS is a *particularly* important problem associated with seed quality, and quality issues for downstream processors can result from different *phenomenology*. For example, rape seed oil content is reduced in harvests containing seeds showing PHS; typically, the oil contains free fatty acids associated with increased cloudiness that can be green due to the presence of chlorophyll. Typically, industries associated with crushing to extract oil will not buy seeds containing greater than 2% free fatty acid.

Seed lots *containing* high levels of hydrolytic enzyme activity associated with sprouted wheat and rye grains (*principally* due to *enzymatic activity* of *alphaamylase*) produce flour that, if used for baking, provides low-quality loaves of bread, with a typically sticky crumb structure and poor loaf volume.

Triticeae PHS is one component of a complex interrelated set of phenotypes *characterized* by inappropriate preharvest alpha-amylase production by grains. It is measured commercially using the Hagberg falling number test (HFN) that provides an indication of starch breakdown by endogenous enzymes.

In the United Kingdom, an HFN score above 250 is usually required for bread-making wheat, and above 225 for grower contracts. Sprouting damage is caused by interaction between development within the maturing seed and prevailing *environmental* conditions. Typically, it occurs when cool, damp conditions are encountered before harvest through excessive moisture remaining in intact ears or as a side effect of increased exposure of ears to ground moisture following lodging. Under these conditions, seeds do not contain enough dormancy to prevent the onset of germination; this phenomenon has also been termed postmaturity sprouting (PoMS).

In addition to physical *environmental* damage, biological damage of wheat seeds by larvae of orange wheat blossom midge (OBM; *Sitodiplosis mosellana)* can also lead to sprouting, although this mechanism is of far less economic significance. This occurs at an earlier stage of seed development than PoMS, when grains still have a high moisture content, and sprouting may be the result of damage to embryo *surrounding* structures, leading to a *reduction* in dormancy capacity. PHS is associated with *production* of low- and high-pI alpha-amylase *(amy2* and amy1, respectively), normally expressed in mature grains of harvested seeds as part of the *germination* process.

From an economical perspective, overexpression of PHS in particular years can cause financial problems for farmers. Throughout the world, where the capacity for cool, damp conditions during the later stages of *maturation* of harvests exists, PHS has been an *important* problem.

In Germany, at least 5 of the last 15 years have seen PHS damage to rye crops; in Poland 6.4% of wheat production and 8% of rye could not be sold for *consumption* due to *sprouting* in 1998. In 2000, up to 45% of wheat production in northern France was used as feed due to low quality as a result of sprouting.

In the United Kingdom, where it is estimated that the average yearly loss due to PHS damage of the wheat crop is ca. GBP 18M, 2004 was a *particularly* bad year due to prevailing U.K. weather *conditions.* In this year almost three *quarters* of group 1 (high quality used for bread making) varieties failed to meet quality standards, with a national average HFN of 231—well below that required for bread making. The *unpredictable* nature of the environmental input to PHS has meant that plant breeders have found it very difficult to provide effective *strategies* for increased genetic resistance.

Field testing of breeding lines and analysis of HFN are not very useful tools because they monitor a highly complex set of *interacting* subtraits. Although PHS is *unpredictable* at the *microscale,* evidence suggests that long-term global weather patterns can predict PHS occurrence.

There is a high degree of correlation between average HFN of the U.K. wheat crop and fluctuation of the *North Atlantic oscillation* (NAO), a *measurement* of the difference in air pressures between the Azores Iberian peninsula region and Arctic Iceland region of the North Atlantic. The NAO has a cycle of 8 years, mirroring the cycling of average U.K. HFN.

Although not helpful in preventing damage to crops, this correlation provides evidence that continued breeding for resistance is important, even during *sustained* periods of low *environmentally* induced damage.

ROLE OF THE MODEL ORGANISM ARABIDOPSIS IN DEFINING GENETIC CONTROL OF DORMACY

Seed dormancy has been studied in many plant species. This happened partly because *researchers* were interested in comparison of species and ecological *significance* of differences found between them. Weed scientists studied species with strong dormancy where this had an impact on the weedy character.

Model species have also been used for study where this has provided important insights to particular *mechanisms.* This has been based on suitability for the analysis of specific phenomena, such as light-induced germination in the case of lettuce (because this species shows a strong response to light). In recent decades, as in other fields of biology, model species have been used to provide genetic variants.

The study of plant hormone mutants *in Arabidopsis* and tomato showed *convincingly* the importance of abscisic acid (ABA) and GA for seed *germination.* In addition, the availability of other types of genetic variants and access to the full genome sequence of *Arabidopsis* has made genetics and molecular biology indispensable to the study of biological processes, including seed dormancy, because these resources allow *identification* of the genes and also processes that control the trait.

In addition, they allow the application of "omic" approaches including full genome gene expression analysis and efficient proteomics. Because the required resources for these studies are presently only available in *Arabidopsis* and rice, it is no surprise that biological research is focusing on these species. Genetic research in other species, such as barley and *Avena fatua,* where genetic variation for dormancy is available will certainly benefit from studies on these models.

Sequence similarity and synteny between gene order in different species will allow the transfer of knowledge from model species to other species. This requires that *mechanisms* involved in a process are similar across species. Although such *similarities* are seen (discussed later), one should not overlook the possibility that some processes or subprocesses can be specific to species (or more likely species group).

This is considered likely for seed dormancy in which qualitatively and quantitatively large differences in dormancy phenotypes (and also possible mechanisms) exist. Although earlier studies on the role of light quality during seed development and during *germination* used *Arabidopsis,* the use of plant hormone mutants has stimulated research on seed biology in *Arabidopsis.*

Much of the work on seed dormancy in *Arabidopsis up* to 2002 has been reviewed previously. Several plant hormones affect seed dormancy. The importance of GA in seed dormancy in *Arabidopsis* was confirmed by the identification of nongerminating mutants that could be restored by the application of GA to the imbibition medium. That application of GA biosynthesis inhibitors such as paclobutrazol (PAC) and uniconazol during *imbibition* inhibited *germination* indicated that de novo synthesis of GA is needed; this has recently been confirmed by measuring GA levels.

The way in which light and cold signals promote the transcription of GA *biosynthesis* genes (especially GA 3-oxidases) and how this signal is transduced to activate genes that affect cell *expansion* are presently some of the best-known parts of hormone signaling related to germination. A dormant genotype resembles a GA-deficient mutant because neither germinates in light on water.

However, because differences between imbibed *nongerminating* GA mutants and wild-type at the protein level are very limited and only become more obvious during radicle protrusion, it seems plausible that GA acts (only) at the stage of radicle *protrusion* when a growth potential should be developed that allows protrusion of the radicle through the *surrounding* envelopes.

GA is not absolutely necessary for germination, which is indicated by the capacity of GAdeficient mutants to germinate when the seed coat is mechanically removed or genetically weakened. The late rise in GA levels after imbibition and the fact that germination can be inhibited rather late by *uniconazol* suggest that the role of GA is relatively late. That dormancy is different from lack of *germination* is suggested by the fact that *dormancy-related* genes are expressed during seed development and that some nondormant mutants such as *fus3* and *delay of germination 1 (dogl)* still require GA for germination.

This would be in agreement with the hypothesis that primary dormancy is induced during seed maturation. Dormancy can only be assessed after imbibition during which much metabolic activity and also gene expression changes take place. Prolonged imbibition in conditions where *germination* does not occur can lead to secondary dormancy, which indicates that dormancy reinduction can take place or that the *developmental* phase of the seed can be reversed to *pregermination* (*maturation* phase) as has been suggested.

This developmental process might be *partially* under epigenetic control as indicated by the *pickle* mutant in which maturation is prevented and that encodes a *chromatin-modeling* protein. Another plant hormone that affects seed dormancy is ABA. The ABA biosynthesis mutants were identified based on the fact that *nondormant* ABA-deficient mutants do not require GA for *germination*.

ABA signal transduction mutants also show a dormancy phenotype similar to that of ABA *biosynthesis* mutants. ABA signal *transduction* mutants that also were characterized by lack of dormancy could be selected directly by their resistance to *germination-inhibiting* concentration of ABA. ABA-deficient mutants in all species studied thus far showed an absence of dormancy.

The role of ABA may be twofold. On one hand, it seems to be required for induction of dormancy during seed development, where *ABI3* is an important *downstream* component. Karssen et al. concluded that ABA induces a dormancy state during seed *maturation*; they observed that ABA levels are high halfway during seed development.

This ABA is partly from maternal origin. Important for dormancy induction was a transient peak of ABA produced at a late stage during seed maturation by the embryo proper. The *conclusion* that ABA was present during imbibition at levels so low that it could not inhibit *germination* did not take into account the possibility that ABA could be resynthesized during *imbibition*.

This was observed later for dormant seeds of the dormant accession Cape Verde Islands (Cvi) and was also suggested by the fact that application of inhibitors of ABA biosynthesis promotes germination. ABA levels in mature seeds also contribute to the inhibition of germination and need to be metabolized before germination takes place. The essential gene for this breakdown appears to be ABA 8'-hydroxylase *(CYP707A)*. Null mutants of this gene show a strongly reduced germination.

The way in which these effects on hormone metabolism are *determined* by the dormancy status of seeds set during maturation and the way in which GA and ABA levels depend on the levels of each other is still not clear. Among the other hormones, clear, although not decisive, roles were found for *brassinosteroids* (BR) and ethylene. Ethylene seems to be required for normal fast germination of seeds, which may also be related to the *sensitivity* of seeds to ABA. BR can promote germination of GA mutants by bypassing the GA *requirement*, which is also an effect of applied ethylene. However, BR mutants germinate normally and BRs do not enhance germination of wild-type seeds.

Phytochrome mutants have allowed dissection of the role of different phytochrome species in seed germination; phytochrome B was especially shown to induce GA *biosynthesis* after exposure to red light. The *importance* of the seed coat or testa could also be demonstrated using genetics because almost all mutants with an altered testa colour or structure showed reduced germination.

Wild-type *Arabidopsis* seeds are brown because of the presence of brown proanthocyanindins (*condensed* tannins) that are flavonoid end products in the inner layer of the inner integument. These compounds affect the structure of the cells in that layer and confer additional resistance to the *protruding* radicle.

It is not clear whether the thick-walled single layer of endosperm (aleurone) in mature seeds plays a significant role in preventing *germination* because no mutants affected *specifically* in this layer have been identified. However, the relevance of this layer, which needs to be weakened to allow germination in species such as tomato and tobacco, is suggested by the expression of cell wall-weakening enzymes in the aleurone layer at the onset of germination. Mutants affected in storage reserve *mobilization* indicate that this process is not required for *germination* but is essential for seedling *establishment*.

However, the *comatose* mutant isolated as a nongerminating mutant has a defect in an ABC* transporter affecting lipid breakdown. This suggests that, differently from other lipid *mobilization* genes, this gene affects germination in *stricto senso*. The application of microarray *technology* and proteomics added another dimension to our *understanding* of seed germination.

The relevance of transcription initiation for germination may be limited because the transcriptional inhibitor α-amanitin does not inhibit *germination*, as does cycloheximide, and hardly affects the levels of major proteins during imbibition. However, for storage *mobilization* and hexose metabolism during establishment, *transcription* certainly plays a role.

Apparently, many changes in transcription observed during imbibition in microarray *experiments* before radicle protrusion could be related to seedling establishment and not to *germination* per se, which seems mainly driven by mRNAs already present in the dry seed. However, a role of newly synthesized transcripts shortly after imbibition cannot be excluded because, at this time point, α-amanitin may not be fully effective. Although progress has been made in understanding of dormancy and germination, especially by using the tools available for *Arabidopsis*, many questions about both processes remain unanswered.

Because progress in the *understanding* of dormancy and germination has been focused on the role of GA and ABA, this emphasis might neglect the importance of other factors. These may be those controlled by genes represented by mutants (e.g., *reduced dormancy, rdol* to *rdo4*) that have not yet been cloned or for which only quantitative trait loci (QTL) positions are known. The importance of maternal factors that are not directly related to the structure of the testa is not well understood.

Indications about the importance of such factors come from the fact that two DOF zinc finger genes, *DAG1* and *DAG2 (DOF affected in germination),* influence germination. These genes *(DAG1* inhibits and *DAG2* promotes germination) are expressed in the vascular tissue of developing seeds but not in mature seeds or during imbibition. Furthermore, the mechanism of afterripening and moist chilling (stratification) is not well understood.

Although the effect of the latter treatment on GA biosynthesis during imbibition is convincingly shown, it seems unlikely that this is the only role of cold because this *treatment* is far more effective than applied GA in breaking dormancy in strongly dormant genotypes. Genetic variation for seed dormancy within species is present between accessions of wild plants and among varieties of *cultivated* plants.

The large environmental effects on the expression of germination characteristics and the involvement of many genes make dormancy genetically a typically quantitative trait. Such traits are now more amenable to genetic analysis because the position of individual QTL and the relative

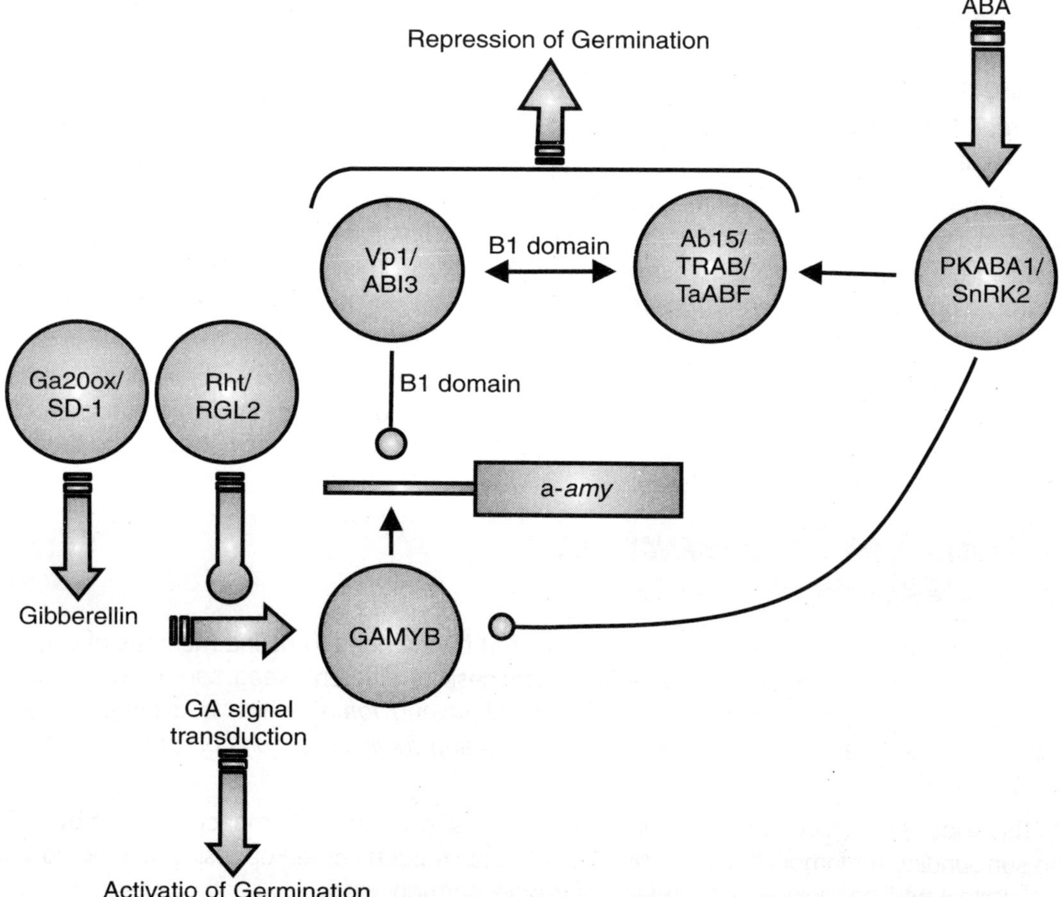

Figure 9.1: Conceptual framework of possible interactions between regulatory molecules involved in ABA and GA control of alpha-amylase synthesis, dormancy, and germination. Solid lines indicate physical interactions, open arrows point to functional consequences. Activation is indicated by arrows, repression by blocked ends.

contribution of these loci can be determined. QTL analysis for seed dormancy requires permanent or immortal mapping populations such as recombinant inbred lines (RILs) because it allows testing a large number of genetically identical seeds per genotype in different environmental conditions. QTL analysis for seed dormancy has been reported *for Arabidopsis.*

This analysis can be followed by study of the individual genes (or chromosome regions containing specific dormancy QTL) by fine mapping and *subsequent* cloning. In this way genes can be identified that control seed dormancy; furthermore, genes that control adaptation to specific *environments* can also be identified. The genes identified in the study of natural variation can be the same as those identified in mutant screens. However, there are several reasons why this is not always the case.

First the parent lines used for mutation experiments can be mutated for specific genes. In the case of *A rabidopsis,* many natural *accessions* show much stronger seed dormancy than the commonly used laboratory accessions Landsberg *erecta (Ler)* and Columbia (Col). Also, mutants that show strong pleiotropic effects such as most ABA mutants will not survive in nature.

Therefore, genes identified by analyzing natural variation are expected to be *ecologically* relevant. The analysis of different RIL populations thus far identified more than twelve regions on the *Arabidopsis* genome where QTLs associated with dormancy are located. Many of these do not colocate with known dormancy loci.

Depending on the parents of the progeny analyzed, the same or different regions are identified. The feasibility to clone such genes has been demonstrated for the *DOGJ locus,* of which *Ler* contains a weak allele and the dormant accession Cvi a strong allele. This gene, which is expressed during seed development and down-regulated during imbibition, encodes a gene of unknown function that is assumed to control genes responsible for dormancy induction.

The cloning of additional QTL and the study of more dormant lines in which regions containing strongly dormant alleles have been introgressed from other *accessions* into an *Ler background* is ongoing and should identify additional genes involved in dormancy induction, dormancy *maintenance,* and dormancy breakage.

MECHANISM OF DORMANCY AND GERMINATION IN CEREALS

Much research into dormancy has been spurred by the need to define methods of control for highly dormant weed species. The possible usefulness of different weed species as models for *investigation* has been reviewed recently. Wild oat *(Avena fatua),* which is a highly pernicious weed, has been used as the major model for *physiological investigations* of grass dormancy for over 100 years.

This species is a good physiological model, *demonstrating* deep dormancy imposed by embryo and surrounding maternal structures. Many studies have been carried out using wild oat to define *biochemical* and physiological changes *controlling* dormancy.

For example, an increase in glycolysis and/or the Krebs cycle has been shown to be an important *determinant* of dormancy breakage in this species. Genetic studies have been attempted using wild oat. Many distinct genetic lines exist with differing dormancy characters and these have provided the material for QTL analyses that have indicated interactions between loci that promote germination or dormancy.

However, because the wild oat genome is hexaploid, with a large genome size (approx. 11,000 Mbp) and very little physical genome information, it is a poor model to use at this level of investigation. Several molecular studies have also used this genetically defined material to study changes in gene expression associated with dormancy and germination phases.

In one study, the *expression* of the wild oat ortholog of maize *Vp1/Arabidopsis* ABI3 was shown to be highly correlated with seed dormancy status. The genetic *components* contributing to PHS have been investigated using a variety of methodologies from whole plant to single seed. Physiological studies of developing wheat caryopses have analyzed the *relationship* between ABA content of seeds and dormancy capacity.

It has been argued that it is important to define the timing of *dormancy* induction during seed development in order to assess the influence of applied hormones and changes in environment. King analyzed the *environmental* influence on in-ear *sprouting* and excised grain germination, comparing cold humid conditions with warm dry.

These analyses established a window of dormancy induction in wheat associated with grain drying and suggest that development favours germination if grain desiccation is prevented at maturity. In addition, studies including this one have shown that dormancy imposition was not related to embryo ABA content. Other studies comparing dormant and nondormant seeds (of varieties or induced mutants) demonstrated large differences in responsiveness of embryos to ABA. Therefore, it is likely that sensitivity to ABA at specific *developmental* periods, in association with the induction of desiccation, are key processes related to dormancy induction and sensitivity to *environmental* conditions that may induce premature *germination.*

The chromosomal locations of wheat embryo sensitivity to ABA and dormancy have been investigated. A cross between wheat cv. Chinese spring (nondormant and ABA insensitive) and line Kitakei- 1354 (dormant, ABA sensitive) was used to *demonstrate* the importance of 4AL and 2D for these characters. At present, the major genetic components used to increase resistance to *sprouting* in wheat are the red grain colour *(R)* loci located on group 5 *chromosomes.*

Dominant alleles promote expression of red pigments (phlobaphenes) within the maternally derived pericarp that tightly surrounds the embryo. It is not known whether the *R* genes provide increased dormancy or are very closely linked to other *dormancy-promoting* loci (for example, *Vp-1,* discussed later). Recently, it has been suggested that the *R* genes encode an myb-like transcription factor that controls expression of genes of the flavonoid pathway within the pericarp during grain development.

Although the *R* loci can be used to provide some resistance to PHS, they also provide characters that reduce the perceived quality of the flour for certain markets. These include colour (red wheat seeds produce discoloured flour that is *perceived negatively* by the *noodle-making* industry), production energy costs, and taste perception.

White wheat types contain no active *R* loci and in general therefore are more prone to PHS. Breeders, particularly in North America and Australia, have concentrated on producing white wheats with increased dormancy levels that should have increased market potential.

POTENTIAL FOR INFORMATION TRANSFER FROM MODEL SYSTEMS TO AGRONOMICALLY IMPROTANT SYSTEMS

Seed dormancy of weeds and sprouting characteristics of crops provide ongoing problems for plant breeders, agrochemical companies, farmers, and downstream processors. Long-term durable solutions are an important target. Genetic alteration of crops to increase resistance to sprouting would provide benefits of quality assurance and *sustainability*.

Because these phases of development are complex, many confounding factors need to be addressed to achieve useful alterations in traits associated with seed *performance*. From the breeding *perspective*, PHS is phenotypically difficult to manipulate because it results from complex physiological and environmental inputs and the *environmental* inputs are difficult or impossible to control and influence.

Multiple genetic loci input into the trait at different levels (morphologically) and stages of seed growth; loci unassociated with seed development per se can influence susceptibility. In addition, mechanisms can be species specific, making it difficult to transfer physiological information from one example to another.

Candidate genes that influence dormancy and *germination*, defined through genetic and molecular approaches, offer important potential tools that could be utilized as highly informative molecular markers for selection and/or genes for *manipulating* development through transgenesis. Genetic approaches to understand PHS in cereals have used QTL and mutant studies to define important regions of the genome, as well as *introgression* studies to add new genetic material with improved trait characteristics.

The Dgenome diploid progenitor *Triticum tauschii* has been used as a donor of embryo and maternal mechanisms for sprouting resistance, via construction of synthetic hexaploid wheat with *T turgidum*. Similarly, *T monococcum* types with high sprouting resistance have been used to introgress this character into triticale. Several QTL studies in wheat and rice have indicated regions of the genome that influence traits of dormancy and sprouting resistance or susceptibility.

In addition, a comparative genetic approach has been used to indicate QTLs from wheat/rice/maize that have *conserved* syntenic relationships (i.e., that lie at the same *chromosomal* positions in the genome). These represent important potential targets for candidate loci having conserved functions associated with dormancy or *germination* traits. Genetic studies in cereal crops suffer the disadvantage of complexity of gene isolation.

This is caused by the extreme difficulty of positional cloning of major gene loci and QTLs-for the most part, because of genome size and polyploidy. Recently, several rice and wheat loci have been cloned through positional methodologies, but in general this remains a long-term and *problematic* approach, especially for QTLs of small effect.

Therefore, the definition of candidate genes defined in simpler model systems provides a complementary approach to identify and characterize regulatory molecules. Models can be utilized at different levels; in the study of mechanisms controlling dormancy these include physiological or biochemical approaches relating biochemical changes to *physiological* characteristics, molecular or biochemical approaches providing mechanistic *understanding* of gene expression regulation and subsequent function, and the use of "simpler" genetic systems. Studies in tomato, *Arabidopsis,* and maize have identified processes and regulator molecules controlling the initiation of *germination* and genetic loci regulating dormancy initiation and the transition to germination.

The alpha-amylase promoter has been used as a model for *transcriptional* regulation in studies of aleurone function and hormone *responsiveness* in association with postgermination events in cereals. Results obtained from all these approaches have been integrated into models for regulation from promoter-transcription factor *interactions* through *developmental* changes.

These models have the potential to define function and usefulness of identified candidate genes in *agriculturally* relevant species and environments. Several examples of candidate genes identified using model systems have been studied and reveal some highly conserved aspects of cell signaling and control of gene expression in flowering plants in relation to post seed-shed development.

However, it is likely that because this phase of development is so complex and *environmental* interactions species specific, many aspects of control may not be shared by all species. An indication of the importance of specific candidate genes can be understood by analysis of conserved function and structure.

Here, the focus is on factors associated with ABA and GA signal *transduction* and synthesis and control of expression of alpha-amylase, a key marker for *germination* and sprouting in cereals. A simple model (based on *information* obtained from *monocot* and dicot species) showing candidate genes and possible interactions between key regulator *molecules* and pathways associated with alpha-amylase production is shown in Figure elsewhere in this chapter, and *components* of the model are discussed next.

Mutants have been isolated in *Arabidopsis,* maize, and rice that exhibit a viviparous phenotype similar to that of PHS. The phenotypes of several mutants *in Arabidopsis,* including severe alleles of *abi3, fus3,* and *lec2,* are outwardly similar to those of mutation at the maize *Vp-1* locus. In each case, dormancy is completely lost and seeds display a viviparous developmental pattern, including the absence of seed *maturation* programs and the premature activation of germination- and *postgermination-associated* development.

These loci have all been cloned and encode highly similar proteins with DNA-binding and transcriptional activation functions, suggesting that they act in one part by repressing gene expression programs associated with *germination.* Conserved structure and function also suggest a central and conserved role in the transition to *germination.*

The phenotypes of these mutants are superficially similar to that of seeds within ears displaying PHS in wheat, although it is also possible that vivipary in these mutants results from disruptions in other developmental pathways. The role of Vp-1 in controlling resistance to PHS and in ABA sensitivity has been investigated.

In this case, it was shown that wheat and closely related progenitors exhibit *significant* alternative splicing of homolog transcripts (derived from each of the single locus *homologous genome positions*) and it was not possible to observe full-length Vp-1 protein in wheat embryo nuclei.

The suggestion that one reason that wheat shows a propensity to sprout is caused by a lack of effective Vp1 activity was tested using transgenic wheat *containing* a correctly spliced *Vp-1* cDNA derived from a highly dormant wild oat ecotype (where *Vp-1 expression* was shown to be highly correlated with dormancy status) under the control of a constitutive promoter.

Transgenic wheat plants showed increased resistance to sprouting in the ear, and isolated seeds also displayed increased sensitivity to applied ABA. Both results indicate, as suggested, that wheat embryos have the capacity to express a higher level of dormancy (and resistance to PHS) through increased activity of "correctly" expressed Vp-1 protein. Others have reported genetic *differences* in ABA *sensitivity* of wheat embryos. One study has shown that a QTL associated with sprouting resistance is present over the chromosomal region containing *Vp-1* on the long arm of 3DL.

The *Vp-1* gene is therefore a good candidate for development of markers associated with reduced splicing of transcripts or increased expression of specific homologs. The Vp-1 protein has been shown to repress expression from a high-pI alphaamylase promoter through the B 1 domain located in the middle of the protein, providing one molecular explanation for Vp-1 repression of germination-associated pathways.

This region has also been shown to interact with another conserved *transcription* factor from *Arabidopsis* and rice (ABI5/TRAB1, respectively) that is highly similar to the wheat protein TaABF. AB13 and AB15 have been studied in detail *in Arabidopsis,* where their roles in ABA-related control of seed development and *germination* are well established.

The TaABF protein interacts with a wheat ABA-induced Ser/Thr SnRK2 protein kinase PKABAI, suggesting that phosphorylation of this transcription factor is an important component of function. Interestingly, AB15 has also been shown to be phosphorylated in an ABAdependant manner in imbibed *Arabidopsis* seeds, mirroring observations in wheat embryos, and TRAB1 is *phosphorylated* in response to ABA treatment in rice. Similar kinases exist in *Arabidopsis* (that contains 10 SnRK2 genes)-two of which exhibit *expression* related to ABA responses-and may be good candidates for functional analysis during seed development and germination.

In tomato, expression of the regulatory subunit of the SnRK1 complex (LeSNF4) is associated with lack of germination in ABA-treated imbibed seeds. These observations provide evidence of a conserved pathway regulating the activation of ABA responses, induction of ABA-regulated molecular interactions, and repression of germination. PKABAI is induced by ABA in wheat and barley and suppresses GAinducible alpha-amylase gene expression in barley. PKABAI has been shown to down-regulate GAMYB-a transcription factor that is part of GA-regulated responses-and is required for activation of expression through a GA response element present in all GA-inducible alpha-amylase promoters.

Rice orthologs of PKABAI appear to exist (SAPK 1 and 2) and expression of SAM was induced by ABA, although the protein is apparently not activated by ABA. GAMYB was originally identified in barley aleurone cells, but has *subsequently* been shown to play a role in rice aleurone by conferring GA responsiveness of amylase production. (These *experiments* used TOS 17 transposon insertion lines to remove expression of the gene.) GAMYB is induced by GA, indicating that this transcription factor is used to integrate signals from GA and ABA signal transduction. Three *Arabidopsis* MYB proteins have been identified with similarity to GAMYB that can substitute for barley GAMYB in *transactivating* the barley alpha-amylase promoter, although their role in controlling germination is not known. A paradigm for *demonstrating* the conservation of function of *agriculturally* important genes is that of the relationship between *Arabidopsis GIBBERELLIN INSENSITIVE (GAI)* and wheat *Reduced height (Rht).*

Both encoded proteins belong to the DELLA subfamily of the GRAS family of plant regulatory proteins. In both cases, dominant mutant alleles produce plants of shortened stature and reduced fertility. Dominant alleles exert their effect by reducing the capacity for GA-induced degradation of the mutant protein; interestingly, alleles *in Arabidopsis* wheat and maize contain deletions of amino acids around a highly conserved section of the protein (the DELLA domain).

These proteins are selected for degradation through the ubiquitin-proteasome pathway via interaction with the F-box protein SLEEPY (SLY); this degradation does not occur with dominant mutant alleles. *Arabidopsis* contains five DELLA protein genes (RGA, GAI, RGL1, RGL2, and RGL3) that are expressed at different levels throughout development, whereas wheat contains only one homolog group on *chromosomes* 4B and 4D.

Of the five *Arabidopsis* genes, *RGL2* has been shown to be the major *determinant* of repression of the initiation of germination; mutant seeds show reduced sensitivity to PAC and enhanced germination potential. Early work with the *Rht3* allele showed that this reduces wheat aleurone amylase expression during *germination* and exogenous GA responsiveness of amylase production by the aleurone.

In addition, this allele has been shown to reduce susceptibility to *prematurity* alphaamylase production that occurs in the absence of sprouting. Genes associated with many of the steps of GA metabolism have been isolated and analyzed from a variety of species.

Individual enzymes may provide useful candidates for *manipulation* of grain GA biosynthesis capacity and hence sensitivity to sprouting. For example, one dwarfing locus in rice (analogous in importance to the *Rht* locus in wheat) *(sd-1)* encodes a GA 20-oxidase gene that catalyses several of the later stages of GA biosynthesis. Recently, genes representing enzymes from this pathway have been mapped using comparative approaches in wheat, rice, and barley.

This approach provides useful information that should allow analysis of the extent to which GA metabolism loci account for chromosomal regions regulating GA-associated phenotypes (including germination) and an indication of the relationship of QTLs associated with GA phenotypes and GA metabolism loci. As the function of candidate genes in *Arabidopsis* and other model systems is revealed, their usefulness for the *manipulation* of agronomically important characters associated with dormancy and *germination* can be tested.

The importance of dormancy induction mechanisms and ABA sensitivity in wheat embryos as determinants of *susceptibility* to PHS means that candidate loci affecting these responses in *Arabidopsis* should be priority targets for further investigation. The recent identification in *Arabidopsis* of the key ABA catabolic enzyme ABA 8'-hydroxylase offers one avenue for analysis of this process in wheat embryos, as mutants display hyperdormancy.

Several other novel regulators influencing ABA sensitivity have recently been described, including SAD1 and *ABH1.* Analysis of crop orthologs of *Arabidopsis* genes associated with QTLs that enhance dormancy potential will also offer new possibilities for *manipulation* of PHS resistance and some indication of pathways important for dormancy in weed species.

Large-scale *bioinformatic* comparisons of *Arabidopsis* gene sequences associated with dormancy and germination with rice genome and wheat/barley EST information should provide candidate orthologous sequences for further analysis. Conserved function allows easy integration of information from model studies to practical application. However, there are several caveats to the use of this approach.

It is notable that QTL studies in *Arabidopsis* have revealed loci previously undetected in mutational screens. Candidate gene studies use conservation as a major tool, and it may be that little variation will be present at loci with highly conserved functions, thereby reducing the usefulness of such genes for marker-assisted breeding. Lastly, it is of course important to identify variation within the species and trait.

Candidate genes and pathways can provide a framework to detect species-specific components; however, in some cases, pathways may be taxon specific and not held in common with those present in distantly related model species. This latter point is exemplified by the identification of vernalization proteins in wheat that do not appear to correspond to those used in *vernalization* pathways in *Arabidopsis*. Several agronomic problems are associated with *germination* and dormancy.

These can be caused by deterioration of quality of seeds in crops or survival of weed seeds in the soil contaminating and competing with subsequent crops. *Understanding* underlying genetic and molecular mechanisms of the different processes that input into dormancy of the embryo and *surrounding* structures is an important component of strategies designed for improving seed quality or controlling seed survival.

Whether via marker-assisted selection or production of transgenic plants, basic research in plant science should provide compelling tools for plant breeders to manipulate traits associated with seed development and subsequent dormancy and *germination*, as well as for agrochemical companies to devise novel chemistries to disrupt the dormancy mechanism of weed seeds.

10

Chapter

MEDICAGO TRUNCATULA

Legumes belong to the *taxonomic* family Fabaceae, containing over 18,000 species divided into the three *subfamilies Mimosoideae, Caesalpinoideae*, and *Papilionoideae*. Legume species have been cultivated for millennia all over the world because of the nutritional value of their seeds.

Nowadays, legumes contribute about 27% of the world's primary crop production; the major single *contributing* species is soybean (*Glycine max*), which is used for multiple applications in the food and feed industries.

Others, such as cowpea (*Vigna unguiculata*), common bean (*Phaseolus vulgaris*), and chickpea (*Cicer arietinum*) contribute significantly to the diets of large numbers of people in Asia, Africa, and South America. The high-quality nutrition of legumes is achieved by the presence of a wealth of secondary metabolites and in the capacity of legumes to live in symbiosis with the nitrogen-fixing bacterium *Rhizobium.*

This symbiosis only occurs under nitrogen limiting conditions and results in formation of complete new organs: the root nodules. Nodules host the *Rhizobium* bacteria, which differentiate in the nodules into symbiotic bacteroids and are the site of catalysis of dinitrogen into ammonia by *nitrogenase.*

As an energy source to achieve N fixation, the bacteria obtain *dicarboxylic* acids from the host plant. By a complex amino-acid cycle, the reduced nitrogen is provided to the plant, where it is accumulated into proteins. The importance of legumes as a protein source for feed and food and their independence of an external nitrogen supply thanks to the symbiosis with *Rhizobium* have encouraged a genomics-led molecular characterization to facilitate applied crop research.

For this purpose, the development of model species has been imperative; *Medicago truncatula* (*Medicago*) and *Lotus japonicus* (*Lotus*) have been selected. In this chapter, we review ongoing *Medicago* research and refer to *Lotus* when it is relevant. We discuss how current progress can already help the characterization of loci in crop species and contribute to the identification of the genes required for critical steps in the establishment of the *legume–Rhizobium* symbiosis.

MEDICAGO TRUNCATULA

Medicago originates from the Mediterranean basin and many accessions have been collected from this region. Phylogenetically, it belongs to the galegoid clade and is closely related to alfalfa (the major world forage legume), lentil, pea, faba bean, and clover.

Unlike these species, *Medicago* has all characteristics of a plant model species: a simple diploid genome, self-fertility, a short generation time (3 to 4 months from seed to seed), and good

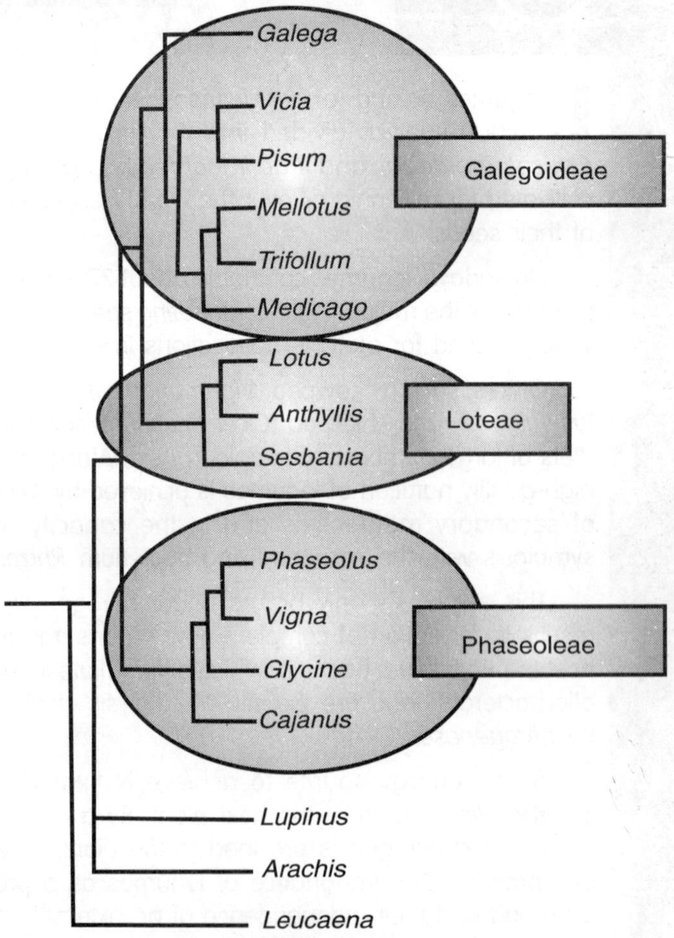

Figure 10.1: Phylogentic relationship of the legume subfamily Papilionoideae. Crop species belong mainly to the Galegoideae or Phaseoleae clade.

genetic transformability. The size of the *Medicago* genome has been estimated to be about 500 Mbp, divided across eight chromosomes and equivalent to four to five times the size of the *Arabidopsis thaliana* (*Arabidopsis*) genome and similar to that of rice (*Oryza sativa*). The *Medicago* genome is more simply organized than that of rice, as visualized during the pachytene stage of meiosis. At this stage, chromosomes are fully

Medicago has a relatively low gene density, based on the number of gene-based molecular markers that can be generated within a partic-ular linkage group. To estimate the size of the gene-rich portion of the *Medicago* genome, two methods have been used:

1. By determining the condensation degree at different positions of euchromatin, the amount of DNA in a euchromatic conformation can be extrapolated. This has resulted in a euchromatin size estimation of 100 to 200 Mb.
2. In the second approach, the difference in genome size between the two accessions Jemalong A17 and R108-1 has been used. To this end the length of pachytene chromosomes and distribution of the hetero/euchromatin were determined.

   The two genotypes do not differ in the length of euchromatin present, but do in the size of the *heterochromatic* pericentromeric blocks. Thus, the difference in genome size between both accessions is contributed by the *heterochromatic* fraction.

The size of the heterochromatic blocks of Jemalong A17 has been estimated to be 320 Mb, leaving the remaining 180 to 230 Mb as euchromatin.

Table 10.1: Genome Size of Crop Legumes.

Common name	Scientific name	Phylogenetic clade	Genome size (Mbp/1 C)
Peanut	*Arachis hypogaea*	Aeschynomeneae	2,813
Cowpea	*Vigna unguiculata*	Phaseoloid	588
Mung bean	*Vigna radiata*	Phaseoloid	515
Common bean	*Phaseolus vulgaris*	Phaseoloid	588
Soybean	*Glycine max*	Phaseoloid	1,103
Pigeon pea	*Cajanus cajan*	Phaseoloid	858
Chickpea	*Cicer arietinum*	Galegoid	931
Lentil	*Lens culinaris*	Galegoid	4,116
Pea	*Pisum sativum*	Galegoid	4,778
Faba bean	*Vicia faba*	Galegoid	26,852
Alfalfa	*Medicago sativa*	Galegoid	1,715
Medicago	*Medicago truncatula*	Galegoid	466
Lotus	*Lotus japonicus*	Loteae	466

Although both calculations are rather indirect, they support the hypothesis that the major part of the *Medicago* genome is composed of repeats within heterochromatic regions. The current *Medicago sequencing* projects are focusing specifically on the *euchromatic* part of the genome. The sequence of this region will be determined in a BAC-by-BAC approach, enabling a full integration of the physical, genetic, and cytogenetic maps.

Functional genomic initiatives have resulted in the generation of >226,000 expressed sequence tags (ESTs) originating from 35 different libraries. These ESTs have been assembled into 18,600 tentative consensus sequences (TCs) of more than 1 EST and 18,200 singleton sequences.

Assuming that *Medicago* has a similar number of genes as *Arabidopsis* does, the vast majority of these are represented by at least one EST. This data set has been used to construct cDNA-based microarrays, as well as oligonucleotide-based chips. In addition, the EST collection has been mined to identify legume-specific genes.

To this end, the *Medicago* ESTs plus similar sets from soybean and *Lotus* were compared to sequence data generated for non-legume plant species, resulting in the identification of 2525 legume-specific EST contigs. Among these are genes specifically induced during *Rhizobium* symbiosis (so-called nodulin genes) and genes encoding for legume-specific seed storage proteins.

However, for the vast majority of genes, the function of the encoded proteins remains largely unknown and probably will need to be elucidated by other methods. To unravel gene function, forward and reverse genetic tools have been applied to *Medicago*. Reverse genetic approaches became possible with the devel-opment of stable transgenic *Medicago* lines, although transformation is significantly less efficient and more time consuming than, for example, for *Arabidopsis*. Of several protocols that have been developed, protocols based on regeneration of transgenic callus have been shown to be the most effective. Lines with increased regeneration efficiency have been selected.

However, crosses between the best line in this *respect—Medicago* R108-1—and lines most widely used for genetic studies result in severe segregation distortion, most likely due to genomic incompatibility. These disadvantages of *Medicago* R108-1 mean that it can be used only in applications that do not require forward genetics.

Genome wide T-DNA and transposon tagging approaches have been initiated for *Medicago*. To effect the latter, the tobacco (*Nicotiana tobaccum*) retrotransposon element Tnt1 was introduced into *Medicago*. The Tnt1 transposition is only activated during tissue culture, during which its copy number remains relatively low. Insertions seem to occur preferentially in the gene-rich portion of the genome, generating gene disruptions.

Currently, a collection of 8000 independent lines representing over 150,000 Tnt1 insertions is being created by an international consortium. In addition to stable transformation, efficient protocols have been developed based on *Agrobacterium* rhizogenes-mediated root transformation. In contrast to stable lines, compound plants are generated in which a nontransformed shoot carries transgenic roots.

The advantage of this system is mainly through a reduction in the time required for acquiring transgenic material; therefore, it is an attractive method to study gene function in roots. Compound plants can be obtained within 4 to 6 weeks and can be nodulated by *Rhizobium* as well as be infected by other symbionts (e.g., *mycorrhizal fungi*) or root pathogens.

Roots generated with this system can be propagated in culture independently of the shoot due to the presence of the *root inducing locus* (*rol*) genes of *A. rhizogenes*. The drawbacks of an A. rhizogenes-mediated root system are:

1. The introduced *rol* genes interfere with plant hormone balance (particularly in overproduction of cytokinin), making this system unsuitable for studies of plant growth regulators.
2. The transformation remains transient because no transgenic offspring are generated.
3. Because the roots are primary transformed tissue, significant variation is observed in expression levels of the introduced transgenes.

In addition, roots can be chimerical because root formation occurs from a group rather than from a single cell. Careful selection is therefore required, and methods based on antibiotic resistance and/or fluorescent markers have been developed.

RNA interference (RNAi) has proven to be a powerful tool to unravel gene function. RNAi can be triggered by generating transgenic lines that express RNAs capable of forming a double-stranded hairpin. For *Medicago* (and legumes in general), the generation of transgenic lines is time consuming, so RNAi has been applied in *A. rhizogenes*—mediated root transformation and shown to be functional.

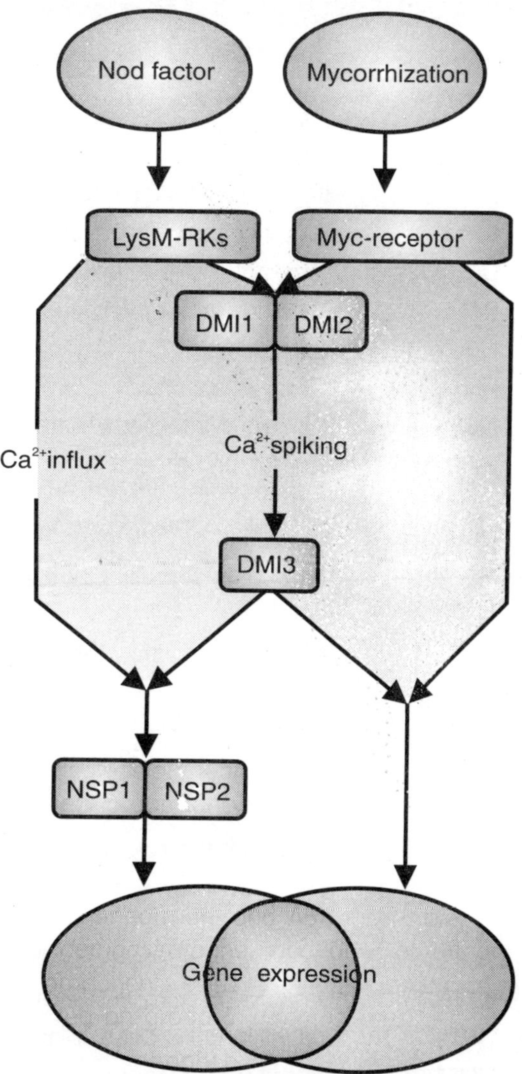

Figure 10.2: A consensus map of Medicago, alfalfa (Medicago sativa), and pea (Pisum sativum). The three species are highly syntenic. Note that Medicago (and alfalfa) linkage group two is represented by pea linkage groups III and VI.

However, systemic spreading of the silencing signal is limited in compound plants. It is transmitted very inefficiently from the transgenic root system to the nontransgenic shoot, and transport from transgenic roots to nontransgenic roots does not occur. Furthermore, because the roots are primary transformed tissue, variation in silencing efficiency tends to occur.

In addition to reverse genetics methods that require transformation, TILLING ("targeted induced local lesions *in* genomes") has been applied in *Medicago* and in *Lotus*. This technique combines

Table 10.2: Medicago Symbiotic Loci Identified by Forward Genetics.

Symbol	Gene name	Protein
	Class I	
BIT	Branching infection threads	Not cloned
DMI1	Does not make infections 1	Putative cation channel
DM12	Does not make infections 2	LRR-RK
DM13	Does not make infections 3	Cat calmodulin kinase
HCL	Hair curling	Not cloned
LIN	Lumpy infections	Not cloned
NIP	Numerous infections with polyphenolics	Not cloned
NFP	Nod factor perception	Likely ortholog of *Lotus* NFR5 encoding a LysM-RK
NSP1	Nodulation signaling pathway 1	Transcription factor
NSP2	Nodulation signaling pathway 2	Transcription factor
PDL	Poodle	Not cloned
RIT	Root hairs infection threads trichomes	Not cloned
SYM1	Symbiosis 1	Not cloned
SYM16	Symbiosis 16	Not cloned
	Class II	
SKI,	Sickle	Not cloned
SUNN	Supernumeric nodules	Likely homolog of *Lotus* HAR encoding an LRR-RK
	Class III	
DNF1	Defective in nitrogen fixation 1	Not cloned
DNF2	Defective in nitrogen fixation 2	Not cloned
DNF3	Defective in nitrogen fixation 3	Not cloned
DNF4	Defective in nitrogen fixation 4	Not cloned
DNF5	Defective in nitrogen fixation 5	Not cloned
DNF6	Defective in nitrogen fixation 6	Not cloned
DNF7	Defective in nitrogen fixation 7	Not cloned
SYM6	Symbiosis 6	Not cloned
SYM17	Symbiosis 17	Not cloned
SYM18	Symbiosis 18	Not cloned
SYM19	Symbiosis 19	Not cloned
SYM20	Symbiosis 20	Not cloned
SYM21	Symbiosis 21	Not cloned

ethyl methanesulfonate (EMS)-induced mutagenesis with the ability to detect base pair changes by heteroduplex analysis by PAGE of Cell digests and generates a range of mutant alleles. The identification of mutant alleles obtained by TILLING is expected in the near future.

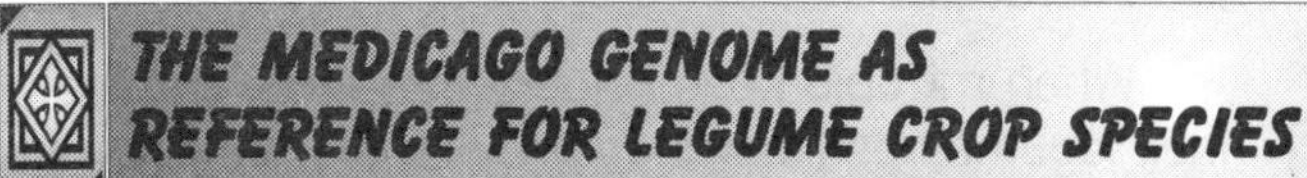

THE MEDICAGO GENOME AS REFERENCE FOR LEGUME CROP SPECIES

To implement knowledge generated from model species for legume crop improvement, comparative genetic maps between model species and economical important crop species need to be constructed. To develop cross-species genetic markers, an *intron-targeted* marker strategy has been shown to represent a powerful approach.

For this purpose, PCR primer pairs are designed to anneal in exon regions conserved between *Medicago* and *Lotus*, soybean, or *Arabidopsis*, designed to amplify across introns. Because introns are significantly more frequently polymorphic than coding regions, these markers are often informative and have been used to integrate the genetic maps of various legume species.

Comparative genetic maps have been created between *Medicago* and the galegoid crop species alfalfa (*Medicago sativa*) and pea, between model species *Medicago* and *Lotus*, and between *Medicago* and the phaseoloid crop species soybean, cowpea, and common bean. A comparison of *Medicago* and alfalfa based on 68 sequence-*characterized* genetic markers indicates that the two *Medicago* genomes are highly similar.

Pea is more differentiated from *Medicago* than is alfalfa. It has a significantly larger genome and contains one chromosome less than both *Medicago* species (seven vs. eight). Despite these differences, a high degree of synteny exists between pea and the *Medicago* species, and only two major chromosomal translocations have been identified.

Medicago linkage group two is distributed over linkage groups III and VI of pea. Similar comparisons have been conducted between *Medicago* and *Lotus*. The *Lotus* genome is about the same size as that of *Medicago* and is divided over six chromosomes. Both species show a significant level of macrosynteny, but several chromosome arm translocations have occurred during evolution.

This synteny is also reflected on the microsynteny scale because the order and orientation of genes have been shown to be conserved significantly. Loteae is a sister group of the Galegoideae, and therefore *Lotus* is more closely related to *Medicago* than to the phaseoloid species.

Alignment of the genetic maps of phaseoloid species and *Medicago* shows significant distortions due to translocations, duplications, and loss of synteny. However, the gene repertoire in orthologous regions in *Medicago* and soybean still displays a degree of conserved gene order.

Based on these studies, it has been concluded that the *Medicago* genome can be effectively used as a reference for the galegoid species. For the phaseoloid species, *Medicago* and *Lotus* can be used, but both show a reduced level of macrosynteny because of translocations.

Rhizobium-Legume Symbiosis

The interaction between *Rhizobium* and legumes has been the subject of many studies over the last century. These have culminated in a detailed description of the steps involved, including infection and primordium formation; a good understanding of the physiology of nitrogen fixation,

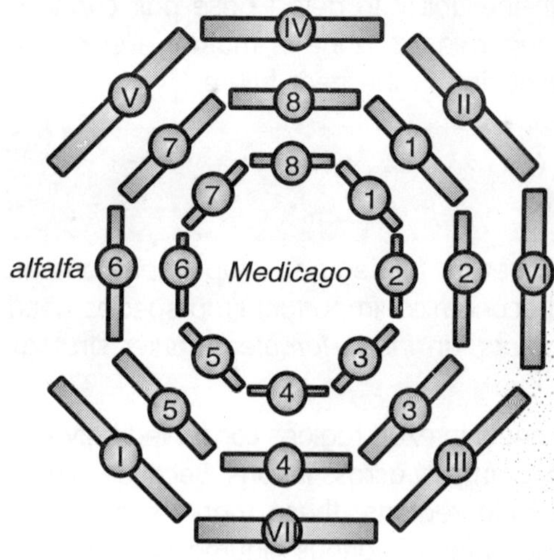

Figure 10.3: Nod factor signaling cascade as identified by forward genetics. Nod factors are perceived by LysM receptor kinases. These activate at least two downstream signaling pathways: one depending on the DMI proteins and a DMI independent pathway for which no specific genes have yet been identified. Both pathways can be discriminated based on distinct Ca^{2+} signals, external Ca^{2+} influx, and perinuclear Ca^{2+} spiking. The DMI pathway is shared mycorrhizal-secreted signal perceived by a hypothetical plant Myc receptor. For mycorrhizal-based signaling, a DMI independent pathway is predicted.

including the requirement for low oxygen tension; the mechanism of nitrogen assimilation; and the identification of plant genes that are regulated during the interaction.

When rhizobia have colonized the root surface of their legume host, they induce morphological changes in root hairs, a phenomenon referred to as root hair deformation.

In some root hairs, rhizobia induce curling as a result of which the bacteria become entrapped in the pocket of the curl. At this point, the plant cell wall is locally modified, the plasma membrane invaginates, and new plant material is deposited. In this way, a tube-like structure, the infection thread, is formed containing the bacteria.

The infection thread grows toward the base of the root hair cell and *subsequently* to the nodule primordium that has simultaneously been formed in the cortex of the root. There, the infection thread ramifies and bacteria are released into primordial cells.

The released bacteria-now called bacteroids-remain surrounded by a membrane of plant origin in a similar fashion as in mitochondria and chloroplasts. Subsequently, bacteroids differentiate and begin to fix nitrogen. Likewise, the nodule primordium *differentiates* into a mature nodule.

Strikingly, genetic analyses have shown that part of the *Rhizobium* symbiosis has evolved from the much older mycorrhizal symbiosis. In contrast to *Rhizobium* symbiosis, which is restricted mainly to legume species, the majority of higher plants have the ability to interact with arbuscular endomycorrhiza, producing a symbiotic association between the plant root and fungi belonging to the order of Glomales.

These fungi grow toward the inner cortical cells of the root, where they differentiate into highly branched structures, the arbuscules. Because the fungus retains hyphae outside the plant it provides the host better access to nutrients such as phosphate.

Genetic approaches in *Rhizobium* have been very successful in identifying the bacterial genes crucial to establishing a proper symbiosis. Many of these genes encode proteins required for production and secretion of a bacterial signaling molecule, the Nod factors.

Nod factors for which the structure has been elucidated all share a β-1,4-linked N-acyl-D-glucosamine backbone of three to six subunits. The nonreducing end of this glucosamine backbone is substituted with a fatty acid of variable structure. Furthermore, at both ends of the backbone, substitutions may be present that include acetyl, sulfuryl, fucosyl, mannosyl, or arabinosyl groups.

Purified Nod factors applied in the nano- to picomolar range are able to induce developmental processes needed for root nodule formation. These responses are provoked at spatially separated sites-specifically, the epidermis, cortical cells, and pericycle. In some species (e.g., alfalfa), Nod factors can even trigger the formation of a complete nodule (lacking bacteria). Just as genetic approaches have led to identifi-cation of genes crucial for signaling organogenesis in *Arabidopsis,* mutants impaired in nodulation have proven to be instrumental in identification of plant genes essential for *Rhizobium-legume* symbiosis and to a better understanding of the underlying mechanisms.

Although several mutants that fail to establish proper symbiosis have been identified for some years in pea and soybean, the identity of the mutated genes has only become possible following establishment of mutagenesis programs in *Medicago* and *Lotus.*

Emerging mutants can be grouped roughly into three classes: class I, non-nodulators (nod-); class II, hypernodulators (nod+++); and class III, nodulators with impaired fixation (nod+, fix-). Here, we will focus on the current knowledge obtained after identification of the genes corresponding to class I and II mutants in *Medicago* and *Lotus.*

Non-Nodulators Enclose Nod Factor Signaling Genes

Various genetic approaches have been used to unravel the Nod factor-signaling cascade. Selection has been made for *Lotus* and *Medicago* mutants that are impaired in the first visible Nod factor-provoked responses.

In a second approach, naturally occurring variation within legumes has been exploited and characterized at the molecular level using *Medicago* in a synteny-based approach. Both methods will be discussed. Among the non-nodulating mutants involving six loci in *Medicago,* only a few are disturbed in most of the Nod factor-induced responses.

Cloning these genes in *Medicago* and *Lotus* has shown that a similar set of genes was mutated, suggesting that mutation screens for impairment in Nod factor signaling are close to saturation. A knock-out mutation in a Nod factor receptor is expected to be impaired in all Nod factor-induced responses.

In *Medicago,* only one mutant displaying such a phenotype has been identified—namely, *nfp;* in *Lotus,* two such loci have been found: *NFR1* and *NFR5.* The other mutants disturbed in Nod factor signaling identified in either species show at least some Nod factor-induced responses in root hair morphology. In *Medicago* there are three *dmi* and two *nsp* mutants. Strikingly, all these genes (including *NFP*) are essential for Nod factor-induced changes in gene expression, as *demonstrated* by *microarray* analysis.

NFR1 and *NFR5* of *Lotus* have been cloned and encode distinct LysM *domaincontaining* receptor kinases (LysM-RK) that, based on their sequence, are localized in the plasma membrane. The putative extracellular regions of both proteins contain LysM domains, which previously have been found in proteins *binding peptidoglycans.*

Thus, these LysM-RKs are good candidates as Nod factor binders because they *contain* an *N*-acetyl *glucosamine* backbone. Because the extracellular domains of NFR1 and NFR5 are markedly different from one another, it seems unlikely that the two *receptors function independently* and recognize the same Nod factor structure.

More probably, a heterodimer involving both receptors is needed for Nod factor perception, a model consistent with the loss of Nod factor responses in both mutants. This is further supported by the atypical serine/threonine kinase in NFR5, which lacks an activation loop that generally regulates kinase activity. Therefore, activation of NFR5-type kinases probably occurs upon phosphorylation by an interacting kinase. *Medicago* NFP is the most likely ortholog of *Lotus* LjNFR5.

The *DMI* and *NSP* genes are positioned downstream of LysM-RK(s) because growth responses in root hairs can be triggered upon Nod factor application. In this respect, the *dmi* mutant root hairs mainly show root hair swelling and only very limited tip growth upon Nod factor perception; *nspl* and *nsp2* mutants show root hair responses more similar to that of the wild-type.

The functioning of the three DMI proteins can be dissected based on Nod factor-induced oscillation of Ca^{2+} concentration. This Ca^{2+} spiking occurs in the perinuclear region of epidermal cells and is induced within a few minutes. Of the *dmi* mutants, only *dmi3* shows this response.

Although the function of this intracellular Ca^{2+} signaling is not yet well understood, pharmacological studies have shown that that it is essential for Nod factorinduced gene expression. DMI1 has a low global similarity to ligand-gated cation channels, whereas DM12 is a receptor kinase in which the putative extracellular region contains three LRR domains.

DMI3 encodes a Ca^{2+} calmodulindependent protein kinase (CCaMK) and is assumed to respond to this Ca^{2+} signal. Genes orthologous to *Medicago* DMI1 and DMI2 have been identified in *Lotus*. Downstream of the DMI-module, NSP1 and NSP2 are functional.

Because all these genes are essential for Nod factor-induced gene expression, it is probable that either of the *NSP* genes encodes a transcription factor that is activated upon Nod factor signaling. Indeed, cloning of NSP1 and NSP2 shows that these genes encode transcription factors belonging to the GRAS family of plant-specific transcription factors.

The three DMI genes are essential not only for *Rhizobium-induced* nodulation, but also for mycorrhizal symbiosis, whereas the putative Nod factor receptors are not. Because the mycorrhizal and *Rhizobium* symbioses only in part trigger expression of a common set of genes, signaling cascades in addition to the DMI module must exist; these (*together* with the *DMI genes*) will be essential to trigger *mycorrhizal* or *Rhizobium* Nod factor specific transcriptional changes.

In the case of Nod factor signaling, the existence of a second such pathway is supported by the different Ca^{2+} response. Apart from the intracellular Ca^{2+} spiking that occurs upon Nod factor perception in a DMI1- and DM12-dependent fashion, an influx of extracellular Ca^{2+} occurs in a DMI-independent manner.

Ca^{2+} influx is one of the first responses in the root epidermis upon Nod factor signaling and is essential for at least some induced *transcriptional* changes. Similarly to the Ca^{2+} influx following Nod factor signaling, it is possible that mycorrhizal fungi also trigger an alternative DMI-independent signaling cascade that, together with the DMI module, is required for *mycorrhizal* specific *transcriptional* changes.

As mentioned earlier, a second strategy based on naturally occurring variation was used to clone a putative Nod factor receptor specifically involved in bacterial infection. In pea accessions originating from the Middle East, the *SYM2* locus was identified as specifically involved in controlling infection thread formation in relation to Nod factor structure.

In the pea accession Afghanistan, this locus inhibits infection by *Rhizobium leguminosarum* bv. *viciae* strains that are unable to add an additional acetate at the reducing end of the sugar backbone of the Nod factor. Thus, the activity of SYM2 depends on the structure of Nod factors secreted by the infecting rhizobia and is part of the mechanism that controls bacterial entry..

Likewise, structure-function relationship studies derived from bacterial genetics have shown that, in *Medicago*, bacterial infection is more dependent on Nod factor structure than are other responses (e.g., nodule primordium formation). To clone a Nod factor receptor essential for *Rhizobium* infection, a synteny-based approach was used to characterize the pea *SYM2* orthologous region in *Medicago*. This region contains several genes encoding LysM-RKs that have been named *LYK*. Knockdown of *LYK3* and *LYK4* by means of *A. rhizogenes*—mediated RNAi has shown that both genes are essential for *Rhizobium* infection in a Nod factor structuredependent manner.

Medicago LYK3 and *LYK4* are homologous to *Lotus* NFR1, although their loss-of-function phenotypes are strikingly different. In addition to *LYK3* and *LYK4* in *Medicago*, the gene *NODULE INCEPTION* (*NIN*) is also essential for infection initiation. *NIN* encodes a protein with homology to transcription factors and was originally cloned in *Lotus* by insertion of an AC transposable element.

The insertion mutant showed excessive root hair deformation and curling, but no infection. In contrast to the genes described previously, *NIN* is induced upon Nod factor perception and therefore cannot be primarily involved in the Nod factor signaling pathway.

Supernodulators

In the 1980s, soybean and pea mutants were identified that formed nodules independent of the nitrogen status of the soil. Apart from this characteristic, these mutant plants formed more nodules than did wild-type plants grown in the absence of nitrogen. As a result, these mutants have been termed supernodulators.

The discovery of supernodulators supports the notions that nodule formation is suppressed by the presence of nitrogen and that legumes have an autoregulatory mechanism that controls the number of nodules formed. Strikingly, supernodulation does not lead to any increase in plant biomass, indicating that nodule formation and nitrogen fixation of the hosted *Rhizobium* are established at the expense of the plant.

The recent cloning of the orthologous genes *HARl* in *Lotus*, *SYM29* in pea, and *NARK* in soybean has allowed for the characterization of an important key regulator of the autoregulatory mechanism. The signature of the protein encoded predicts that it functions as a receptor kinase because of the presence of extracellular leucine-rich repeats (LRRs) and an intracellular serine/threonine kinase domain.

Based on its homology to *Arabidopsis* CLAVATA1 and the observation that the number of lateral roots in the *Lotus harl* mutant is affected after inoculation with *Rhizobium*, it has been suggested that this protein has a role in mediating control over root organ formation, including lateral roots as well as nodules.

PERSPECTIVE

The capacity to form root nodules in which bacteria convert nitrogen into ammonia allows the seed of a legume crop to accumulate high protein content. This is a unique feature among plants. To keep pace with the growing world demand for protein-rich food, *Rhizobium* symbiosis must be exploited to its limits.

This need has led to development of the legume model species *Lotus* and *Medicago*, of which the latter is presented in this review in more detail. In addition, nonsymbiotic traits, like tolerance to biotic and abiotic stress, seed quality, plant architecture, and flowering behaviour, are important for legume crop performance. Therefore, these traits are important targets for legume breeders.

Series of gene-based genetic markers that can be used across legume species have been developed and used to integrate genetic linkage maps. The possibility to exploit the synteny between model and crop legumes will certainly be instrumental in future legume breeding. This has been recognized; several well-funded programs covering many aspects of legume biology have been implemented worldwide.

This will ultimately lead to full integration of the *Medicago* and *Lotus* genome sequence with high-density genetic maps of crop legumes. Development of model systems has speeded up identification of genes encoding key players in *Rhizobium* symbiosis.

Initially, these studies have been focused on genes involved in Nod factor signaling because perception of this bacterial signal molecule by the plant forms the main trigger for root nodule development. Strikingly, the number of genes essential for Nod factor signaling that can be identified genetically is low and is conserved among all legume species studied so far.

Furthermore, mutations in these genes mainly affect symbiosis, *suggesting* that they do not play important roles in other plant *processes*. It can be expected that in the near future the link between the Nod factor signaling network and common cellular processes will be *elucidated* and thereby will make available *knowledge* of how plants rewire processes for organ formation.

Homologs of genes affected in *non-nodulating* mutants such as Nod factor receptor LysM-RKs or *DMI* genes are present also in non-legumes. This indicates that the processes needed for nodule formation could be, in part, already present in nonlegume species and suggests that *Rhizobium* has recruited genes involved in general plant development for nodule formation.

The observation that *dmi* mutants are also impaired in the *interaction* with arbuscular mycorrhiza has led to the hypothesis that nodule formation evolved from this more widespread symbiosis. This suggests that non-legumes may lack a spectrum of the genes that enable the establishment of a symbiotic relationship with rhizobia. The longstanding dream of nodulated riceor other important non-legume crop species-might be feasible.

However, *transferring* the capacity of nitrogen fixation to non-legume species will be a difficult task that will depend largely on how much processes needed for nodule development are present in non-legumes and can be geared to each other in the way in which the process occurs in legumes. This implies that we need to know how many additional *components* must be transferred to non-legumes.

An in-depth analysis of the 2500+ legume-specific genes identified could provide clues in this direction. A major challenge is to uncover the extent to which non-legume homologs of genes such as LysM-RKs and *DMI* are able to complement *corresponding* legume mutants. Such studies would give insight into the extent to which the functioning of Nod factor signaling genes are unique to legumes. The availability of model species will definitely prove their value in finding answers to these *exciting* questions.

NITRATE ASSIMILATION IN PLANTS

Model species are generally first selected because they allow novel insights into processes that are poorly understood; they are then maintained and accepted because of their experimental amenability and usefulness in understanding further processes.

Plants represent an *extraordinarily* diverse group of living beings at the molecular, genetic, biochemical, and physiological levels, and their great biodiversity reflects the evolution of complex genetic and *biochemical* networks.

Unicellular organisms lack the complexity and sophistication of higher plant systems. However they remain useful as a tool to understand aspects of *fundamental* plant biology at the cell level that have not yet been elucidated.

In particular, conditions for efficient uptake, metabolism, and regulation of nutrient acquisition, as well as for *accumulation* of assimilate-derived products are relevant for determining the efficiency of growth at whole plant and thus at crop levels.

For this reason, many *research* groups have focused their attention on understanding key processes in model plant systems. It is becoming clear that present model systems are not as *representative* as had been hoped, so, for a proper understanding, each system needs to be studied in itself.

At the same time, it seems that, to dissect the major processes at the molecular level, it is necessary to work in a system simple enough to control the influence of extraneous factors. Interestingly, the molecular dissection of metabolic steps (e.g., nitrate assimilation) has shown that the *differences* between *unicellular* and land plants can be smaller than feared at the level of complexity of gene and protein families and with respect to cellular strategies adopted to achieve particular end results.

An ideal model system needs a set of basic physiological, *biochemical*, and molecular techniques, along with developed resources in genetics, genomics, and transgenesis. These will allow for the ready genetic dissection of mutant phenotype, together with a *straightforward* correlation of phenotype to genotype. The *Chlamydomonas* system fulfils most of these requirements, as summarized in Table elsewhere in this chapter.

The use of *Chlamydomonas* as an amenable biological system was first described in Harris's book. The advantages of *Chlamydomonas* as a model unicellular system were further presented by Rochaix et al. and were recently compared to those of *Arabidopsis*.

Decoding the nuclear and chloroplast genomes of *Chlamydomonas* and developing molecular tools have strengthened the position of this organism for study of important plant cell processes such as photosynthesis, chloroplast inheritance and biology, mitochondrial genetics, nutrient deficiency, carbon metabolism, and nitrate assimilation.

THE NITRATE ASSIMILATION PATHWAY AND ITS KEY POINTS

Ammonium and nitrate are the primary inorganic nitrogen sources for plant growth. Though many species use ammonium in preference to nitrate, the majority of plants, algae, and microorganisms is able to use nitrate efficiently because ammonium is about 10 to 1000 times less abundant than nitrate in natural soils, except in a few ecosystems such as coniferous forests.

Table 11.1: Advantages of the Chlamydomonas System in Molecular Plant Biology.

Property	*Description*
Genome similar to *Arabidopsis*	Haploid organism with a 10^8-bp genome
Genome sequencing project	Mostly sequenced; third assembly released
EST and microarrays	Above 200,000
Genetics	Excellent classical genetics, with standard tetrad and complementation analysis
Transformation	Nuclear, chloroplast, and mitochondrial genomes transformed
Markers	Array of selectable markers including antibiotic resistance available
Interfering gene expression	Antisense and RNAi methodologies
Mutant library	A 22,000 mutant library with mostly single insertions

In addition, nitrate provides an efficient signal for modulating metabolic processes and plant architecture. The assimilation of ammonium ion has a lower energy cost than that of nitrate and

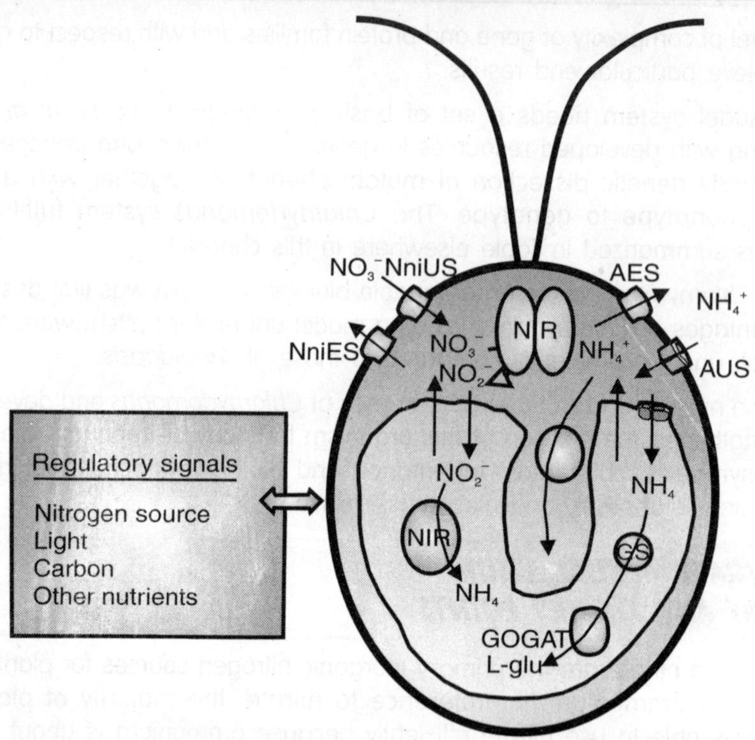

Figure 11.1: The nitrate assimilation pathway in Chlamydomonas. NniUS = nitrate/nitrite uptake influx systems; NniES = nitrate/nitrite efflux systems; AUS = ammonium uptake influx systems; AES = ammonium efflux systems.

many genes involved in nitrate assimilation are strongly repressed in the presence of ammonium. However, many plants, especially herbaceous crop plants, utilize nitrate or a combination of both ions as the preferred nitrogen form for growth. Thus, about 75% of the nitrogen in proteins consumed by man is *ultimately* derived from nitrate assimilated by plants. The *fundamental* role of the nitrate assimilation pathway in plant nutrition has been the object of intensive studies for many years and has been reviewed by many authors.

This chapter represents an update of information about nitrate assimilation in *Chlamydomonas* as a source of information for the equivalent process in higher plants.

The basic steps for nitrate utilization in a single photosynthetic eukaryotic cell are as follow:

1. The entry of nitrate into the cell by means of specific transport systems
2. A first reduction step from nitrate to nitrite, which occurs in the cytosol and is catalyzed by the nitrate reductase (NR) enzyme
3. Nitrite transport to the chloroplast
4. A second reduction step that occurs in the chloroplast, where nitrite reductase (NiR) catalyses nitrite reduction to ammonium
5. Finally, ammonium incorporation into carbon skeletons by the glutamine synthetase/ glutamate synthase cycle (GS/GOGAT)

Table 11.2: Major Elements for Nitrate and Ammonium Assimilation in Chlamydomonas.

Gene product	*Gene/name*	*Scaffold/ linkage group*	*Protein (aa residues)*	*Possible and subcellular localization predictions*
Nitrate reductase	*Nia1* (also Nit1) (C_520041)	52/IX	?	Cyt
Nitrite reductase	*Nii1* (C-520008)	52/IX	589	Chl
Glutamine synthetase	*GS1* (C-20337)	2/II	382	Cyt
	GS2 (also GLN2) (C_380043)	38/XII/XIII	380	Chl
	GS3 (C_380117)	38/XII/XIII		Chl
Glutamate synthase	*NADH-GOGAT* (C_1440026)	144		
	Fd-GOGAT (C-160008)	16/XII–XIII	847	Chl
Nitrate transporter NRT1 family	*Nrt1.1* (C_40176)	4/IV		
Nitrate transporter NRT2 family	*Nrt2.1* (also Nar3) (C_520006)	52/IX	547	Pm
	Nrt2.2 (also Nar4) (C_520007)	52/IX		Pm
	Nrt2.3 (C_330081)	33/IX	572	Pm
	Nrt2.4-5 159/III (C_1590030-1)			Pm
	Nrt2.6 (C-20370)	2/II		Pm
Nitrate transporter component NAR2	*Nar2* (C_520042)	52/IX		Pm
Nitrite transporter NAR1 family	*Nar1.1* (C_520040)	52/IX		Chl
	Nar1.2 (also LciA) (C_90197)	9/VI	336	Chl
	Nar1.3 (C_8440001)	844,100		
	Nar1.4 (C_720018)	72/VII		

(Table Contd.)

Gene product	Gene/name	Scaffold/ linkage group	Protein (aa residues)	Possible and subcellular localization predictions
	Nar1.5 (C_70011)	7/XII-XIII		Chl
	Nar1.6 (C280009)	28/I		
Ammonium transporter AMT1 family	*Amt1.1* (C_110147)	11 / III	539	Pm
	Amt1.2 (C_4560001)	456	542	Chl
	Amt1.3 (Amt3) (C_2680003)	268	579	Pm
	Amt1.4 *(Amt4)* (C_930017)	93	498	Chl
	Amt1.5 *(Amt5)* (C_220054)	22/IX	610	Pm
	CrAmt1.6 *(Amt6)* (C_980024)	98		Er,Prx
	Amt1.7 *(Amt7)* (C_20186)	2/ II	a: 411 b: 487	a: Mit b: Pm
	Amt1.8 *(Amt8)* (C_380121)	38/ XII–XIII	481	Pm, Chl
Positive regulator NIT2	*Nit2* (C_860001)	86/III		Nuc
MoCo carrier protein	Mcp1 (C_700030)	70	165	
Alternative oxydase	Aox1 (C_330029)	33/IX	360	Mit
NADP+-malate dehydrogenase	*NADP-Mdh* (C_520009)	52/IX	415	Chl

Chlamydomonas does not use intracellular compartments to store nitrate or ammonium. Thus, under conditions where the cell assimilation capability is exceeded, extrusion systems come into play, thereby avoiding any toxic effects of excessive intracellular ammonium or nitrite ions.

The nitrate assimilation pathway is governed by specific regulatory genes mediating positive and negative responses against nitrate and ammonium, respectively. In addition, control of this pathway is connected to a network of stimuli needed to coordinate nitrate assimilation with that of carbon, sulfur, and other nutrients (phosphorous, potassium, etc.).

Light is also essential for inorganic nutrient assimilation in photosynthetic organisms. A complex network underlying the process of nitrate assimilation has been proposed on the basis of transcriptome analyses in response to nitrate, potassium, etc..

The *Chlamydomonas* genes involved in nitrate assimilation are shown in Table elsewhere in this chapter. Some of the direct gene products have been studied and characterized; others have been taken from the recently released genome sequence database and will require further study.

As shown, structural genes responsible for nitrate and nitrite reduction are single copy, while those for ammonium incorporation correspond to plastidial (GS2) and cytosolic (GS1) forms. The large number of genes involved in the transport of nitrate, nitrite, and ammonium is particularly notable.

On *Chlamydomonas* chromosome IX, two gene clusters contain most of the nitrate assimilation genes, the majority of which are under control of the regulatory gene *Nit2*. One occurs in a region of about 36 kb, where the structural genes *Nia1* (encoding *NR*), *Nii1*(*NiR*), *Nar1.1* (a plastidial nitrite transporter), *Nar2* (a component of some nitrate transporters), and *Nrt2.1* and *Nrt2.2* (nitrate transporter components) are found.

In this cluster, a plastidial malate dehydrogenase gene, *NMdh*, not regulated by nitrate is also present and has been proposed to play an important role in the supply of reducing power for nitrate reduction. The second cluster (about 10 kb in size) contains two nitrate-regulated genes, *Nrt2.3* (a nitrite transporter) and Aox1 (a mitochondrial alternative oxidase).

Clustering of nitrate assimilation genes has also been reported in *Aspergillus nidulans* and in *Hansenula polymorpha*, so this may well represent a cellular strategy to optimize the regulation of this pathway. In *Aspergillus*, the intergenic region between the divergently transcribed *niiA* (NiR) and *niaD* (NR) genes contains multiple *NirA* (pathway-specific positive regulator)-binding sites, which act bidirectionally.

Interestingly, genes encoding GS2, GS3, fd-GOGAT, and AMT1.8 are located in chromosome XII and chromosome XIII. None of these genes are regulated by nitrate, but they are related to ammonium assimilation. *GS2* and *GS3* are bidirectional clustered genes.

THE USEFULNESS OF CHLORATE-RESISTANT MUTANTS TO STUDY OF NITRATE ASSIMILATION

The isolation and characterization of mutants deficient in the nitrate assimilation pathway, mostly through chlorate resistance, have been powerful tools in defining structural and regulatory elements of nitrate metabolism, as well as in understanding the function of each component. As an analog of nitrate, chlorate can be reduced by NR to produce chlorite, which is cell toxic. Nevertheless, chlorate becomes toxic and causes mutagenesis in the cells by a process dependent on its transport.

Thus, mutants incapable of taking up chlorate and those deficient in NR activity could be selected in chlorate media. This strategy has been widely used in fungi, algae, and plants; the mutants generated in this way have been used to study the nitrate assimilation pathway.

The characterization of *Chlamydomonas* chlorate-resistant mutants has led to identification of

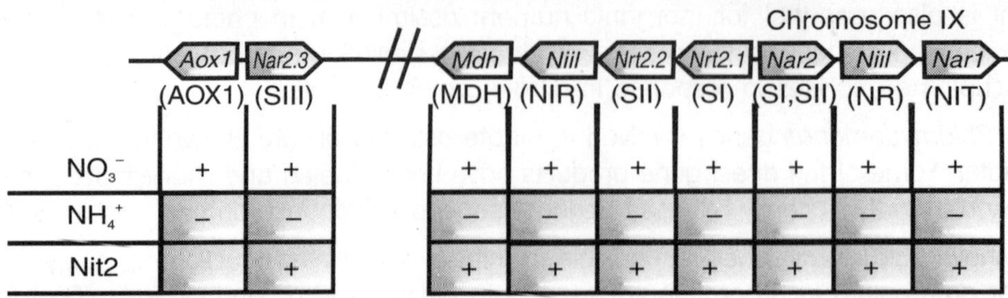

	AOX1	SIII	MDH	NIR	SII	SI	SI,SII	NR	NIT
NO_3^-	+	+	+	+	+	+	+	+	+
NH_4^+	–	–	–	–	–	–	–	–	–
Nit2		+	+	+	+	+	+	+	+

Figure 11.2: Nitrate assimilation genes clustered in chromosome IX. Plus and minus symbols refer to positive and negative acting effectors, respectively, on the gene shown above.

several loci involved in nitrate assimilation. The function of genes encoding nitrate transporters, nitrate reductase, nitrite reductase, the plastidial nitrite transporter NAR1, and the regulatory gene *Nit2*, was demonstrated by characterization of chlorate-resistant mutants.

A chlorate-resistant mutant related to light regulation of nitrate assimilation has also been characterized, though the locus affected has yet to be identified. This mutant shares several characteristics with the *Arabidopsis* CR88 mutant that carries a lesion in a gene encoding a chloroplast targeted Hsp90 protein and shows a pleiotropic phenotype.

NITRATE AND NITRITE REDUCITON AND ITS REGULATION

Nitrate reductases from photosynthetic eukaryotes are homodimeric proteins that use pyridine nucleotides as electron donors. Each monomer is a 100- to 120-kDa polypeptide containing three prosthetic groups—flavin adenin dinucleotide (FAD), heme b557, and molybdenum cofactor (Moco)—that are present in three functional domains spaced by two short hinge regions.

NR sequences from different eukaryotic organisms show a high level of conservation; however, they differ from the cyanobacterial NR, which uses ferredoxin as an electron donor, as do a number of other prokaryotic NRs. One or two genes can be identified for NR in different organisms. Barley and *Arabidopsis* contain two structural loci (*Nia1* and *Nia2*), one of which (*Nia1*) encodes the most abundant isoform (NADH-dependent).

In other organisms such as *Nicotiana plumbaginifolia, Lotus japonicus,* or *Chlamydomonas reinhardtii,* only one gene is responsible for NR activity. Molybdenum is a micronutrient essential for nitrate reduction. Its presence in a *molybdopterin* cofactor (Moco) was first genetically identified in *Aspergillus nidulans* as a cofactor common to NR and xanthine dehydrogenase. Afterwards, Moco was also found to be associated with aldehyde oxidase and sulfite oxidase.

Thus, Moco is *essential* in key metabolic processes: nitrate *assimilation,* purine catabolism, biosynthesis of phytohormone abscisic acid, and *detoxification* of sulfite. Moco consists of the molybdopterin (MPT), an incompletely alkylated aromatic pterin complexing one Mo atom via a dithiolene group to its four-carbon side chain.

The pathway of Moco biosynthesis has been dissected in plants and its steps determined from GTP to Moco. For this purpose, the isolation and *characterization* of Moco mutants using chlorate *resistance* was *particularly* important. The Moco mutants have been *classified* into six different *complementation* groups (*CnxA-CnxF*). Biochemical and genetic *characterization* of *Chlamydomonas* NR-deficient mutants has allowed for identification of seven loci (*Nit3* to *Nit7, Nit10*, and *Nit11*) related to Moco biosynthesis.

Single mutants defective at the *Nit5* or *Nit6* genes show a wild phenotype, whereas the double mutant *Nit5–Nit6 lacks* Moco and molybdate uptake activity. Molybdate uptake by *Chlamydomonas* cells is thought to be mediated by specific transporters: a high-capacity system related to the *Nit5* gene function and another system with less capacity. Moco-carrier protein (CP) activity was first identified in *Chlamydomonas* and its presence could also be demonstrated in *Vicia faba* seeds.

The corresponding gene, *Mcp1*, was isolated and functionally characterized. Both pure Moco-CP from *Chlamydomonas* and the recombinant protein were able to protect Moco from inactivation by oxygen very efficiently. Moco-CP is proposed to participate directly in transfer of the prosthetic

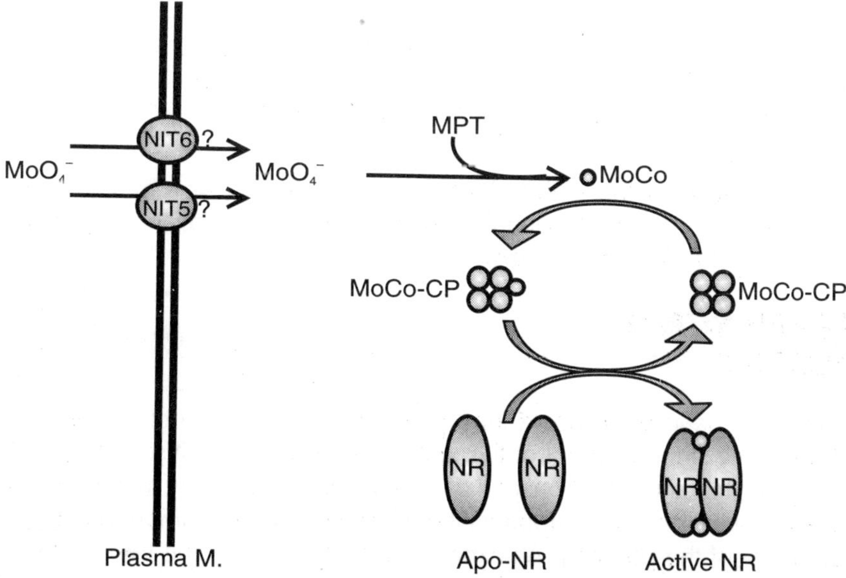

Figure 11.3: Model for molybdate transport and Moco-carrier protein function in Chlamydomonas. MPT = molybdopterin.

group Moco to the apoNR. Orthologs of *Mcp1* have been found in prokaryotic organisms, but *identification* in eukaryotes is not easily deduced from sequence homology data. NR activity is tightly regulated by environmental factors such as light, nitrogen, and carbon availability.

In plants, NR and GS have been reported to bind 14-3-3 proteins. The mechanism of plant NR reversible inactivation depends on phosphorylation and binding of 14-3-3. The interaction of 14-3-3 proteins with a number of metabolic *enzymes* is sequence specific and suggests *involvement* in regulation of complicated daily rhythms in sugar metabolism in coordination with photosynthesis, ATP production, and nitrate reduction.

Chlamydomonas NR is not affected by regulatory interactions with 14-3-3. Nevertheless, GS1 is *phosphorylated* and binds 14-3-3. The function of the 14-3-3 regulatory mechanism is not clear and has been related to protein turnover in the cells. The *Chlamydomonas* NR is subject to a redox interconversion regulatory mechanism.

Thus, in the absence of nitrate, NR becomes over-reduced and inactivated. This mechanism has been shown to be operative *in vivo* and *in vitro;* it can be reversed *in vivo* by resupply of nitrate and *in vitro* by ferricyanide oxidation. Inactivation of NR results in a decrease in the enzyme's half-life. Nitrite reduction is a six-electron step catalyzed by NiR, which uses reduced ferredoxin (fd) as electron donor.

In photosynthetic eukaryotic organisms, NiR is located at the chloroplast stroma and is also present in the plastids of nonphotosynthetic tissues. The holoenzyme is encoded by a nuclear gene. The 63-kDa protein contains two redox centers: a siroheme and a [4Fe-4S] cluster. The N- and C-terminal parts of the protein are proposed to bind ferredoxin and the [4Fe-4S] redox center, respectively.

NiR is encoded by one gene in *Chlamydomonas* as in *Arabidopsis,* but other plants such as tobacco may contain as many as four genes. The *Chlamydomonas* NiR shows a similar regulation pattern as NR and requires absence of ammonium and presence of light and nitrate for maximum expression.

However, no regulation at the activity level has been shown, so any changes in enzyme amounts appear to be due to transcriptional regulation. Post-transcriptional regulation has been demonstrated in *Nicotiana* and *Arabidopsis.*

NITRATE AND NITRITE TRANSPORTERS

Much attention has been paid over a long period to the reduction of nitrate because this is considered to be the key step in control of the pathway. However, the first transport step through the plasma membrane, together with transport at the chloroplast envelope membrane, has become the focus of present interest because it appears that these steps play an important role in regulation of the overall efficiency of the pathway. As expected for this key role, the transporters are subject to fine regulatory control.

Transport Systems

As indicated in Figure elsewhere in this chapter, nitrate transport systems should work in influx as well as efflux processes to provide sufficient nitrate under different nutritional and environmental conditions to satisfy the total demand of nitrogen. These transporters should be operative at the level of the plasma and the plastidial membranes.

Nitrate/nitrite transporters have been classified on the basis of their substrate affinity, specificity and requirements for induction into: constitutive high-affinity nitrate transport systems (cHANTS), inducible high-affinity nitrate transport systems (iHANTS), constitutive low-affinity nitrate transport systems (cLANTS), and inducible low-affinity nitrate transport systems (iLANTS).

Studies on nitrate transport suggest that in plants, algae, and fungi, nitrate uptake is electrogenic and driven by proton cotransport. The proton gradient is maintained by a H+-ATPase. Thus, the use of H+-ATPase inhibitors and alkalization of external medium inhibit nitrate uptake.

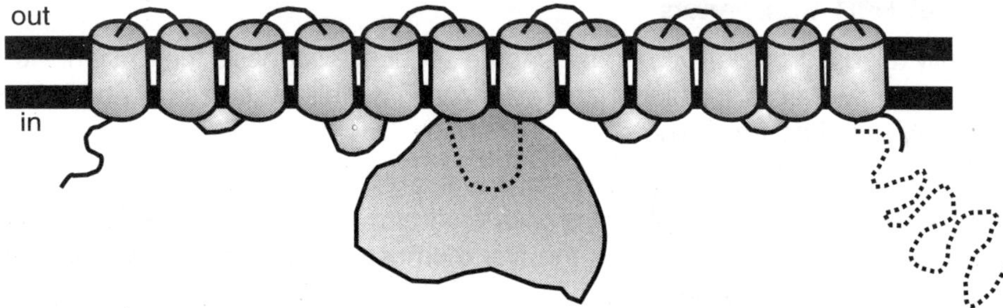

Figure 11.4: Models for NRT2 proteins. Central loop and C terminal region are indicated in continuous line for the fungal NRT2 transporter and in discontinuous line for algal/plant transporters.

Other nitrate transport mechanisms have also been proposed. A Na^+/NO_3^- symport system was suggested in the cyanobacteria *Anacystis nidulans R2* and an active ATP-dependent transport system (ABC transporter) in *Synechoccocus* sp. PCC7942. In *Escherichia coli*, the nitrate transporters NarK and NarU might be able to catalyze nitrate uptake or nitrate–nitrite antiport.

Voltagedependent chloride channels might also participate in nitrate transport. In *Arabidopsis*, AtCLC-a is induced by nitrate and its involvement in the control of the intracellular nitrate status has been suggested. cHANTS have been described in higher plants such as *Nicotiana*, barley, and *Arabidopsis* and are characterized by low values of Km and Vmax (typically 6 to 20 *μM* and 0.3 to 0.8 μmol $g^{-1}.h^{-1}$, respectively).

The cHATS provides high-affinity, low-capacity activity for NO_3^- entry in uninduced plants. Nevertheless, cHATS activity is upregulated (approximately threefold) by exposure to NO_3^-. The iHANTS have been identified and well characterized in higher plants (*Arabidopsis, Nicotiana*, barley, etc.), algae (*Chlamydomonas, Chlorella*), yeasts (*Hansenula polymorpha*), and fungi (*Aspergillus, Neurospora*). The iHANTSs provide high affinity and capacity (Km 20 to *100 μM* and Vmax 3–8 μmol $g^{-1}.h^{-1}$).

These transporters require NO_3^- or NO_2^- to be induced and are subject to repression by nitrogen metabolites such as ammonium and glutamine. The cLANTS and iLANTSs have been identified in higher plants and can significantly contribute to nitrate uptake at millimolar nitrate concentrations. The specificity for the nitrate ion has also been used to name the transporter. In *Chlamydomonas*, physiological studies with mutant strains carrying particular transporters have demonstrated their ability to distinguish between NO_3^- and NO_2^-.

Thus, the nitrate and nitrite transporters in this alga can be classified into nitrate specific, nitrite specific, and nitrate/nitrite bispecific. In addition, some plant nitrate transporters have been shown to transport amino acids. The biochemical characteristics for nitrate transport activities seem to be as complex as the picture for nitrate transporter genes.

This gene complexity is observed from comparison of *Arabidopsis* and *Chlamydomonas* genomes. Three families of nitrate/nitrite transporters, *Nrt1, Nrt2*, and *Nar1*, operate in photosynthetic eukaryotic organisms. For these families, there exist 51 Nrt1, *7 Nrt2*, and no Nar1-like genes in *Arabidopsis*, whereas in *Chlamydomonas* there are *1* Nrt1, *5 Nrt2*, and *6* Nar1 genes. The precise role of each transporter in terms of substrate specificity, capacity, localization, and participation in nitrate assimilation and its efficiency is a challenge that is now starting to be addressed.

Functionality of NRT1 Transporters

The recently released *Chlamydomonas* genome sequence suggests the existence of a putative NRT1 transporter. However, a functional characterization of this system is needed. The following section refers only to data from plants.

The NRT1 transporters (also named PTR transporters) belong to the POT family, which includes numbers of H+-dependent oligopeptide transporters from mammals, plants, fungi, and bacteria. The *Arabidopsis* AtNrt1.1 (*CHL1*) gene was the first member of the NRT1 family identified and was cloned on the basis of its function.

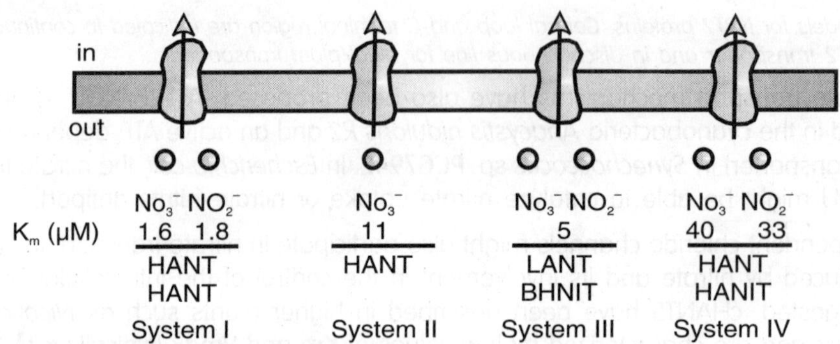

Figure 11.5: Single- and two-component nitrate transport systems in Chlamydomonas.

Mutants in the AtNrt1.1 gene (*chl1* mutants) were originally isolated in the early 1970s from screens based on ClO3 resistance and later shown to be defective in ClO_3^- and NO_3^- uptake. The AtNrt1.1 gene was cloned by T-DNA tagging and found to encode a hydrophobic 65-kDa protein with the characteristic features of a typical membrane transporter.

NRT1 proteins are predicted to have 12 transmembrane domains, with a long loop containing many charged residues separating the first six transmembrane domains from the second six, and short N- and C-terminal ends. The Nand C-terminal domains are quite short. The *Arabidopsis chl1* mutant was the primary source of information about the function of the AtNRT1.1 transporter. CHL1 was initially described as a NO_3^- inducible low-affinity transporter.

The present picture concerning NRT1.1 is complex and shows how a single transporter is involved in regulating multiple functions:

NRT1.1 is now considered as a dual affinity transporter, both HATS and LATS. The phosphorylation of NRT1.1, triggered by limited external

NO_3^- availability, is responsible for the shift from low to high affinity, thus adapting the functional properties of the transporter to the resource level in the root environment.

NRT1.1 is strongly expressed in nascent organs of root and shoot (root tips, emerging lateral roots, and nascent leaves) and plays a crucial role in early phases of development of these young organs. In particular, NRT1.1 mutants display altered root architecture in some conditions, with reduced growth of primary and secondary roots, even in the absence of added NO_3^- in the external medium. This suggests an alternative function for NRT1.1, independent of NO_3^- transport.

It has been reported recently that the mutation of NRT1.1 also leads to lower sensitivity to drought, related to a reduced stomatal opening because of impaired NO_3^- transport in stomata guard cells.

Clearly, the view that NRT1.1 behaves only as a transporter in charge of the NO_3^- uptake from the external medium is an oversimplification. This protein appears to fulfill multiple physiological functions, which are becoming evident more than 30 years after identification of the first NRT1.1 mutant.

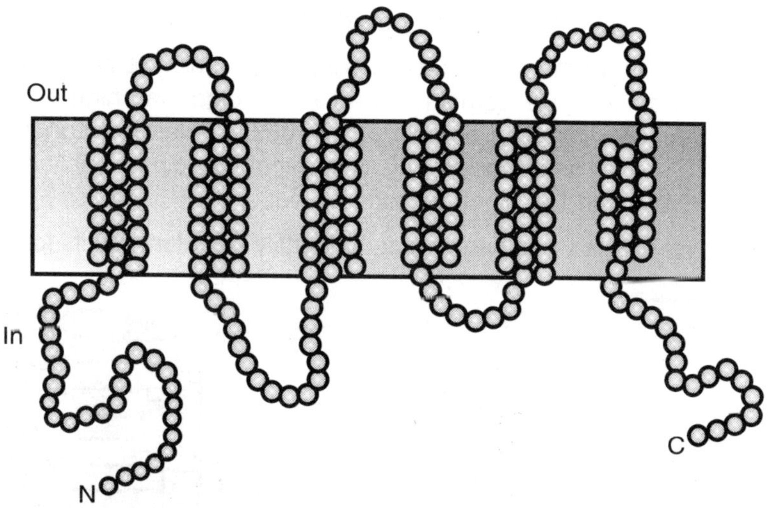

Figure 11.6: Model for the NAR1.1 protein showing the six transmembrane domains. Residues mostly conserved among FNT proteins are shown in red.

One candidate suggested as LATS is the *AtNRT1.2* gene product. This gene is constitutively expressed in the absence of nitrate, and its functional analysis in *Xenopus* oocytes shows specificity for nitrate as a substrate (Km 5.9 *mM*), but it is not able to transport dipeptides or histidine. Transgenic plants containing an antisense *AtNrt1.2* also confirm its role as a LANT. Recently, the *Arabidopsis* NRT1.4 was shown to be a LANT related to nitrate homeostasis in the leaf petiole, so defects in this gene alter leaf development.

Functional studies have been performed over other plant species to demonstrate the functionality of NRT1 transporters. For example, *Brassica napus* NRT1.2 was confirmed to be a LANT. However, in this case the Km for nitrate was voltage dependent, increasing from 4 *mM* at a membrane potential of 40 mV to 14 *mM* at 180mV. In addition to its NO_3^- transport activity, BnNRT1.2 was also found to be able to transport L-histidine, generating even larger currents than with NO_3^-.

Functionality of NRT2 Transporters

The NRT2 transporters, also named NNP (for nitrate–nitrite porter), belong to the major facilitator superfamily (MFS), which includes sugar transporters from mam-mals, plants, yeast, and bacteria. MFS is a divergent group of proteins that are typically 500 to 600 amino acids in length and have a characteristic membrane topology of 12 transmembrane domains arranged as two sets of six, connected by a cytosolic loop. The first eukaryotic member from this family was cloned from *Aspergillus* (*Emericella nidulans, crn*A). A mutation in *crnA* conferred resistance to ClO_3^- and a partial defect in NO_3^- uptake. Subsequently, two homologous genes were discovered in *Chlamydomonas*. These transporters were located in the nitrate assim-ilation gene cluster. Later, a second cluster was shown to contain a third *Nrt2* gene in the alga.

Nrt2 genes have been cloned from a wide range of plant species, fungi, algae, yeast, and bacteria. On the basis of their structural features, members of the NNP family have been classified into several groups. Transporters (from bacteria) are the smallest members of this protein family and have a minimal amount of sequence outside the 12 transmembrane domains.

The fungal members of the family have a large hydrophilic central loop of 90 amino acids located between transmembrane domains 6 and 7. The algal and higher plant members of the family have an extended C-terminal domain of ~70 amino acids that can include an N-terminal sequence extension of ~20 amino acids. This N-terminal domain is highly conserved among the NRT2 family but is absent in the algal and barley sequences.

In silico structural analysis has been made to identify signature motifs for the NRT2 family and

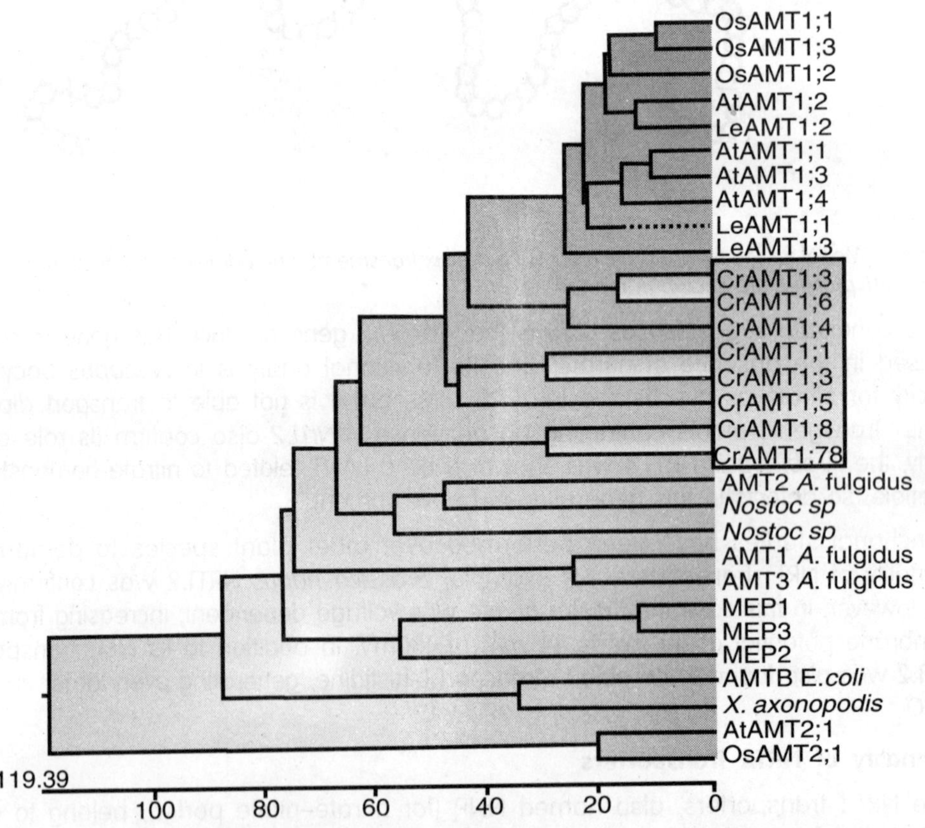

Figure 11.7: Phylogenetic tree of the AMT1 proteins. Alignment was performed with Clust-alW and the tree with the DNAStar package. Chlamydomonas AMT1 are included in a box. Other details are given in the text.

possible substrate recognition. Recently, it has been shown experimentally in *Aspergillus crnA* that many conserved glycine residues throughout the protein sequence have a structural role, while certain conserved charged or polar residues within transmembrane domains (for example, arginine residues conserved within transmembrane domains 2 and 8) are involved in a nitrate-binding function.

NRT2 transporters seem to involve one or two components. In *Chlamydomonas*, a combination of mutant analysis and oocyte expression experi-ments indicates that some NRT2 members (NRT2.1 and NRT2.2) require an addi-tional protein (NAR2) for their functionality, and are thus two-component systems. Other NRT2 proteins, such as NRT2.3 from *Chlamydomonas* and CrnA from *Aspergillus*, are single-component systems.

Until now, NRT2 proteins from plants have not been shown to be functional as single proteins, and they might require a protein homolog to *Chlamydomonas* NAR2, as recently discovered in plants. Nine *Nar2*-type genes have been identified in *Hordeum vulgare* and two in *Arabidopsis*. NAR2 proteins appear to have a single transmembrane domain and may interact with NRT2.1 to modify its function. The amino acid sequence KX2 KX2 LCYX2 SX3 RXWRX3 DX4 DK between amino acids 140 and 180 seems to be characteristic of the higher plant family of NAR2 proteins.

The use of the *Xenopus* oocytes expression system has shown that an NAR2 mRNA (*HvNAR2.3*) was able to reconstitute high-affinity NO_3^- transport activity when co-injected with mRNA for the otherwise inactive *HvNrt2.1*. This result provides strong evidence for the utility of the *Chlamydomonas* model for higher plants.

Chlamydomonas mutant strains defective in several of the nitrate gene clusters have allowed identification of four high-affinity nitrate/nitrite transporters. System I corresponds to a bispecific HANT/HANiT encoded by *Nrt2.1/Nar2*; system II to a monospecific HANT encoded by *Nrt2.2/Nar2*; system III to a bispecific HANiT and LANT, probably encoded by *Nrt2.3*; and system IV to a bispecific HANT/HANiT for which the gene responsible has yet to be identified.

The pH dependence of the nitrate-elicited currents by NRT2.1–NAR2 expressed in oocytes is consistent with an H^+ -cotransport mechanism. These transport systems are differentially regulated by the carbon and nitrogen source. Systems I, II, and III are optimally expressed at high CO_2, and their activity is blocked by ammonium; system IV is expressed optimally under limiting CO_2 and its activity is not inhibited by ammonium.

In contrast to systems I, II, and III, system IV is inhibited by CO_2. Concerning the function for each of these systems, mutants deleted in systems I and II and carrying functional systems III and IV are unable to grow efficiently in nitrate media. Thus, under sufficient CO_2, systems I and II have a primary function in the provision of nitrate for growth and systems I and III in nitrite entry.

NAR1 Transporters

The nitrite transport step into the chloroplast is not well documented in plants, probably because of the lack of any molecular evidence and of the long-standing assumption that nitrite can diffuse freely as nitrous acid into the chloroplast.

However, nitrite uptake into intact pea chloroplasts shows saturation kinetics, pref-erence for alkaline pH, and sensitivity to protein modifiers; this favours the existence of a nitrite-mediated channel or transporter vs. the permeation of nitrous acid.

Nitrite transport into chloroplast inner envelope vesicles from pea has been evaluated by Shingles et al.. These authors propose that nitrite rapidly diffuses across the plastid membrane depending on a proton gradient, so the proton-linked NO_2^- transport should be bidirectional.

Nitrite concentrations change *significantly* in roots of barley seedlings, depending on the nitrate availability in the environment, and in spinach leaves during the light–dark transitions. Thus, the need for a plastidial nitrite *transporter* is important for two reasons: (1) to avoid cellular toxicity of nitrite; and (2) to increase efficiency of the nitrite step.

Studies with *Chlamydomonas Nar1.1* have provided the first molecular evidence that nitrite transport to the chloroplast is a regulated process mediated by specific transporters rather than the result of diffusion. In spite of identification of six members of the NAR1 family in *Chlamydomonas*, further studies will be required to know the role of each NAR1 protein.

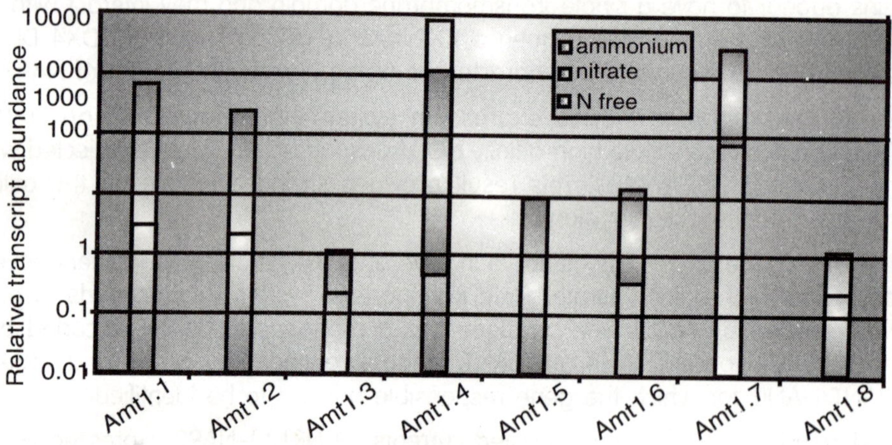

Figure 11.8: Expression of Amt1 transcripts in different nitrogenous media. The relative abundance is referred to that of ubiquitin ligase transcript used as a control.

Available data show that the *Nar1* gene family may be closely associated with carbon and nitrogen metabolism because *Nar*1.1 is nitrate upregulated and under control of the nitrate-pathway-specific regu-latory gene *Nit*2, whereas *Nar*1.2 (*Lci*A) is upregulated by low CO_2 and under control of the carbon-pathway-specific regulatory gene *ccm1*.

In plants, NAR1 proteins cannot be identified on the basis of sequence homology, but the role of NAR1.1 is so fundamental in *Chlamydomonas* that its function needs to be carried out by another protein family in plants. The precise function for *Nar1*.1, as a chloroplast nitrite transporter allowing for efficient nitrate utilization, has been deduced on the basis of the following data:

Nar1.1 is located in the nitrate cluster and coregulated with the other clustered genes (Figure 7.2; [41]). Thus, *Nar1.1* is expressed in nitrate but not in ammonium media.

NAR1.1 corresponds to an integral membrane protein predicted to have six spanning-membrane domains, a plastidial localization, and significant iden-tity to formate and nitrite transporters from bacteria. The nitrite uptake activity by intact chloroplasts isolated from Nar1;1 + and Nar1;1 – strains supports the notion that NAR1;1 is a plastidial nitrite trans-porter with an apparent Ks for nitrite of about 5 μ*M*.

Nar1.1 allows nitrate utilization when this nutrient is limiting for the cells. This limitation of nitrate takes place in strains lacking HANT systems I and II at nitrate millimolar concentrations, or in strains having HANT with nitrate micromolar concentrations in the medium.

Nar1.1 improves nitrate use efficiency for growth under light/dark cycles and low CO_2 environments. Under such conditions, strains lacking *Nar1.1* uncouple nitrate reduction from the cells' capability to assimilate the ammonium produced because of a significant deregulation in expression of enzymes and transporters for nitrate assimilation including GS1.

AMMONIUM ASSIMILATION

Incorporation of ammonium is a basic process shared by nitrate assimilation and other alternative nitrogen source pathways. The GS/GOGAT cycle is the major step for ammonium incorporation into carbon skeletons by photosynthetic organisms. GS catalyses the formation of glutamine from ammonium and glutamate in an ATP-dependent reaction.

In plants, GS isozymes are encoded by multigene families and some of their members show cytosolic localization, while others have a plastidial or nodular localization. The *Chlamydomonas* genome sequence shows three *GS* genes; *GS1* encodes a cytosolic form and *GS2* and *GS3* are proposed to encode plastidial isozymes.

Although a *GS3* expression pattern is not documented, the major transcript level corresponds to that of *GS2*, which is constitutively expressed with respect to nitrogen supply. *GS1* transcripts increase in cells grown in nitrate and decrease in cells grown in ammonium. GS1 is proposed to play an active role in nitrate assimilation.

GOGAT catalyses the transfer of the amide group from glutamine to 2-oxoglu-tarate in a reaction that requires reducing equivalents. Two molecules of glutamate, the substrate of GS, are produced. Two GOGAT isoenzymes—one specific for reduced ferredoxin as electron donor and another specific for NADH—have been characterized in plants.

Fd-GOGAT is a 130- to 150-kDa monomer, with a [3Fe-4S] center located in plastids and roots; NADH-GOGAT is a 158- to 240-kDa monomer with the same prosthetic group, located in the plastids. GOGAT gene number per genome differs between species. In *Chlamydomonas*, single genes encod-ing the NADH- and the ferredoxin-GOGAT are present. These enzymes have been characterized in detail.

AMMONIUM TRANSPORT GENES

As schematized in Figure elsewhere in this chapter, ammonium transporters are expected to operate in different cell localizations (plasma membrane, chloroplast, and mitochondria) mediating influx and efflux. Because ammonium is a strong negative signal of nitrate assimilation, it is important to know the different transporters that could mediate its effects.

Knowledge of the mechanisms regulating ammonium transport systems is also essential for

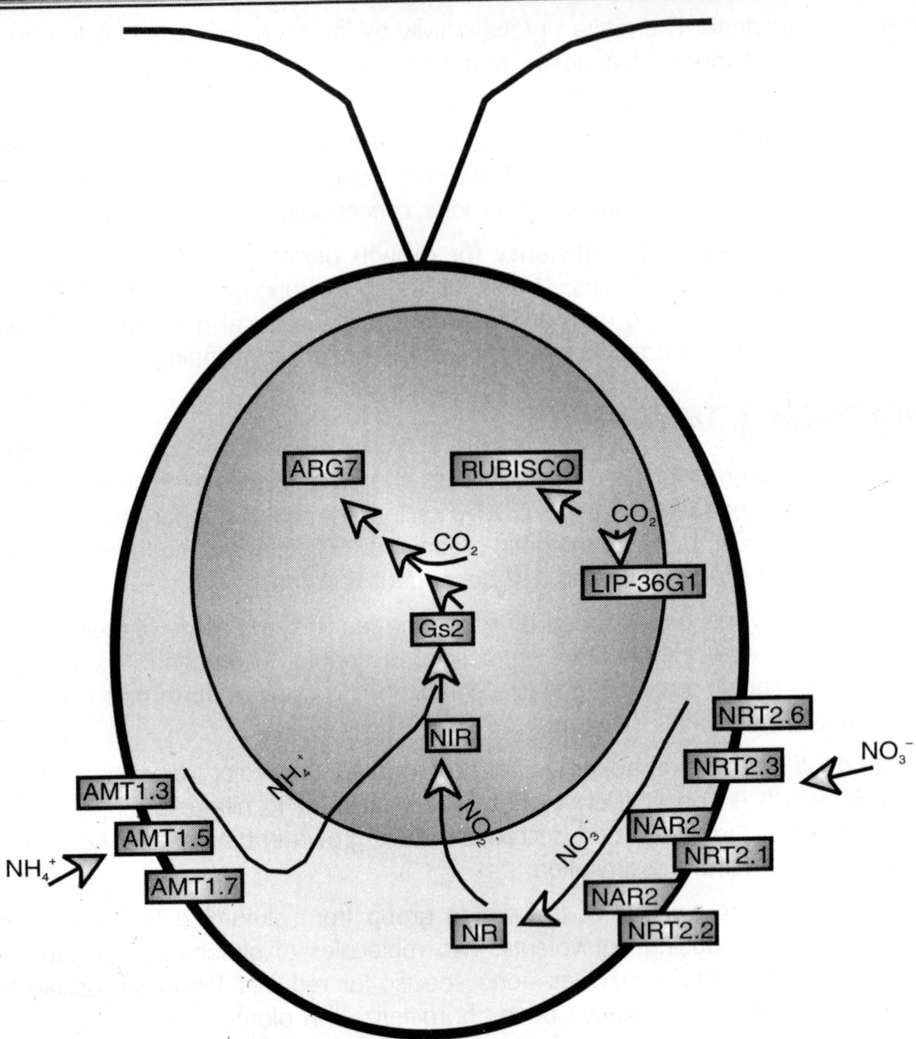

Figure 11.9: Gene targets of Chlamy1 protein, modulat-ing circadian rhythm. The scheme includes genes proposed to be a target of chlamy1 by Waltenberger et al. (pink boxes), plus other putative target genes proposed on the basis of their sequence containing UG repeats.

better understanding of nitrogen metabolism and plant growth. At the physiological level, biphasic kinetics for ammonium uptake in several species have been assigned to two different systems of transport: low-affinity trans-port systems (LATS), which are related to passive K^+ channels, and high-affinity transport systems (HATS) that mediate active transport by coupling NH^+_4 influx to a H^+ gradient.

Similarly, at the physiological level, in *Chlamydomonas* cells, two transport systems subject to circadian rhythm are thought to participate in ammonium uptake: a constitutively expressed, low-affinity version and a high-affinity version negatively regulated by ammonium. Although LATSs have not yet been described at the molecular level, HATSs have been widely described in diverse organisms like plants, yeasts, bacteria, fungi, and animals, and form the AMT/MEP family.

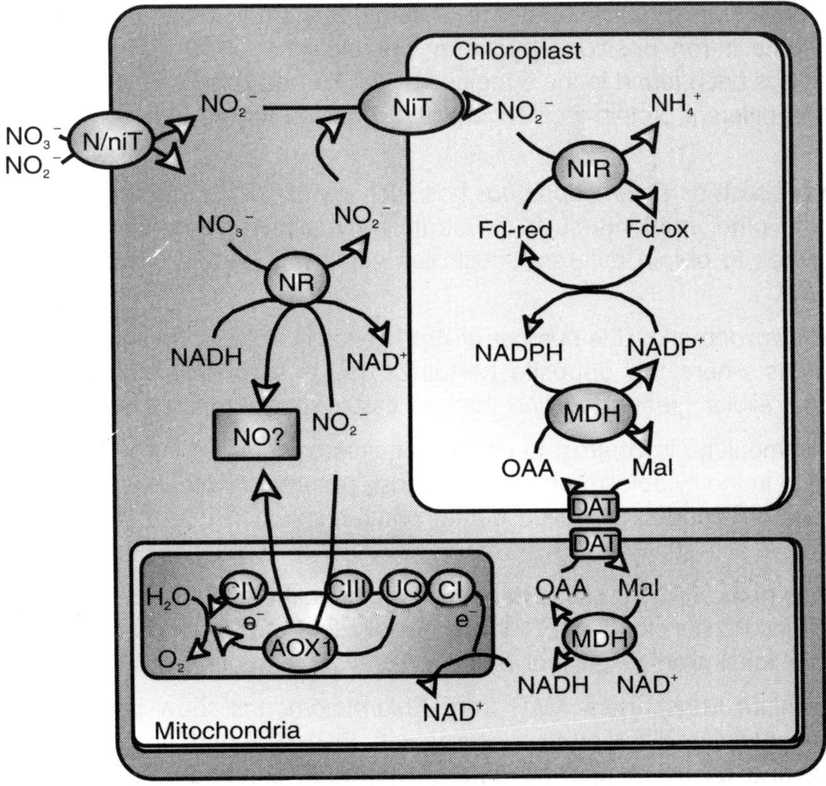

Figure 11.10: The redox valves and the nitrate assimilation pathway.

This family has also been related by homology with the human rhesus (Rh) blood proteins. *Chlamydomo-nas* is one of the few organisms that have genes from both families: *Amt* and *Rh*. A possible role of the *Chlamydomonas* Rh proteins is as a bidirectional channel for ammonia and CO_2.

In plants, most AMTs have been included in a large family designated AMT1, with the exception of some AMTs identified in *Arabidopsis* and *Lotus japonicus* that are included in AMT2, a new family of transporters whose sequence identity is closer to yeast and bacteria than to plants. Five members of the AMT1 family have been described in *Arabidopsis* and three in tomato; its existence is also known in other species such as rice, *Brassicanapus*, and *Lotus japonicus*.

Eight members of the *Amt1* gene family (*Amt*1.(1 to 8) genes) have been iden-tified in *Chlamydomonas*, representing the largest family thus far described in any organism. On the other hand, no *Amt2* genes have been found in *Chlamydomonas*, suggesting that the proposed role of AMT2 transporters in plants could be undertaken by some of the AMT1 proteins in the alga.

In Figure elsewhere in this chapter, the phylogenetic relationships among these transporters and those from other organisms are shown. AMT1.1 to 6 are closely related to the plant AMT proteins, whereas AMT7 and 8 are more distantly related to plants.

The *Chlamydomonas Amt1* genes show interesting evolutionary relationships because a particular feature present in all is the existence of some very small exons (less than 40 bp) and conservation of some intron positions that may be related to its transcriptional regulation. Alternative splicing has been found in the 5 regions of *Amt1.5* and *Amt1.7*. This complex regulation might even result in different protein localization, as suggested for the two isoforms deduced for AMT1.7.

That a single cell such as *Chlamydomonas* has such a wide set of putative transport systems for ammonium and other nutrients such as nitrate may reflect the need for complementary affinities and activities to obtain different substrates effi-ciently under changing environmental conditions.

However, in *Chlamydomonas*, the number of *Amt1* genes is higher than that of the *Nrt2* genes, in contrast to plants where the opposite pertains. This is surprising if *Chlamydomonas* is considered as a unicellular green alga and thus no tissue specialization exists.

In any case, ammonium, in contrast to nitrate, requires a regulated intracellular flux because it can be assimilated in the cytoplasm or the chloroplast, accumulated in the vacuole, or excreted to the medium, and ammonium produced during photo-respiration needs to be exported from the mitochondria.

The localizations of the AMT1 proteins remain unknown, although several predictions suggest a clear chloroplast localization for AMT1.2 and especially for AMT1.4, whose hypothetical signal peptide has substantial similarity with that of the *Chlamydomonas* chloroplast protein NiR.

Putative ammonium transporters AMT1 from *Chlamydomonas* show particular expression patterns in media containing different nitrogen sources: (1) *Amt1.4* and *Amt1.7* have the highest expression levels; and (2) by comparing different nitrogen conditions, putative ammonium transporters 1, 2, 4, and 5 have maximum expression in an N-free medium; the ammonium transporters 3 and 8 show maximum expression in nitrate medium, whereas 6 and 7 are enhanced in ammonium medium.

The expression of the *Chlamydomonas Amt1;1* has a complex regulatory mechanism responding to repression by ammonium, ammonium derivatives, and nitrate in a process mediated by the regulatory gene *Nit2*, so NIT2 provides a signal for the presence of a usable nitrogen form—nitrate—connecting the pathway of ammonium uptake with that of nitrate assimilation.

This means that NIT2 has a dual role in gene expression: the well-known positive one on nitrate assimilation and a novel negative one on *Amt1;1* expression. *Chlamydomonas Amt1;1* has an expression pattern similar to that of the *Arabidopsis* gene *Amt1;1*, which encodes the major system responsible for ammonium uptake.

NITRATE ASSIMILATION AND LIGHT/CIRCADIAN RHYTHM

Light positively regulates most of the nitrate assimilation pathway genes and activities. It is known that NR activity and mRNA levels of plants grown in a day/night cycle fluctuate between a maximal level at the beginning of the day period and an almost undetectable level at the end of the day.

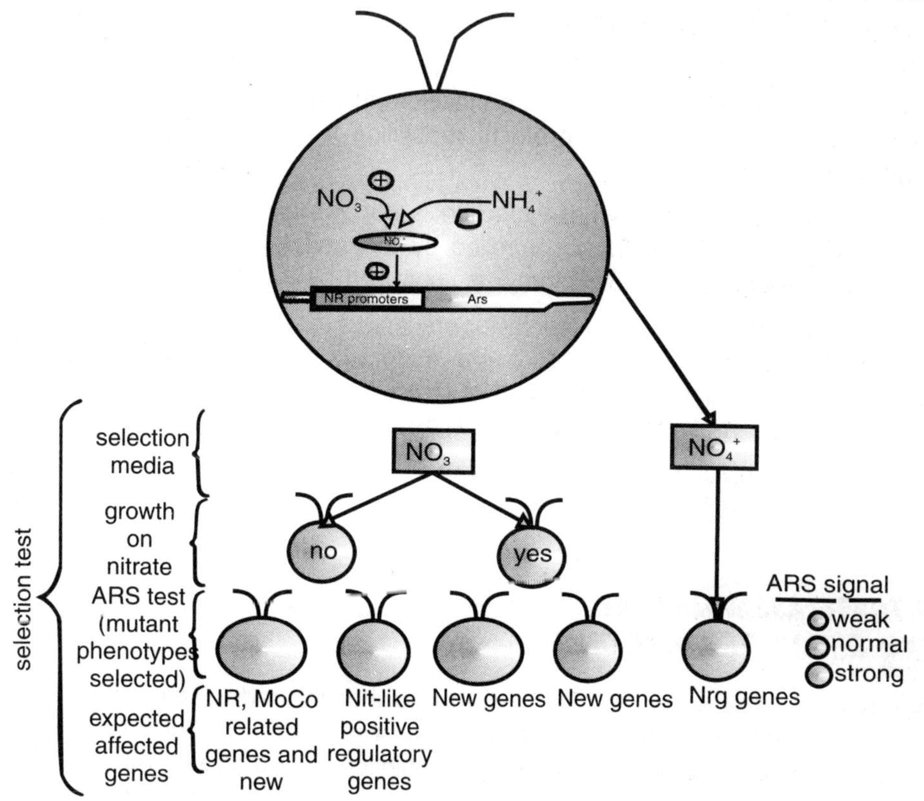

Figure 11.11: The isolation of nitrate assimilation mutants defective in positive and negative signaling. ARS = arylsulfatase activity.

Further evidence of this light regulation are the diurnal changes of nitrate and nitrite uptake that generally peak during the light period and reach a minimum in the dark. An endogenous clock regulates the temporal expression of genes/mRNAs involved in the circadian pathway. In *Chlamydomonas*, a recently identified, clock-controlled RNA-binding protein (Chlamy 1) represents an analog of the circadian trans-acting factor, CCTR, from the phylogenetically diverse algal species *Gon-yaulax polyedra*.

Chlamy1 protein could act as a translational repressor, pre-venting translation of UG-repeats-bearing mRNAs from the end of the light until the end of the dark phase. The strength of the interaction between Chlamy1 and transcripts may be influenced by the number of UG units in a region.

Chlamy1 has been cloned and is composed of two subunits: C1 is involved in protein–protein interaction and C3 bears three RNA recognition motif domains. Chlamy1 was reported to bind mRNAs whose products are key components of nitrogen and CO_2 metabolism. Some of the genes under Chlamy1 control are involved in uptake of nitrite (*Nrt2*.3), its reduction to ammonium (*Nii1*), fixation of ammonium as glutamine (*GS2*), and arginine biosynthesis (*ARG7*).

Other proteins encoded by these RNAs are related to photosynthesis and the CO_2 shuttle into the chloroplast (LIP36-G1), and CO_2 fixation as ribulose-1,5-bisphosphate carboxylase (RBCS1, the small subunit of RUBISCO).

Another is the light-dependent NADPH-protochlorophyllide-oxidoreductase (L-POR), one of two PORs that catalyze the reduction of protochlorophyllide to chlorophyllide, a regulatory step in chlorophyll biosynthesis.

Finally, YPTC4 is a G-protein whose function is not yet fully understood. These genes, plus others containing UG units putatively controlled by Chlamy1, are shown in Figure elsewhere in this chapter. The binding activity of Chlamy1 is controlled by the circadian clock.

Thus, the regulatory properties of this factor are limited to a certain time window, which begins at the end of the day phase and ends in the middle of the night phase. During this period, binding activity of Chlamy1 is high, and at other time points it is low. Clock-controlled RNA-binding proteins have also been identified in other species.

In *Arabidopsis*, a glycine-rich RNA-binding protein, GRP7, was characterized. Both transcripts and the protein are components of a negative feedback circuit capable of generating a stable oscillation.

NITRATE ASSIMILATION AND REDOX REGULATION

It has been recently shown in *Chlamydomonas* that the redox state of the plasto-quinone (PQ) pool plays a key role in *Nia1* gene expression. A reduced PQ pool is needed as a positive signal to allow *Nia1* expression so that chemical inhibitors or mutations preventing loading of PQ with electrons have a negative effect on expression.

These results are related to others reported in tobacco, *Arabidopsis*, and *Lemna*, where signals from the photosynthetic electron flow are controlling *Nia1* gene expression. Nonetheless, in plants, an oxidized state of an electron transport component appears to activate expression.

This apparent disagreement may reflect differences in the physiology and biochemistry of nitrate reduction in plants and algae. Nitrate and nitrite reduction requires reducing power. In photosynthetic cells, this reducing power is supplied by the photosynthetic electron transfer chain (medi-ated by NADPH-ferredoxin oxido-reductase) generating reduced ferredoxin for nitrite reduction and redox valves, which feed the cytosol with NAD(P)H for nitrate reduction.

Some participation of the Calvin cycle steps has been demonstrated: glucose-6-phosphate dehydrogenase and Calvin cycle activities are increased in nitrate media in *Chlamydomonas* and *Arabidopsis*. In *Chlamydomonas*, NMDH and AOX1, which are closely related to their coun-terparts in plants, are involved in processes that control the fine-tuning of redox metabolism in chloroplast and mitochondrion.

NMDH partici-pates in the export of reducing equivalents from the chloroplast to the cytosol by the malate–oxalacetate shuttle. Activity of this "malate valve" is regulated by the redox switch of NMDH, which is turned on and off by light signaling mediated by the thioredoxin–ferredoxin system.

AOX is a mitochondrial terminal oxidase that funnels electrons to oxygen from reduced ubiquinone. It bypasses the respiratory pathway through complexes III and IV of the mitochondrial electron transfer chain and the concomitant generation of electrochemical potential; thus, synthesis of the amount of ATP per electron pair is diminished.

AOX could play a role as a redox valve to balance carbon metabolism and mitochondrial electron transport. Clustering of *Nmdh1* and *Aox1* genes with nitrate assimilation genes in *Chlamydomonas* provides a direct molecular link between NADH and ATP pool regulation and nitrate assimilation. NMDH provides reducing power for cytosolic nitrate reduction, and its activity level is compromised by competition for reduced ferredoxin, the electron donor for nitrite reductase.

Specific induction of *Aox1* by nitrate and not by ammonium might indicate the requirement to adapt mitochondrial electron flow and thus ATP synthesis during assimilation of these nitrogen sources. This adaptation is especially critical under stress conditions, such as high concentrations of nitrite in the cytosol that might act as substrate of NR and be converted to nitric oxide or in the mitochondria where NO has been shown to be produced by the alternative oxidase in *Chlorella*.

NO in turn inhibits the mitochondrial electron transport chain, so alternative oxidase would be essential for ATP production under NO presence. Whether or not NO production is the result of stress or a metabolic condition that triggers a signaling cascade will require the identification of intermediates in this cascade and the final target of action for the cell response.

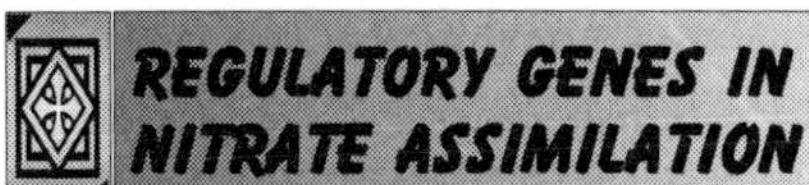

REGULATORY GENES IN NITRATE ASSIMILATION

Membrane proteins, such as pumps, ion channels, transporters, and receptors, are responsible for controlling the flow of nutrients and other solutes across the plasma membrane and cellular organelles. They also provide the appropriate signaling metabolites for modulating the cell response to changing needs according to nutrient availability and developmental or physiological status.

Nitrate is sensed, and this up- or down-regulates the expression of a large number of genes, some of which are specific for nitrate assimilation. Other genes link this route to different metabolic pathways. Studies with *Arabidopsis* have found that by using transcriptome analysis in short-term treatments with low levels of nitrate, more than 1000 genes are significantly induced or repressed within 20 min of treatment. In roots, 270 genes were differentially expressed in media containing nitrate as opposed to ammonium nitrate.

In *Chlamydomonas*, nitrate signaling occurs intracellularly and directly depends on the activity of the nitrate transport systems. Thus, in *Chlamydomonas* the presence of particular high-affinity nitrate transport systems has an important regulatory role in expression of nitrate assimilation genes. According to their affinity, specificity, and capacity for nitrate, these systems are responsible for differential nitrate signaling by regulating intracellular concentrations of nitrate.

Recently, serial analysis of gene expression (SAGE) transcriptome analysis in the nitrate transporter Chl1 mutant of *Arabidopsis* has revealed how a single trans-porter affects the expression of hundreds of genes. Because of the marked deregu-lation of *Nrt2.1* in the mutant, it was suggested that *Nrt1.1* plays a direct signaling role in regulating other nitrate transporters.

Ammonium transporters have been suggested to participate in the sensing of ammonium. The yeast ammonium transporter MEP2, dispensable for cell growth, has been implicated in the sensing of ammonium. In *Chlamy-domonas*, the ammonium-uptake defective mutant strain 2170 is also relieved from the negative effect of ammonium or methylammonium on the nitrate pathway.

The regulatory genes for nitrate assimilation are well defined in fungi and yeast. Two positively acting regulatory genes are required for expression of nitrate transporter and nitrate reduction genes in fungi. *NirA/Nit4/YNA1* genes from *Aspergillus, Neurospora,* and *Hansenula,* respectively, are pathway-specific genes involved in nitrate induction and correspond to GAL4-like Cys6/Zn2 binuclear zinc cluster. AREA/NIT2 from *Aspergillus* and *Neurospora* are major regula-tory proteins mediating nitrogen repression from readily usable nitrogen sources such as ammonium and glutamine and correspond to GATA-binding transcription factors.

The mechanism proposed for the NIT2 negative function in *Neu-rospora* is that NMR1, a negatively acting protein, interacts with NIT2 in the presence of catabolic repressors and prevents the binding of NIT2 to DNA for transcription of target genes.

Various attempts have been made to identify regulatory genes from plants. However, the relationship of the isolated genes to regulation of the nitrate pathway has not been shown, even though a GATA motif was found in the promoter of the spinach NiR gene and *in silico* analysis revealed a large number of GATA-family transcription factors in *Arabidopsis* and rice.

The regulatory fungal model does not seem to fit in photosynthetic eukaryotes and clues for regulatory genes may therefore come from systems such as *Chlamy-domonas*. It has been hypothesized that several genes act to mediate the positive effects of nitrate and the negative ones of ammonium. Their deficiency may lead to partial phenotypes that can explain the difficulties experienced in the genetic dis-section of regulation in algae and plants.

In *Chlamydomonas*, the expression of nitrate assimilation genes is co-regulated. These genes are subject to repression by ammonium, induction by nitrate, and the control of the positive-acting regulatory gene *Nit2*. This gene was cloned by transposon tagging from chlorate-resistant mutants. Although struc-tural analysis of the deduced NIT2 protein has not been performed, the *Chlamydomo-nas* genome database reveals that NIT2 is a transcription factor containing an RWPXRK box present in plant proteins involved in N control and the *Chlamydomonas* mid protein involved in minus mating type dominance.

Nit2 is itself subject to ammonium repression, which implies an additional level of control. A functional genomics approach to identify regulatory genes for nitrate assim-ilation has been recently performed in *Chlamydomonas*. By taking advantage of deletion events that occur during integration of a heterologous marker in *Chlamy-domonas* transformation, along with the *Chlamydomonas* genome sequence, anordered mutant library of 22,000 strains has recently been obtained.

Assuming that *Chlamydomonas* contains about 17,000 genes and deletions affect wide genomic regions, it has been proposed that such a number of mutants will be sufficient to cover most of the *Chlamydomonas* genome.

The arylsulfatase reporter gene under the control of the *Nia1* gene promoter has served as a sensor for identifying regulatory mutants. A forward screening of the library allowed for the selection of 145 mutants defective at putative genes related to the positive signal of nitrate or the negative signal of ammonium.

The ammonium-insensitive mutants were found to be defective at genomic regions bearing putative new genes related to regulatory functions such as guanylate cyclase, protein kinase, peptidyl-prolyl isomerase, or DNA binding, although no direct evidence correlating these genes with ammonium repression exists. In addition, some insertions in ammonium-insensitive mutants map to unknown regions or genes in the genome.

From these and previous results, it is becoming evident that a complex network of signaling proteins mediates the effects of ammonium and its derivatives on the nitrate assimilation pathway. Now, it will be possible to explore additional regulatory pathways by direct or reverse genetics—for example, those involving TOR or PII proteins reported in other organisms and considered to be integrators of nutrient availability (amino acids and energy) and as key players of nutrient-mediated signal transduction.

CONCLUSION

As the most important pathway for nitrogen acquisition by crop plants, nitrate assimilation needs to be understood in its most basic molecular aspects. In spite of its simplicity, this pathway has many intriguing questions still to solve. Model systems can provide invaluable information in the search for key genes and functions in plants and, finally, to the optimization of nitrogen use efficiency. In the age of genomics, comparative biology and mutant collections, the work carried out in *Chlamydomonas* should progress further in this direction.

PLANT TRANS PROTEIN MICRO-ARRAYS

P*osttranslational modification* of proteins by *phosphorylation* is the most abundant type of cellular *regulation* affecting *essentially* every intracellular process of eukaryotes. Phosphorylation of a protein can cause changes in its structure, stability, *enzymatic* activity, the ability to interact with other molecules, or its subcellular localization.

Protein kinases catalyze the reversible phosphorylation of protein substrates at serine, threonine or tyrosine residues. In *Arabidopsis, e.g.,* serin/*threonine kinases* represent about *4%* of the proteome, but their *biological* function is not well understood. Therefore, *high-throughput* proteomics methods for global analysis of protein kinase function are needed to identify downstream substrates.

Various techniques for *determining* consensus phosphorylation site sequences, including peptide libraries and peptide arrays led to the identification of kinase substrates. *Furthermore,* a solid-phase *phosphorylation* screening of X phage cDNA *expression* libraries as well as different protein-protein interaction screening methods, such as overlay methods and the yeast two-hybrid system have been applied in this respect.

More recently, protein kinases were engineered to accept unnatural adenosine triphosphate (cyclopentyl ATP) analogs and have been used to identify specific substrates. Furthermore, preliminary studies have demonstrated the suitability of protein microarrays for the study of *phosphorylation* by kinases.

To elucidate the sites of phosphorylation, *antibodies* against phosphorylated protein epitopes may be used for detection on protein microarrays. Furthermore, peptide arrays or mass spectrometry (MS)-based methods may be applied in this respect. We describe here a protein microarray-based method for phosphorylation screening of proteins and the identification of potential kinase substrates in a high-throughput manner.

We successfully used this screening tool to identify novel targets for the barley casein kinase 2α (CK2α) as well as for different *Arabidopsis thaliana* mitogen-activated protein (MAP) kinases. Our method utilizes plant protein microarrays generated as described in detail in the previous chapter in this book.

Then microarrays have to be incubated with the soluble and active kinase in the presence of radioactive [γ^{33}-P] ATP. Radioactive signals detected by phosphor imager or X-ray film were then evaluated for the selection of potential substrates.

We verified the potential substrates by on-blot *phosphorylation* in vitro. As a whole, our approach allows *shortlisting* candidate substrates of plant protein kinases for further analysis. Follow-up in vivo experiments are essential to evaluate their physiological relevance.

MATERIALS

Purification of Recombinant Kinases Under Native Conditions

1. Medium for culturing cells. 2YT or LB containing 100 µg/mL ampicillin, 15 µg/mL kanamycin, and 2% glucose.
2. Isopropyl-β-D-thiogalactopyranosid (IPTG).
3. Native lysis buffer: 300 mM NaCl, 50 mM NaH_2PO_4, 10 mM imidazole, pH 8.0.
4. Native wash buffer: 300 mM NaCl, 50 mM NaH_2PO_4, 20 mM imidazole, pH 8.0.
5. Native elution buffer: 300 mM NaCl, 50 mM NaH_2PO_4, 250 mM imidazole, pH 8.0.
6. Lysozyme.
7. Phenylmethylsulfonyl fluoride (PMSF).
8. Ultrasonic homogenizer (Branson Ultrasonic, Danbury, CT).
9. NiNTA-agarose (NTA:nickel-nitrilotri-acetic acid, Qiagen).
10. 1 mL Polypropylene columns (Qiagen, Hilden, Germany).
11. Bradford reagent (Bio-Rad, Munich, Germany).

Kinase Assay on Protein Microarrays

1. TBS + Tween (TBST): 10 mM Tris-HCl, pH 7.5, 150 mM NaCl, 0.1% (v/v) Tween.
2. Blocking solution: 2% bovine serum albumin (BSA; Sigma, St. Louis, MO) in TBST.
3. [γ^{33}-P]ATP, 250 µCi/mL (Amersham Pharmacia Biotech Europe, Freiburg, Germany)
4. CK2a buffer: 25 mM Tris-HCl pH 8.5, 10 mM $MgCl_2$, 1 mM dithiothreitol (DTT).
5. FASTTM-slides (Whatman Schleicher & Schuell, Dassel, Germany).
6. Cover slide (Carl Roth, Karlsruhe, Germany).
7. X-ray cassette (Hypercassette, Amersham Pharmacia, Freiburg, Germany).

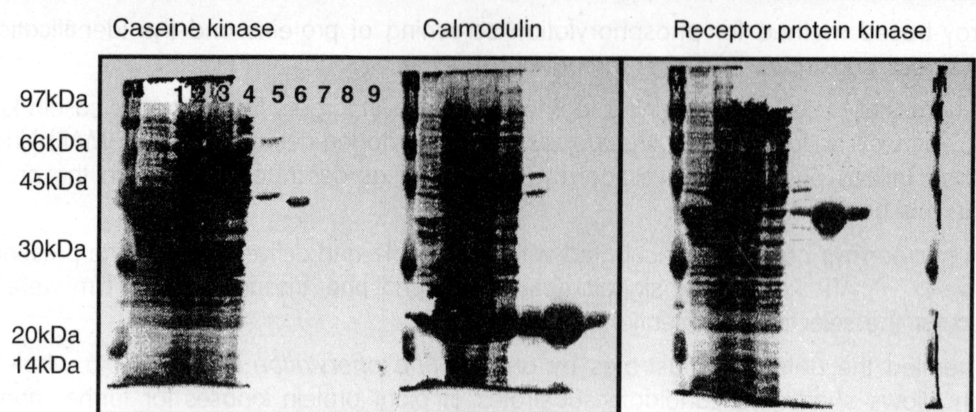

Figure 12.1: SDS-PAGE (15%) of two native purified kinases and calmodulin (Coomassiestained). Nine samples of each purification step were separated: 1, pellet (insoluble fraction after centrifugation); 2, supernatant (soluble fraction after centrifugation); 3, first column flowthrough; 4, 5, first and second washing fraction; 6–9, eluates. M, molecular weight marker. Approximate sizes of the protein markers are shown at the left in kDa.

8. X-ray film (Kodak, Stuttgart, Germany).
9. Imaging screen (BAS-SR 2025, Fujifilm, Japan).
10. Phosphor imager (BAS-Reader-5000, Fujifilm).

METHODS

Purification of Recombinant Kinases Under Native Conditions

For phosphorylation screening, the kinase has to be soluble, active, and pure, which means that no other kinase should be present. To achieve this, largescale production and purification of recombinant His-tagged protein kinases can be carried out from cDNA expression libraries, e.g., a barley expression library constructed in the *E. coli* vector pQE30NST (accession number: AF074376; 22) or an *Arabidopsis* expression library in vector pQE30NASTattB. To express and purify protein kinases from these libraries, we apply the following protocol:

For expression:

1. Falcon tubes (50 mL) were filled with 10 mL of medium for culturing cells.
2. Cultures were inoculated with bacteria expressing the His-tagged protein kinase. They were taken from 384-well plates, which were stored at –80°C.
3. After overnight growth at 37°C with vigorous shaking, the cultures were transferred to 300-mL Erlenmeyer flasks, and 90 mL of prewarmed 2YT medium, supplemented with 100 μg/mL of ampicillin, and 15 μg/mL of kanamycin were added. The incubation was continued until an OD of 0.7 was reached.
4. To induce protein expression, IPTG was added to a final concentration of 1 mM, and incubation was continued for 4 h at 37°C.
5. Cells were transferred into 50-mL Falcon tubes (two for each culture), harvested by

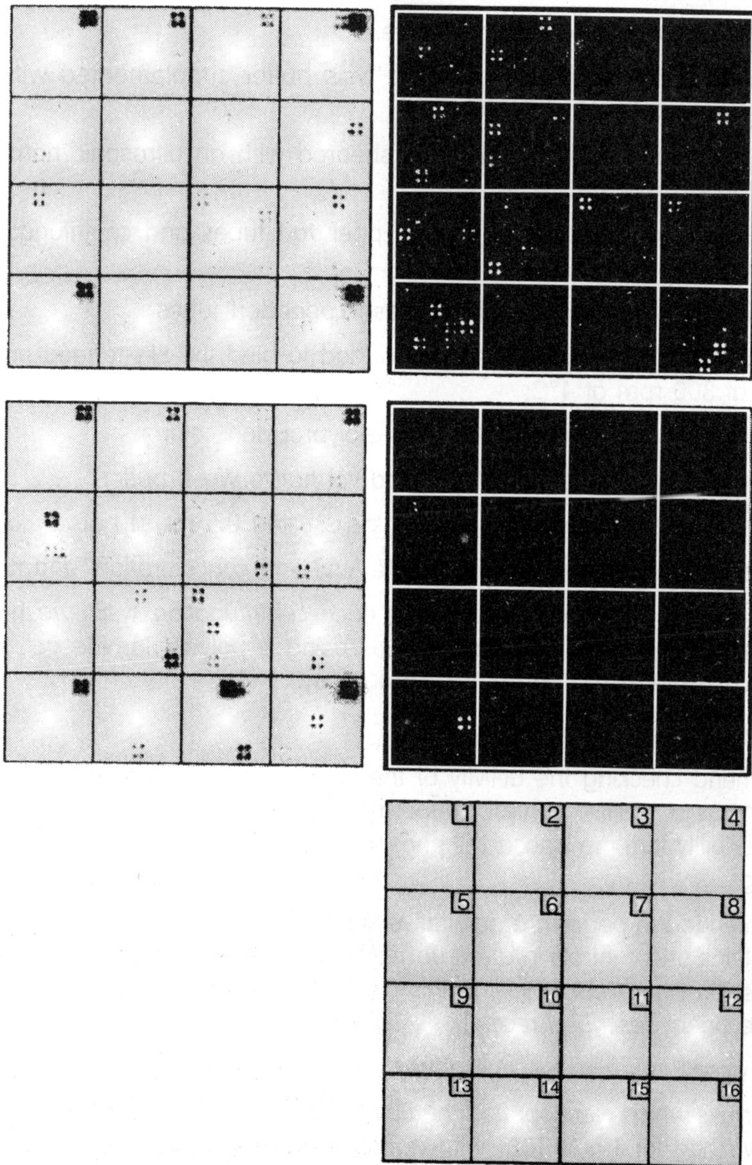

Figure 12.2: Kinase and immunoassay with 768 different barley proteins, immobilized on FAST-slides (384 proteins in each field). Left: X-ray film image with the results of a representative CK2α assay. Right: equal microarrays screened with an anti-RGSHis6 antibody. The map at the bottom shows the positions of the 16 controls. The purified proteins and the controls were spotted in quadruplicate. Controls: 1, 4, 13, and 16, high-mobility group (HMG)(1), 0.29 mg/mL; 2, HMG(2), 0.76 mg/mL..

centrifugation at 1900g for 10 min, and stored immediately at –80°C.

For native protein purification (all the following steps were performed at 0–4°C):

1. The frozen pellets were thawed on ice.
2. Cells were lysed in 0.5 mL of native lysis buffer supplemented with 0.25 mg/mL of lysozyme and 0.1 mM of PMSF.
3. Lysates were pooled, and DNA was sheared with an ultrasonic homogenizer for 3X 1 min at 50% power on ice.
4. Lysates were transferred into 1.5-mL Eppendorf tubes and centrifuged at 20,000g, 4°C for 30 min.
5. Supernatants were transferred into fresh Eppendorf tubes.
6. Then 250 µL of NiNTA-agarose were added to bind the His-tagged proteins by shaking for 1 h at 300 rpm at 4°C.
7. The mixtures were transferred to 1-mL polypropylen columns.
8. The columns were washed with 10 bed vol native wash buffer.
9. Proteins were eluted with four elution steps using 0.5 mL of native elution buffer each.
10. Proteins were mixed with glycerol (20% [v/v] end concentration) and stored at –80°C.

We recommend collecting an aliquot from each *centrifugation*, lysis, wash, and elution step to control the efficiency of the purification steps using a polyacrylamide gel, e.g., 15%. Protein *concentration* was determined by the Bradford assay.

Kinase Assay on Protein Microarrays

We recommend checking the activity of the kinase prior to microarray *experiments* using a known *substrate* as a positive control. Furthermore, prior to microarray experiments, one has to exclude the possibility that the kinase will phosphorylate 3' or 5' tags of the *recombinant* proteins, such as His-tags.

Here, we intended to perform a qualitative signal evaluation after the kinase assay. Therefore we spotted proteins and controls in *quadruplicates* on FAST-slides using a 10 × 10 spotting pattern, as *demonstrated* in Figure elsewhere in this chapter. Spotting patterns for following quantitative evaluation have been *described* recently.

For kinase assay:

1. Microarrays which were spotted the day before and stored at 4°C in a closed slide holder, were washed for 1 h in TBST with vigorous shaking at room temperature.
2. Washed microarrays were blocked for 1 h at room temperature with 2% BSA/TBST.
3. Blocked microarrays were placed on Whatman paper soaked with TBST to avoid drying of the microarrays during the subsequent kinase reaction.
4. Microarrays were incubated with 250 tL kinase solution containing 13 µg/mL of CK2α and 25 µCi/mL of radioactive labeled [γ^{33}-P]ATP in CK2α buffer for 1 h at room temperature. This incubation was performed underneath a cover slide.

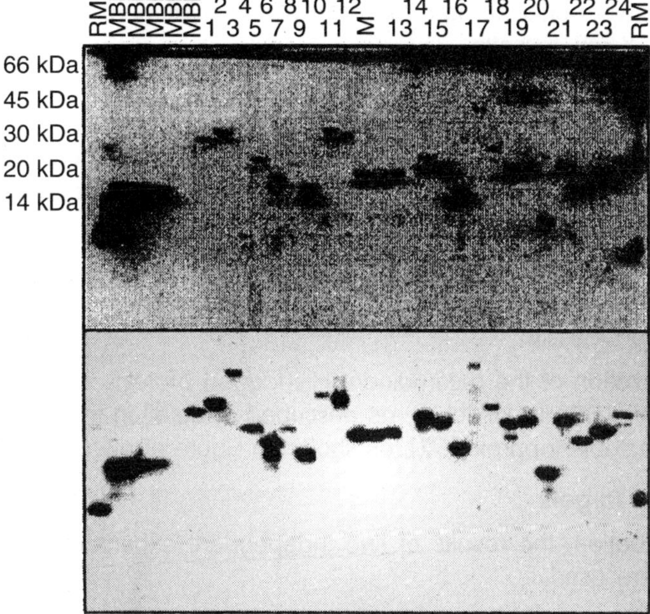

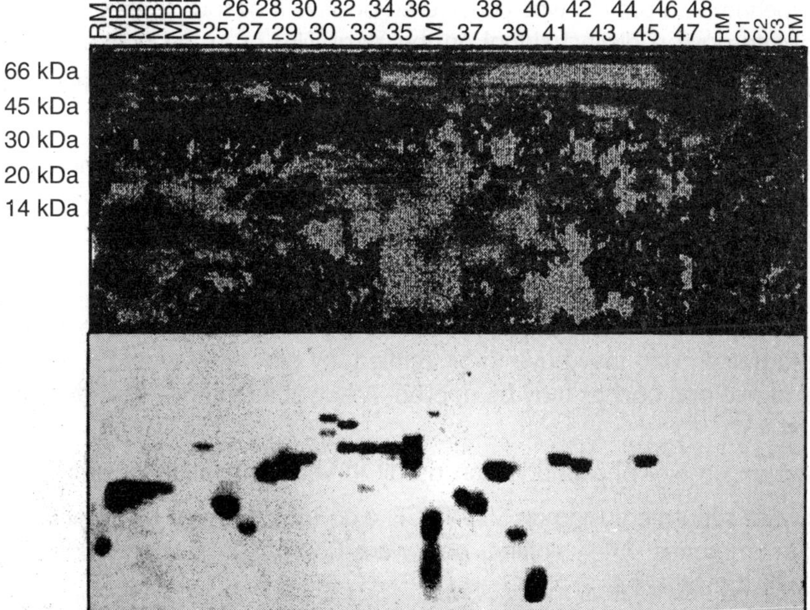

Figure 12.3: Verification of kinase substrates. SDS-PAGE (15%) gels of potential kinase substrates (upper panel) and autoradiograms of these proteins after blotting and kinase reaction (lower panel). Gels were Coomassie stained after electrophoresis and blotting.

5. Six wash steps of 30 min each were performed in TBST.
6. Finally, the microarrays were dried using a microtiter plate centrifuge or by manual fanning and *transferred* to an X-ray cassette.
7. Signal detection was performed by means of an X-ray film. The film was laid on the microarrays, and the cassette was stored for a suitable time at –80°C. Afterwards, the film was developed in a dark chamber. Alternatively, the *microarrays* could be exposed to an imaging screen that would be screened afterward by a phosphor imager for signal detection.

Figure elsewhere in this chapter shows a typical result of a phosphorylation screening experiment with CK2a after exposure of the microarrays to X-ray film. Quadruples of several proteins gave distinct signals on the film.

To determine the immobilization of the recombinant His-tagged proteins on the microarrays we screened them with an anti-RGS-His_6 antibody, as described in detail in this book. Nearly all the spotted proteins were detectable (approx 95%), as shown in Figure elsewhere in this chapter.

Selection of Potential Kinase Targets

For qualitative signal evaluation, the results of two independent experiments with different preparations of the kinase were used.

Criteria for selection of the potential kinase targets were:

1. In one experiment, a protein was considered positive if all four spots of the corresponding quadruplicate gave a distinct signal on the X-ray film or in the phosphor imager.
2. A protein was regarded as a potential target protein if it attained positive results in both experiments.

Thus, for example, 21 potential CK2a target proteins were identified out of nearly 800 different barley proteins. The selection of potential targets using a threshold-based *quantification* system after phosphorylation on *microarrays* has been described recently.

Verification of Potential Phosphorylation Targets

As this microarray-based phosphorylation method is a screening tool to identify potential phosphorylation targets in vitro they have to be verified, by other methods. For such verifications, various in vitro or in vivo approaches may be applied. A few of them have already been mentioned in the introduction.

We performed an on-blot phosphorylation assay to verify potential substrates in vitro:

1. Proteins were separated using an SDS–PAGE, e.g., 15%, followed by blotting of the proteins on PVDF membranes. After blotting, gels were Coomassie stained to check the transfer efficiency of the proteins.
2. The phosphorylation reaction was carried out on the blot membrane using reaction conditions as described for the microarray-based assay in an appropriate volume.

Figure elsewhere in this chapter shows a typical example of such an on-blot phosphorylation experiment. Forty-eight proteins, which were identified in both microarray *experiments* as potential substrates using a MAP kinase, were verified with this method. Nearly all were confirmed, as shown in Fig. 3. We used the same positive controls as in the microarray experiments: different concentrations of myelin basic protein (MBP; artificial substrate of MAP kinases). For negative controls proteins were used that had been tested in the microarray-based kinase assay and had not been identified as potential substrates.

Other in vitro verification methods are phosphorylation assays performed in solution with subsequent detection of the phosphorylated proteins by SDS electrophoresis and autoradiography. Such an assay has been used by Fukunaga and Hunter to verify candidate kinase substrates identified by phosphorylation screening of an expression phage cDNA library.

Notes

1. We recommend storing kinases at –80°C by using glycerol (end concentration 20%) or lyophilizing the kinases in the appropriate buffer and storing them afterward at –80°C. The lyophilized kinases need to be dissolved in ddH_2O shortly before use.
2. Prior to performing microarray-based kinase experiments, we recommend testing the activity of the prepared kinase by in-gel assays or kinase assays in solution using a known substrate of the respective kinase. Optimal conditions for kinase activation (e.g., optimal buffer conditions, kinase concentration, and incubation time of the kinase) must be determined experimentally for each individual kinase, if they are not known from the literature.
3. The use of a positive control is *recommended* (e.g., a known substrate of the kinase). In studies with barley CK2α, we used as positive controls different barley proteins, which share a strong homology with different plant HMG (high mobility group) proteins. These proteins are well-known targets of CK2α. In another study, we analyzed different *Arabidopsis* MAP kinases. In this case a known artificial substrate, MBP, was used as a positive control.
4. Recombinant proteins often contain 3' or 5' tags, which are encoded by the cloning vector in addition to the coding sequence of the respective protein. *Phosphorylation* of these tags may result in false-positive results and should therefore be excluded prior to microarray experiments. One possibility is to perform test assays in solution with the active kinase using some recombinant proteins that are known not to be substrates of the respective kinases and that are expressed the same manner (same tags) as the recombinant proteins used for the *microarray experiments*. These selected proteins should yield negative results in the test.
5. With respect to surface coating, we tried to phosphorylate *immobilized* denatured proteins on FAST-slides (coated with a nitrocellulose-derived polymer) and epoxy slides. We detected phosphorylation only on FAST-slides, although Histagged proteins were detectable with anti-RGS-His6 antibody on both surfaces. In other studies, nanowells, BSA-NHS (BSA-N-hydroxysuccinimide) slides, or SAM2 (streptavidin-coated membrane) slides have been used as surfaces in *microarray-based* kinase assays. We reviewed different micorarray surfaces for *different microarray* applications including phosphorylation studies.

6. The maximum spot density, which at these individual spots can still be differentiated in a phosphorylation assay, is highly dependent on the resolution of the phosphor imager used to scan the imaging plates. For determination of this density, we recommend the use of *commercially* available protein kinase (e.g., PKA from New England Biolabs). When using a scanner with a resolution of 10 μm (e.g., BioImager FLA 8,000, Fujifilm, Japan) in combination with spotting patterns up to 11 × 11 (spot distance: 410 μm), we received sufficient resolution to evaluate the radioactive signals. Applications of higher spotting patterns (14 × 14-pattern with a spot distance of 321 μm; 15 × 15 pattern with a spot distance of 300 μm) were not suitable for radioactive detection, as signals of intense spots interfered with neighboring spot signals. This led to faulty analysis of the signal *intensity* of *neighboring* proteins as well as faulty *background* corrections.
7. This washing step in TBST was performed to remove urea from the microarrays, as we recognized that urea reduces the activity of the kinase in the microarray assay.
8. For detection of radioactive signals with X-ray film, the film was laid onto the microarrays, which were covered with cling film. The exposure time depends on the signal intensity and can vary between 30 min to a few days. The sensitivity of the X-ray films can cause problems, as blackening of the films is only in a very limited area that is linear and *proportional* to signal intensity. For detection with a *phosphor* imager, the imaging plates are laid onto the covered *microarrays* and then scanned. Detection by *phosphor* imager is 10 to 100 times more sensitive than by X-ray films, and therefore it is possible to get faster results and/or detect poor signals, *respectively*. Furthermore, the system has a higher dynamic range compared with the X-ray film. Strong signals as well as poor signals are detected within one *exposure* and the *linear signal-to-intensity* range has a dimension over 5 (100,000:1). This shows that the phosphor imager is better for quantification of radioactive signals than X-ray films.

GENERATION OF PLANT PROTEIN MICROARRAYS

Protein microarrays consist of hundreds or even thousands of addressable protein samples, which are arranged in a systematic order in high density on coated glass slides . They allow parallel, fast, and easy analysis of proteins for expression and modification as well as for their molecular interactions with antibodies, other proteins, DNA, or small molecules.

Using protein antigen *microarrays,* several studies have been performed to *characterize* specific antibodies or to screen sera from patients suffering from diverse diseases. As an ideal for- mat to characterize a specific antibody, proteome arrays are envisaged. To analyze several antibodies on a single microarray, different multiplexing approaches may be applied.

Compared with Western blotting, the application of protein microarrays for antibody characterization has several advantages including a higher sensitivity of protein detection. Proteomics methods, e.g., 2D electrophoresis/mass *spectrometry* methods or protein microarray technology, are finding increasing applications in the plant field.

Here we describe a method to generate plant protein *microarrays* and to utilize them for the investigation of antigen-antibody interactions. We successfully applied the method to generate the first plant protein microarrays containing 96 *Arabidopsis* proteins, which were detectable with a limit of approx 2 to 4 fmol per spot with an anti-RGS-His_6 antibody on FAST™-slides.

Utilizing these arrays, we were able to show that a monoclonal anti-TCP1 antibody and polyclonal anti-MYB6 and anti-DOF11 sera recognized their respective antigens on the microarrays and did not cross-react with the other *immobilized* proteins including other DOF and MYB transcription factors. For this method, RGS-His_6-tagged plant proteins are expressed and purified in high throughput using characterized cDNA expression clones. Purified proteins are robotically arrayed onto FAST-slides.

The generated protein microarrays are then incubated with the respective antibodies, followed by fluorescently labeled secondary antibodies. Signal detection and evaluation allows one to identify proteins interacting with the tested antibody.

2. MATERIALS

High-Throughput Protein Expression and Purification

1. Media for culturing cells: 2 YT or LB medium containing 100 µg/mL ampicillin, 15 µg/mL kanamycin, and 2% glucose.
2. 96-Well microtiter plates.
3. 384-Well microtiter plates.
4. 96-Deep-well plates: 96-well microtiter plates with 2-mL cavities.

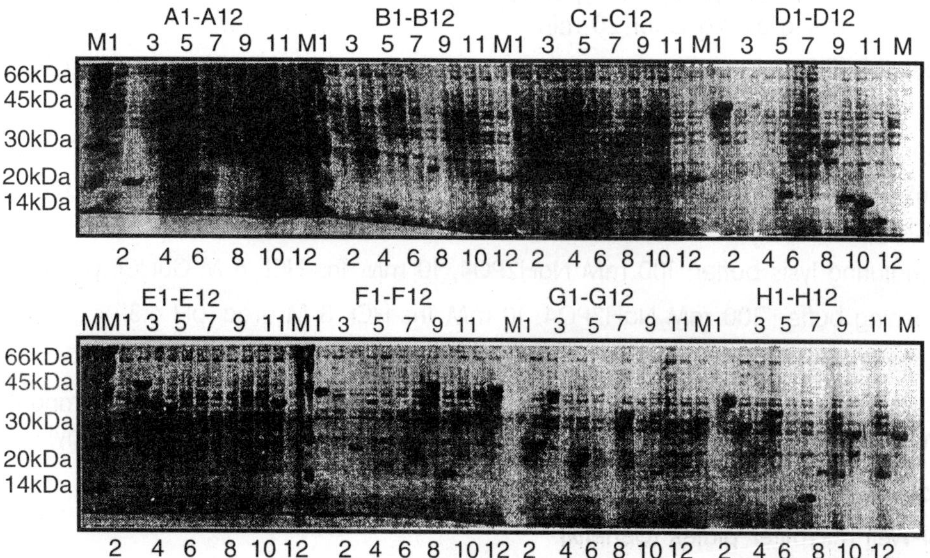

Figure 12.4: SDS-Page (15%) of the lysates of 96 Arabidopsis expression clones after induction of expression. Expression clones were obtained from an Arabidopsis cDNA expression library. Lysates were separated using a 15% SDS-PAGE and then Coomassie stained. A1–H12, coordinates of the 96-well plate. M, marker. Approximate sizes of the protein markers are shown at the left in kDa.

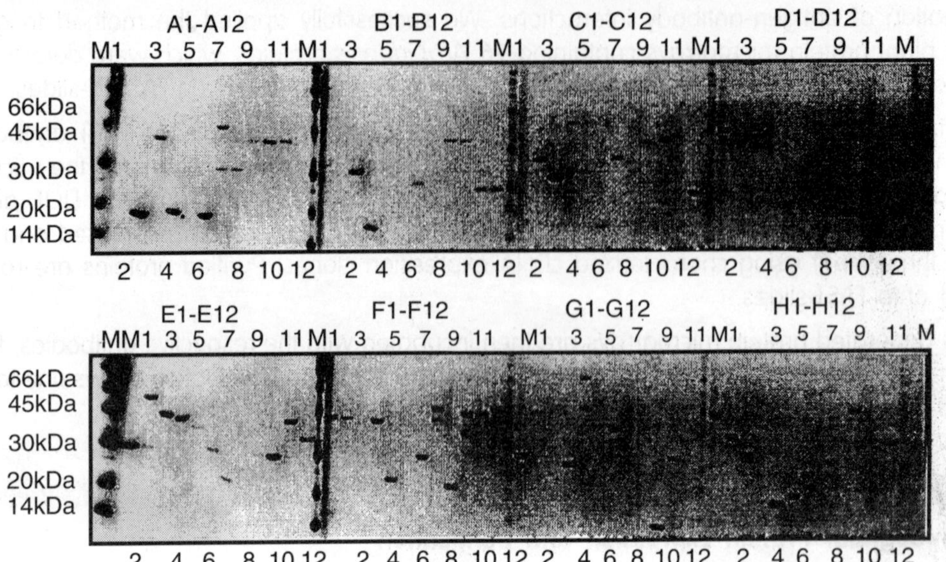

Figure 12.5: SDS-Page (15%) of the 96 purified His-tagged Arabidopsis proteins. Proteins were expressed and purified from an Arabidopsis cDNA expression library. Proteins were separated using a 15% SDS-PAGE and then Coomassie stained. A1–H12, coordinates of the 96-well plate. M, Marker. Approximate sizes of the protein markers are shown at the left in kDa.

5. Replicator with 96 pins (Nunc, Wiesbaden, Germany).
6. 4X SB medium: 48 g/L bacto-tryptone, 96 g/L yeast extract, 0.8% (v/v) glycerol, sterilized by autoclaving at 120°C for 20 min.
7. 20X PP buffer : 17 mM KH2PO4, *72 mM* K2HPO4, sterilized by filtration through a 0.2-μm pore size filter. 8. Thiamin.
9. Isopropyl-(3-D-thiogalactopyranosid (IPTG; MBI Fermentas, St. Leon-Rot, Germany).
10. MultiScreenHTS Vacuum Manifold (Millipore, Eschborn, Germany). 11. NiNTA-agarose (NTA: nickel-nitrilotriacetic acid; Qiagen).
12. Denaturing lysis buffer: 100 mM NaH2PO4, 10 mM Tris-HCl, 6 M GuHCl, pH 8.0.
13. Washing buffer: 100 mM NaH2PO4, 10 mM Tris-HCl, 8 M urea, pH 6.3.
14. Elution buffer: 100 mM NaH2PO4, 10 mM Tris-HCl, 8 M urea, pH 4.5.
15. 96-Well filter plate with a non-protein-binding 0.65-μm pore size PVDF membrane (Millipore Multiscreen MADVN 6550). 16. Bradford reagent (Bio-Rad, Munich, Germany).

Generation of Plant Protein Microarrays

1. 384-Well microtiter plates (Genetix).
2. FAST-slides (Whatman Schleicher & Schuell, Dassel, Germany). 3. QArray system (Genetix, New Milton, UK).
4. Mouse, rat, and rabbit IgG antibodies (Santa Cruz, Biotechnology, Santa Cruz, CA).

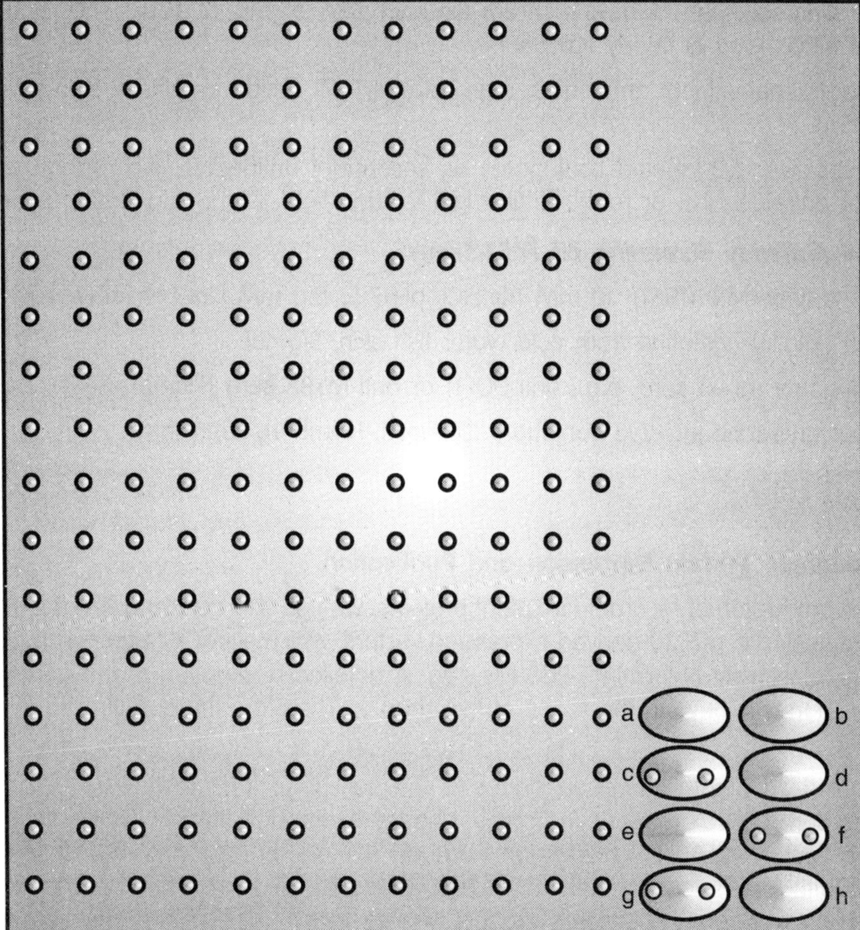

Figure 12.6: Detection of 96 His-tagged Arabidopsis proteins on FAST-slides using an anti-RGS-His_6 antibody. Ninety six proteins were expressed and purified from an Arabidopsis cDNA expression library (33). The proteins and several controls (a–h) were spotted as horizontal duplicates in a 4 × 4 spotting pattern on FAST-slides. The microarrays were screened with anti-RGS-His_6 antibody. Controls: a, elution buffer; b, PBS; c, rabbit antimouse IgG-Cy3 conjugate, diluted 1:25 in PBS; d, BSA, 20 pmol/ µL in PBS; e, normal rabbit IgG, diluted 1:10 in PBS; f, normal rat IgG, diluted 1:10 in PBS; g, mouse anti-RGS-His_6 antibody, diluted 1:10 in PBS; h, normal mouse IgG, diluted 1:10 in PBS.

Antibody Screening on Protein Microarrays

1. Arrayscanner: 428 Arrayscanner System (Affimetrix, Palo Alto, CA) or ScanArray 4000 (PerkinElmer Life Science, Cologne, Germany). 2. Cover slide (Carl Roth, Karlsruhe, Germany).

3. GenePixPro4.0 software.

Monoclonal Antibody Screening on FAST-Slides

1. TBS + Tween-20 (TBST): 10 mM Tris-HCl, pH 7.5, 150 mM NaCl, 0,1% (v/v) Tween-20.

2. Blocking solution: 2% bovine serum albumin (BSA; Sigma, St. Louis, MO) in TBST. 3. Mouse anti-RGS-His6 antibody (Qiagen).
4. Specific monoclonal antibodies, e.g., rat anti-TCP1 antibody (Affinity Bioreagents, Cholden, USA).
5. Respective Cy3-labeled conjugates as secondary antibodies, e.g., rabbit antimouse IgG for anti-RGS-His6 or rabbit antirat IgG for anti-TCP1 (Dianova, Hamburg, Germany).

Polyclonal Antibody Screening on FAST-Slides

1. TBS + Tween-20 (TBST): 10 mM Tris-HCl, pH 7.5, 150 mM NaCl, 0,1% (v/v) Tween-20.
2. Fish gelatine (gelatine from cold water fish skin; Sigma).
3. Polyclonal rabbit sera, e.g., anti-DOF11 or anti-MYB6 sera (Pineda, Berlin, Germany).
4. Goat antirabbit IgG-Cy3 conjugate (Dianova, Hamburg, Germany).

3. METHODS

High-Throughput Protein Expression and Purification

For the production of recombinant plant proteins, we use characterized *E. coli* cDNA expression clones constructed in pQE30-derived expression vectors, which allow for expression of recombinant plant proteins with an N-terminal RGS-His_6-tag. A detailed description of the generation of such expression clones is beyond the scope of this chapter. Therefore the authors provide only a short summary.

Characterized plant expression clones in the described format may be obtained from the following ordered cDNA expression libraries: barley expression library constructed in the E. *coli* expression vector pQE30NST (accession number: AF074376) and *Arabidopsis* library in vector pQE30NASTattB (accession number: AY386205). These libraries have been constructed according to the protocol of Konrad Buessow's group using deoxythymidine oligonucleotides for priming to produce recombinant proteins with their complete N-terminus.

As another source of characterized plant protein expression clones, we used full-length *Arabidopsis* clones generated by directional cloning of open reading frames via restriction-dependent cloning using gene-specific primers or via recombination dependent cloning utilizing the GATEWAY cloning technology. To express and purify plant proteins from these resources in high through put and small scale we apply the following protocols: For expression:

1. 96-Deep-well plates with 2-mL cavities were filled with 100 tL 2YT medium supplemented with 2% glucose, 100 µg/mL ampicillin, and 15 tg/mL kanamycin.
2. Cultures of recombinant clones were inoculated from 384-well or from 96-well microtiter plates that had been stored at –80°C. For inoculation, replicators carrying 96 pins were used.
3. After 16 h of growth at 37°C with vigorous shaking, 900 tL of prewarmed medium (1X SB medium, 1X PP buffer supplemented with 100 µg/mL ampicillin, 15 tg/mL kanamycin, 20 tg/mL thiamin) were added, and the incubation was continued for 2 h.

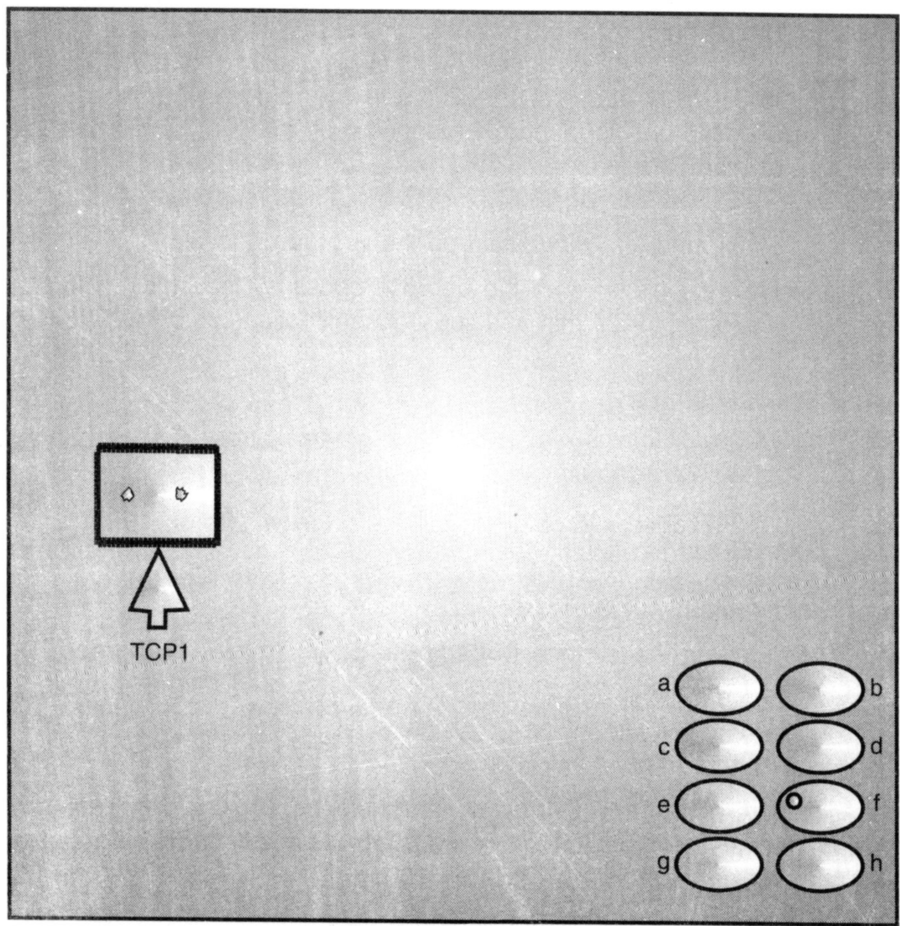

Figure 12.7: Screening of an Arabidopsis protein microarray containing 96 proteins with an anti-TCP1 antibody. Proteins were expressed and purified from Arabidopsis fulllength cDNA expression clones. The proteins a... several controls (a–h) were spotted as horizontal duplicates in a 4 × 4 spotting pattern on FAST-slides. The microarrays were screened with anti-TCP1 antibody.

4. To induce protein expression, IPTG was added to a final concentration of 1 mM, and incubation was continued for another 4 to 5 h.

5. Cells were harvested by centrifugation at 1500g for 10 min at 4°C, and the pellets were stored at –80°C.

6. Aliquots of the lysates of harvested cells can be used to control the efficiency of the expression on a 15% polyacrylamide gel.

For protein purification:

1. First, 150 tL of denaturing lysis buffer were added to the thawed pellets. Pellets were resolved by extensive vortexing and afterward incubated for 30 min at room temperature on a shaker.

2. The lysated cells were centrifuged at 1900g. The supernatant was transferred to a 96-

well filter plate and immediately drawn into a fresh filter plate by using a vacuum manifold.

3. Afterwards 30 μL NiNTA-agarose (1:2 diluted in lysis buffer) were added to each well, the plate was sealed with tape, and His-tagged proteins were bound by shaking for 1 h at 300 rpm at room temperature.
4. The agarose beads were washed three times by resuspending in 100 μL washing buffer, shaking for 5 min, and removing liquid on the vacuum manifold.
5. Finally, proteins were eluted from the agarose beads by incubation with 80 μL elution buffer without shaking for 20 min and subsequent filtration into a fresh 96-well microtiter plate.
6. Proteins were stored at 4°C.
7. Purified proteins can be separated, e.g., on a 15% polyacrylamide gel, and protein concentrations can be determined by the Bradford assay.

Using the methods described, we obtained purified proteins with an average concentration of approximately 100 μg/mL.

Generation of Plant Protein Microarrays

We used FAST-slides as the surface to generate plant protein microarrays for antibody characterization or for phosphorylation studies.

1. Prior to microarray production, 20 μL of the purified plant proteins and of the controls were transferred to 384-well microtiter plates.
2. FAST-slides were placed in a QArray system equipped with humidity control (65 to 70%) and 16-blunt end stainless steel print pins with a tip diameter of 150 tm and a pin distance of 4.5 mm. We used a spotting depth of 120 μm.
3. Each sample was loaded once, transferring a volume of 0.6 nL per spot (40).
4. After each transfer the spotting gadget was washed in ddH_2O (6 s), dried with a heat fan (2 s), washed in 80% ethanol (6 s), and dried again (3 s) to prevent cross contamination.
5. Protein microarrays were stored at 4°C in a closed slide holder.

Proteins can be spotted in different spotting patterns, e.g., in 4 × 4, 8 × 8, 10 × 10 patterns, and so on. Also, the protein can be spotted in duplicates, quadruplicates, and so forth. For antibody-antigen interaction studies, we recommend patterns with horizontal duplicates spotted in two identical fields.

Plant protein microarrays produced by this protocol can be utilized for antibody characterization, giving reproducible results for at least 3 wk when stored at 4°C. Furthermore, such plant protein microarrays may be applied for phosphorylation studies with protein kinases, as also described in detail in this book.

For detection of the immobilized recombinant plant proteins, slides are incubated with an anti-RGS-His_6 antibody according to the following protocol. As an example, Figure elsewhere in this chapter shows a microarray containing 96 *Arabidopsis* proteins, which we expressed and purified from a recently described ordered cDNA expression library.

Proteins were spotted in a 4 × 4 spotting pattern with horizontally positioned duplicates in two identical fields. All the recombinant proteins gave a signal indicating that the efficiencies of protein expression, purification, and transfer were sufficient for detection of the tested recombinant proteins on the microarrays.

Antibody Screening on Protein Microarrays

Monoclonal Antibody Screening on FAST-Slides

1. Plant protein microarrays were blocked for 1 h at room temperature with 2% BSA/TBST.
2. Mouse anti-RGS-His6 (1:2000 dilution) or plant-specific antibodies (e.g., rat anti TCP1 antibody; 1:1000 dilution) were diluted in blocking solution and then applied onto the arrays for 1 h at room temperature.
3. Two 10-min wash steps with TBST were performed.
4. The slides were further incubated for 1 h at room temperature with the respective Cy3-labeled secondary antibody (rabbit antimouse IgG for anti-RGS-His6 primary antibody; rabbit antirat IgG conjugate for anti-TCP1 antibody), which were applied at a 1:800 dilution in blocking solution.
5. Then three wash steps of 30 min each were performed in TBST.
6. Prior to scanning, we dried the microarrays using a microtiter plate centrifuge or by manual fanning.
7. Signal detection was performed by means of a 428 Arrayscanner System or a ScanArray 4,000.

All antibody incubation steps were carried out in a 200-μL volume underneath a cover slide.

To exclude nonspecific binding of the secondary antibody, we performed a parallel incubation of the protein microarrays with blocking solution without primary antibody followed by incubation with the respective secondary antibody, including all washing steps just described.

Figure elsewhere in this chapter shows an example of the fluorescence image after incubation of an Arabidopsis protein microarray with a monoclonal anti-TCP1 antibody. Ninety six Arabidopsis proteins obtained from full-length cDNA expression clones were immobilized on FAST-slides using a 4 × 4 spotting pattern with horizontally positioned duplicates in two identical fields. The image shows that the anti-TCP1 antibody specifically recognizes only the duplicate of TCP1 protein spots and does not cross-react with the other immobilized Arabidopsis proteins.

Polyclonal Antibody Screening on FAST-Slides

1. The slides were blocked for 1 h at room temperature in fish gelatine (10% in TBST).
2. Rabbit sera were diluted in blocking solution (e.g., 1:1000 for anti-DOF11 or 1:500 for anti-MYB6 serum and then applied to the arrays for 1 h at room temperature.
3. Two 10-min wash steps with TBST were performed.

4. The slides were further incubated for 1 h at room temperature with goat antirabbit IgG-Cy3 conjugate, which was applied at a 1:800 dilution in blocking solution (*see Note 13*).
5. Then three wash steps of 30 min each were performed in TBST.
6. Prior to scanning, we dried the microarrays using a microtiter plate centrifuge or by manual fanning.
7. Signal detection was performed by means of a 428 Arrayscanner System or a ScanArray 4000.

All antibody incubation steps were carried out in a 200-µL volume underneath a cover slide.

Image Analysis

The median spot intensity data (local background subtracted) were obtained with GenePixPro4.0; for further comparison, the average values of the intensities obtained from both spots of the duplicates were calculated.

Notes

1. pH values of the buffers have to be adjusted prior to every protein purification.
2. To yield extracts with higher content of specific recombinant proteins (e.g., for low-copy plasmids) for the subsequent purification, one can generate protein lysates from the pellets of two parallel 1-mL cultures using 100 µL lysis buffer each, which are then combined to 200 µL for the subsequent centrifugation.
3. Take off the supernatant very carefully. Only transfer clear supernatant to the filter plate to ensure that the plate will not occlude during the subsequent filtration.
4. Use a vacuum pump that is suitable for microtiter plates.
5. Ensure that no drops are retained underneath the filter plate after filtration; if so, they have to be transferred manually. To avoid this problem, the filtration should be carried out with enough pump power and as fast as possible. (The vacuum has to build up *quickly* and should not be held up longer then *necessary*, because the filter plates *become* too *leaky otherwise*.)
6. Another possibility for performing high-throughput protein purification is to apply a Qiagen BioRobot 8,000 and the Ni-NTA Superflow 96 BioRobotKit (Qiagen) using the same buffers for lysis, washing, and elution as in the case of manual *purification*. Disadvantages of the robot method compared with the manual method are higher costs for consumables and the fact that the elution volume may not be adjusted to lower than 350 µL. Using manual purification, elution volumes from 30 to 80 µL may be used. Small elution volumes may be *beneficial* if rarely expressed proteins are purified.
7. FAST-slides are coated with a nitrocellulose-derived polymer. In terms of surface structure, they belong to the 3D microarrays, which have a higher *immobilization* capacity than 2D slides. For antigen-antibody interaction studies using monoclonal antibodies, we utilized self-made PAA slides in addition to FASTslides. PAA slides have been prepared according to published protocols. Angenendt and co-workers compared different surfaces for antigenantibody interaction studies. We reviewed different microarray surfaces for different

microarray applications including antigen-antibody interaction studies.

8. Solutions without protein (elution buffer, PBS) as well as an unrelated protein without an RGS-His6-tag (BSA) served as negative controls. As positive controls, we spotted the Cy3-labeled secondary antibodies (e.g., mouse IgG-Cy3 conjugate). To test the binding quality of the secondary antibodies we spotted respective *monoclonal* primary antibodies as well as IgG of the same species as the primary antibody (e.g., mouse IgG in the screening *experiments* with mouse anti-RGS-His6 antibody or rabbit IgG in the *screening experiments* with *polyclonal* sera) as additional positive controls.

9. The spotting pattern used for antigen-antibody interaction studies on microarrays depends on the number of proteins and replicates to be analyzed. The maximum spot density at which individual spots can still be differentiated with fluorescence detection is highly dependent on the spotting method and the resolution of the microarray scanner.

   When we used our spotting protocol, a scanner with a resolution of 10 μm, and spotting patterns of 14 × 14 (spot distance of 321 μm) or 15 × 15 (spot distance of 300 μm), we obtained sufficient resolution to evaluate the fluorescence signals (data not shown). We have not tested higher spotting patterns up to now. Lueking and co-workers successfully applied a 13 × 13 spotting pattern for antibody-antigen interaction studies with fluorescence detection using a similar spotting protocol on PAA slides but loading each sample five times to create one spot. (We loaded each sample once.)

   Michaud and colleagues used yeast proteome arrays spotted at even higher densities (16 × 18) to characterize antibodies utilizing fluorescence detection. We reviewed different antigen-antibody interaction studies using protein *microarrays* besides other applications.

10. In our proof-of-principle study to characterize plant antibodies, we used a 4 × 4 horizontal duplicate spotting pattern with a spot spacing of 1050 μm. Two identical fields were spotted on each slide.

11. The optimal antibody dilution, if not known from literature, may be determined prior to microarray studies by Western blotting *experiments*.

12. To prevent drying of the microarrays during antibody incubation, we put them on Whatman paper soaked with TBST.

13. Owing to fluorescence labeling of the secondary antibody, this incubation step and all subsequent steps have to be performed in the dark.

14. Fluorescently labeled conjugates are stored at 4°C or, after addition of glycerol (end concentration of 50%), at –20°C.

15. Alternatively to this protocol for screening of monoclonal antibodies on FASTslides, we applied the following protocol: the FAST-slides were blocked with 1X PBS/0.5% (v/v) Tween-20 solution for 2 h at room temperature. A suitable dilution of monoclonal antibodies was then applied onto the microarrays for 2 h at room temperature, followed by two 10-min wash steps with blocking solution.

    The slides were further incubated with the respective secondary Cy3labeled conjugate (species specific with regard to primary antibody) for 1 h at room temperature. The slides were washed two times with 1X PBS/0.5% (v/v) Tween-20 for 30 min and then two times with 1X PBS for 20 min at room temperature before the signal detection was performed.

In the case of the monoclonal anti-PhyB antibody Pea-25, only the application of this protocol led to the detection of PhyB on FAST-slides.

13

Chapter

PLANT DEFENCE RESPONSE

The yeast two-hybrid approach has been widely used for library screening to identify and isolate genes encoding proteins that interact with a favourite protein. It has lhe *capability* of *identifying* and isolating desired clones ln a relatively short time by allowing one to easily screen through tens of millions of clones. There are many two-hybrid systems available.

Unfortunately, many of them may not be reliable; unacceptable backgrounds and *troublesome* falsepositive results often are seen with many two-hybrid systems. Here, we describe a reliable GAL4-based yeast two-hybrid system that includes a bait plasmid vector derived from plasmid pPC86, which is based on a *CEN/ARS* DNA *replication* system rather than the 2 μ replication origin.

This twohybrid system is compatible with most two-hybrid libraries and gives extremely low background and few false-positive clones, making it ideal for library *screening* purposes.

We have used it not only to isolate *Arabidopsis* NPR1-interactors from rice but also to pull out the rice ortholog of NPR1 by using one of the NPR1-interactors as bait to back-screen the rice library. The unique elements contributing to the successful use of this system are discussed.

MATERIALS

1. Plasmid vector pMC86: The construction of this plasmid bas been described previously.
2. Yeast strain HF7c.
3. YPD agar plates and liquid medium.

4. Synthetic dropout (SD) media. SD-Trp, SD-Trp-Leu, and SD-Trp-Leu-His + 10 mM 3-AT plates. SD-Trp liquid medium.
5. 1X TE/LiAc. Prepare from 10X LiAc (1N) and 10X TE (0.1 M Tris-HCl, 10 mM ethylene diamine tetraacetic acid, pH 7.5). For storage of yeast competent cells, add sterile glycerol to approx 25%.
6. Polyethylene glycol (PEG)/LiAc solution containing 40% PEG. Make fresh by mixing 1 v of 10X TE, 1 v of 10X LiAc, and 8 v of 50% PEG stock. Do not subject 50% PEG to prolonged autoclaving; autoclave briefly or filtrate.
7. Denatured, sheared herring testes carrier DNA (10 mg/mL).
8. Dimethyl sulfoxide.
9. Z buffer: $Na_2HPO_47H_2O$ (16.1 g/L), NaH2PO4H2O (5.5 g/L), KCl (0.75 g/L), $MgSO_47H_2O$ (0.246 g/L), pH 7.0 and autoclave.
10. X-gal/Z buffer: Mix 66.8 μL X-gal stock solution (20 mg/mL in dimethyl formamide) and 10.8 μL of β-mercaptoethanol with 4 mL of Z buffer.
11. Yeast lysis solution: 2% Triton X-100, 100 mM NaCl, 1% sodium dodecyl sulfate, 1 mM ethylene diamine tetraacetic acid, and 10 mM Tris-HCl, pH 8.0.

METHODS

The following protocols are derived mainly from Clontech yeast protocols.

Preparation of the Bait Plasmid and Yeast Competent Cells

1. Construct bait plasmid coding for a fusion protein of the GAL4 DNA binding domain and your favourite protein by using the multiple cloning sites available in pMC86. The pMC86 plasmid, carrying the GAL4 DNA binding domain and the Trp1 selection marker, is compatible with vectors with the *Leu2* selection, such as pAD-GAL4 from Stratagene, for library construction. For library vectors that carry the Trp1 selection, the pPC97 plasmid can be used for bait construction instead.
2. Inoculate a fresh colony of yeast HF7c in 3 mL of YPD liquid medium in the morning and grow the culture at 30°C with vigorous shaking. Inoculate 50 mL of YPD liquid medium with this 3 mL of HF7c culture at the end of the day and grow overnight.
3. Centrifuge yeast cells the next morning at 3000g at room temperature for 5 min, remove the supernatant, resuspend cells in 30 mL of sterile water, and spin down again. To prepare competent cells for storage, resuspend the yeast cells in sterile 1X TE/LiAc containing 25% glycerol. The competent cells can be stored at –80°C for at least 1 yr; however, transformation efficiency will gradually decrease. As an alternative, the yeast transformation kit from Zymo Research (Orange, CA) works fairly well.
4. Transform HF7c competent cells with the bait plasmid: Mix 1 μL of pMC86 derived bait plasmid DNA (approx 0.1 μg) and 5 μL of denatured herring testes carrier DNA (10 mg/mL) with 50 μL of the competent yeast cells. Add 300 μL of PEG/LiAc solution to the cells, mix, and incubate at 30°C with shaking for 30 to 60 min. Spread the mixture directly on a SD-

Trp plate. Yeast colonies will appear in 2 to 3 d. Streak out individual colonies and make glycerol stocks in YPD medium plus 25% glycerol.

Librar-Scale Transformation With Library DNA (Prey Plasmid)

1. Purify library DNA by using large-scale plasmid purification columns, such as Qiagen Maxi plasmid columns. Normally, at least several hundred micrograms of library DNA is needed.
2. On the morning of day 0, inoculate 3 mL of SD-Trp medium with several fresh colonies of HF7c containing the bait and grow at 30°C with vigorous shaking. At the end of the day, inoculate 100 mL of SD-Trp medium with this 3-mL seed culture and grow overnight. The culture should almost reach stationary phase at the end of day one. Inoculate 1000 mL of SD-Trp medium with the whole 100-mL culture and grow overnight. In the morning of day 2, the culture should reach the late log phase. Spin down the yeast cells at 5000g for 5 min at room temperature. Remove the supernatant. Resuspend cells in 500 mL of sterile water and spin down cells again.
3. Resuspend the cells in 8 mL of 1X TE/LiAc. Mix 2 mL of denatured herring testes carrier DNA (10 mg/mL) and 100 to 500 µg of library DNA (prey plasmid) with the cells. Add this mixture to 60 mL of PEG/LiAc solution. Mix well.
4. Incubate at 30°C for 30 min with shaking.
5. Add 7 mL of dimethyl sulfoxide. Mix well with swirling.
6. Heat shock for 15 min at 42°C with occasional swirling. Chill cells on ice for 1 to 2 min.
7. Spin at 5000g for 5 min at room temperature to pellet cells. Remove the supernatant.
8. Resuspend the cells in 10 mL of 1X TE, pH 7.5.
9. Plate out cells on SD-Trp-Leu-His medium containing 10 mM 3-AT. This procedure requires approx 50 150×15-mm plates. Also plate out a small aliquot of cells on SD-Trp-Leu medium to estimate the total amount of yeast transformants. Incubate the plates at 30°C for up to 2 wk. Seal plates after a few days of incubation to slow down plate drying.
10. This protocol typically yields several to 20 million yeast transformants that grow on SD-Trp-Leu medium.

Perform β-Galactosidase Filter Assay to Select Positie Clones

1. Yeast colonies of putative interactors would start to appear in 5 d. Some yeast colonies may not show up until 2 wk after transformation.
2. Streak good colonies to new SD-Trp-Leu-His plates containing 10 mM 3-AT. The cells should grow into patches in 2 to 3 d.
3. Scrape up half of the cell mass of each clone from plate and patch on a sterile 3-mm filter circle with sterile toothpicks. Place the filter circle on a SD-Trp-Leu plate and incubate at 30°C overnight.
4. Lift the filter circle and air-dry it. Dip it in liquid N_2 for 10 s to permeabilize yeast cells. Thaw the filter at room temperature for 1 to 2 min.

5. At this time, add approx 1.9 mL of X-gal/Z buffer solution to a filter circle in a plate (100×15 mm) to prepare an X-gal-saturated filter circle.
6. Overlay the cells/filter on the X-gal-saturated filter in the plate.
7. Incubate at 30°C or room temperature until blue colours develop. This may take an hour to overnight incubation depending on the strength of interaction.

Isolation of Plasmid DNA From Yeast and Retransformation of Yeast Cells for Confirmation of Positie Interaction

1. Scrape up yeast cells from plates and resuspend them in 200 µL yeast lysis solution. Add 200 µL of phenol/chloroform and 200 mg of acid-washed glass beads.
2. Vortex for 2 min to break cells. Spin at 14,000 rpm for 5 min at room temperature.
3. Transfer the supernatant to a clean microcentrifuge tube. Add two volumes of ethanol and 1/10 volume of 3 M NaOAc to precipitate the DNA.
4. Spin down DNA and rinse the pellet with 70% ethanol. Dry the pellet.
5. Resuspend the DNA pellet in 20 µL of Tris-HCl or TE buffer, pH 8.0.
6. Transform *Escherichia coli* cells with 1 µL of the DNA by electroporation.
7. Pick two transformed *E. coli* colonies, grow in 2 mL of luria broth medium with carbenicillin or ampicillin, and extract plasmid DNA by miniprep.
8. Cut DNA with enzymes and run on a gel to confirm the presence of the prey plasmid and the insert. Based on the restriction patterns, the isolated clones can often be divided into groups.
9. Transform HF7c yeast cells with the isolated prey plasmid and the bait plasmid simultaneously. Plate out on SD-Trp-Leu medium.
10. Transfer several colonies for each putative clone to a SD-Trp-Leu-His plate containing 10 mM 3-AT to test for growth. Also perform β-galactosidase assay to confirm the interaction.

Notes

1. The pMC86 plasmid, derived from pPC86 and pPC97, is based on the *CEN6/ ARSH4* replication system, different from the 2-µ replication origin. In contrast to the high copy number of 2-µ-based plasmids, the *CEN6/ARSH4* replication system-based pMC86 has a low copy number. We have noticed that when a 2-µ-based plasmid is used as bait, the number of false-positive clones tends to be higher, possibly as a result of the higher DNA recombination events between the plasmid DNA and the genome. In addition to lowering false-positive clones, this feature of pMC86 also makes easier the recovery of the prey plasmid like pAD-GAL4, carrying the 2-µ replication origin because most of the plasmid DNA recovered from yeast would be the prey plasmid.
2. The HF7c yeast strain, carrying Trp1, *Leu2,* and *His3* selectable markers and the *LacZ* reporter gene, has an extremely low background when plated on SD-TrpLeu-His medium containing 10 mM 3-AT. When streaked directly on plate, HF7c does not require addition of 3-AT to suppress its growth. This feature is critical in library screening. On the contrary,

many other commonly used yeast strains, such as PJ69-4A, carry significant leaky *His3* activity and require much higher 3-AT concentrations to suppress their growth, often give high backgrounds during library screening. It is possible to lower the concentration of 3-AT in medium for screening when using HF7c. In general, the level of background is proportional to the cell mass spread on each plate. The more cell mass on each plate, the higher concentration of 3-AT is needed.

3. We have noticed that the larger the protein encoded by the bait, the higher the number of false-positive clones, which may be attributable to the fact that yeast contains in its genome many protein sequences that can serve as a transcription activation domain and that DNA recombination rates are high in yeast. Any recombination event that creates a fusion protein between the bait and a transcription activation domain will generate a false-positive clone. In general, it is a good practice to keep the coding sequence of the bait less than 2 kb to avoid high number of false-positive clones.
4. However, HF7c does carry a disadvantageous feature; it often does not grow very vigorously in the SD-Trp medium and requires more time to grow to the needed cell mass compared to some other strains. This feature may contribute to lower transformation efficiency sometimes.
5. Some bait constructs may result in slower growth of the HF7c cells. This would usually give rise to lower transformation efficiency. To boost transformation efficiency, one can grow the HF7c cells in YPD medium for 2 h after spinning down cells from the 1000 mL SD-Trp culture before proceeding to transformation. After the heat shock treatment, the cells can also be cultured in YPD medium for a few hours before spun down for plating in order to help the yeast cells recover from the treatments.

THE USE OF PROTOPLASTS TO STUDY INNATE IMMUNE RESPONSES

Introduction

Plants rely on innate immune responses to launch *inducible* defense *against* bacterial, fungal, and viral pathogens upon *recognition* of diverse *pathogenderived* elicitors. The *elicitors* are either *conserved* among several microbial species (*pathogen-associated molecular patterns* [PAMPs]) or specific to some races of a pathogen species (avirulence [Avr] or type III effectors). The recognition of PAMPs is likely mediated by *receptor-like* kinases with *extracellular* leucine-rich repeats.

In *Arabidopsis, FLS2* encodes an leucine-rich-repeatreceptor-like kinase as the receptor for bacterial flagellin. Avr or type III effectors are recognized by plant resistance proteins to trigger gene-for-gene resistance. R proteins are associated with plasma membrane or *localized* in the *intracellular* cytosol or nucleus to directly or indirectly interact with *avr* gene products that are secreted and translocated by bacterial type III secretion system into plant cells.

So far, more than 40 R genes have been identified in diverse plant species, but the signal transduction pathways activated by R proteins are still poorly *understood.* Extensive genetic screens have led to the isolation of many important *components* in gene-for-gene resistance and PAMP-mediated basal resistance. However, their biochemical functions and molecular actions in defense responses are largely unknown. It has been widely assumed that PAMP and Avr trigger mostly convergent innate immune responses, including calcium influx, kinase activation, *oxidative* signaling, *transcription reprogramming* and, in some cases, programmed cell death.

Recently, analyses of global gene expression profiles have suggested that similar defense gene expression programs are shared by compatible (disease caused by virulent bacteria) and incompatible (resistance to avirulent bacteria) plant-pathogen interactions at the genome level.

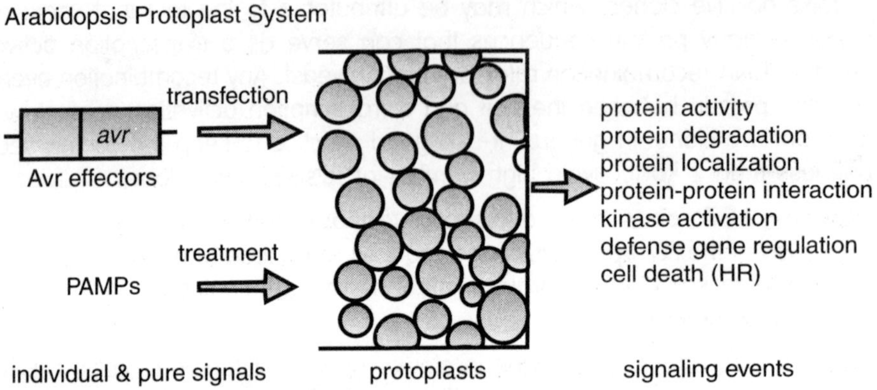

Figure 13.1: The use of protoplast transient assays to study early signaling events mediated by Avr and pathogen-associated molecular pattern. 35S is the constitutive promoter derived from cauliflower mosaic virus. HR, hypersensitive response.

However, because the whole plant–pathogen interactions display complex responses stimulated simultaneously by a large array of *extracellular* and intracellular pathogen elicitors, the traditional approach provided limited resolution in dissecting the molecular mechanisms of early defense ignaling events at the cellular level. The use of transient gene *expression* in a cell-based system has facilitated the rapid discoveries of signal transduction pathways in many multicellular organisms.

The freshly isolated *Arabidopsis* mesophyll protoplasts display *physiological* and cell-autonomous responses to a broad spectrum of signals, including light, sugar, auxin, *cytokinin*, abscisic acid, hydrogen peroxide, and stresses, similar to those found in intact tissues and plants. These protoplasts also have been used to investigate cell death induced by a fungal elicitor *fumonisin* B1 and a type III effector AvrRpt2. Notably, *Arabidopsis* mesophyll protoplasts have been developed to study plant innate immune responses, including activation of mitogen-activated protein kinase cascades and WRKY transcription factors triggered by flagellin.

Future applications of the protoplast transient expression system could facilitate the dissection and comparison of different types of immune responses triggered by individual pathogen-derived elicitors at the cellular and molecular level. The protoplast system provides unique *opportunities* to explore the elusive early signaling events in plant disease resistance.

We have demonstrated that protoplasts could be transfected with bacterial *avr* genes under the control of a *constitutive* or inducible promoter, or treated with different PAMPs to study cell death, defense gene regulation, protein *degradation* and interaction, and kinase activation. The same approach could be used to study the functions of R proteins and other signaling *components* in the defense network. In combination with genetic, genomic, proteomic, and computational tools, this powerful cell-based system will broaden our understanding the signal transduction mechanisms of plant innate immunity.

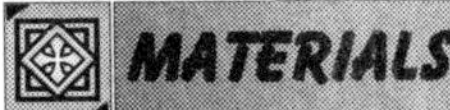

MATERIALS

Construction of the Plant Epression Plasmids

1. Effector constructs: clone the desired coding region of *avr* genes, R genes, or other signaling genes into a plant expression vector behind the constitutive 35S promoter or an inducible promoter.
2. Reporter constructs: fuse the promoter of various target genes with a reporter gene, such as the *LUC* (firefly luciferase), *GFP* (green fluorescent protein), or *GUS* ((-glucuronidase) genes.

Protoplast Isolation and Transfection

1. Plant material: 4-wk-old *Arabidopsis* plants (Col-0 or Ler) grown in soil in the greenhouse or growth chamber (30–65% relative humidity, 20–25°C, 50–100 μ mol/m^{-2}/s light, 10- to 13-h photoperiod).
2. Enzyme solution: 1.5% cellulase R10, 0.4% macerozyme R10, 0.4 M mannitol, 20 mM KCl, 20 mM MES, pH 5.7
3. 0.45-μm Filter.
4. Razor blades.
5. Desiccator.
6. 35- to 75-μm nylon mesh.
7. 30-mL Round-bottom polypropylene tubes.
8. Hemacytometer.
9. 2-mL Round-bottom tubes.
10. 40% (w/v) polyethylene glycol (PEG) solution: To make 10 mL of PEG solution, add 4 g of PEG4000 into 3 mL of H2O, 2.5 mL of 0.8 M mannitol, and 1 mL of 1 M CaCl2.
11. W5 solution: 154 mM NaCl, 125 mM $CaCl_2$, 5 mM KCl, 2 mM MES pH 5.7.
12. MMg solution: 0.4 M mannitol, 15 mM $MgCl_2$, 4 mM MES pH 5.7.
13. WI solution: 0.5 M mannitol, 20 mM KCl, 4 mM MES pH 5.7
14. Tissue culture plates (6-well, 12-well, or 24-well).
15. PAMP: Flg22, the conserved 22 amino acids of flagellin, chemically synthesized according to the published peptide sequence.

Immune Response Assas

1. Light microscope.
2. Evans blue (Sigma).

3. Fluorescent microscope.
4. YO-PRO-1 (Molecular Probes, Y-3603).
5. 4-Methylumbelliferyl-(-D-glucuronide (MUG).
6. Fluorometer.
7. Cell lysis buffer: 25 mM Tris-phosphate pH 7.8, 2 mM 1, 2-diaminocyclohexane*N,N,N,N-tetraacetic* acid, 10% glycerol, 1% Triton X-100, 2 mM dithiothreitol (DTT).
8. Luciferase assay substrate (Promega, E1501).
9. Luminometer (Monolight™ 3010, BD Bioscience).
10. TRIzol Reagent (Invitrogen).
11. Oligo(dT) (500 ng/µL; Invitrogen).
12. dNTP (Mix of dATP, dTTP, dGTP, and dCTP; New England Biolabs).
13. 5X first strand buffer (Invitrogen).
14. 0.1 M DTT (Invitrogen).
15. RNase inhibitor (40 U/µL, Invitrogen).
16. M-MLV reverse transcriptase (200 U/µL, Invitrogen). 17. RNase-free DNase I (Invitrogen).

METHODS

Protoplast Isolation

1. Prepare enzyme solution.
2. Heat the enzyme solution at 55°C for 10 min to inactivate proteases and enhance enzyme solubility.
3. Cool the solution to room temperature before adding 10 mM $CaCl_2$ and 0.1% bovine serum albumin.
4. Pass the solution through a 0.45-µm filter into a Petri dish.
5. Cut well-expanded *Arabidopsis* leaves (usually the middle section of the third or fourth pair of true leaves approx 1–1.5 cm in length) into 0.5-mm strips with fresh razor blades and digest the leaf strips in the enzyme solution in a Petri dish.
6. Cover the Petri dish with the foil and apply vacuum infiltration by using a desiccator for 30 min.
7. Continue the digestion without vacuum or shaking for another 2.5–3 h. The digestion time may vary depending on the material and experimental goals.
8. Release the protoplasts by gently shaking the Petri dish by hand or use a shaker at 80 rpm for 1 min. Be gentle with the protoplasts. Some leaves now turn transparent and the enzyme solution becomes green.
9. Add equal volume of W5 solution to facilitate protoplast centrifugation.

10. Filter the enzyme solution containing protoplasts with a 35- to 75-μm nylon mesh into a 30 mL round-bottom tube.
11. Pellet the protoplasts by spinning for 2 min at 100g or speed 3 using an IEC clinical centrifuge.
12. Resuspend the protoplasts in 0.5 mL of W5 solution by gently shaking.
13. Count protoplasts using a hemacytometer under the light microscope and adjust the protoplasts in the W5 solution to a density of 2×10^5 /mL.
14. Keep the protoplasts on ice for at least 30 min in the W5 solution to allow recovery from isolation stress.
15. The protoplasts should settle to the bottom of the tube in 5–10 min. Before PEGCa^{2+} transfection, pipet the W5 solution out and resuspend the protoplasts in MMg solution at a density of 2×10^5/mL.

PEG Transfection, PP Treatment, and Incubation

1. Prepare 40% (w/v) PEG solution with 0.2 M mannitol and 100 mM CaCl2.
2. Take out the plasmid DNA from the –20°C freezer and thaw it completely.
3. Add 20 μL (20–40 μg) of the mixed effector and reporter DNA into a roundbottom 2 mL tube.
4. Add 200 μL of protoplasts in MMg solution prepared from elsewhere in this chapter. into the tube.
5. After adding protoplasts, immediately add 220 μL of 40% PEG into the tube and mix well gently.
6. Incubate at room temperature (23°C) for 5–30 min.
7. Stop the transfection by adding 0.8 mL W5 solution and mix well.
8. Spin at 100g for 2 min and remove PEG.
9. Resuspend the protoplasts gently with 100 μL WI.
10. Add the protoplasts into a six-well tissue culture plate with 1 mL of WI.
11. Treat the protoplasts with PAMPs.
11. Incubate the protoplasts under desirable conditions.
12. After incubation for 2 to 16 h, protoplasts could be investigated immediately for cell death, GFP expression, protein localization, or gene expression.
13. Alternatively, harvest protoplasts by centrifugation at 100g for 2 min and remove the supernatant. Freeze and store the samples at –80°C until ready for diverse assays.

Immune Response Assas

Cell Death Assays

ENS BLUE STINING

1. Add Evans blue dye to the protoplasts in WI solution to a final concentration of 0.04%.

2. Incubate for 10 min at room temperature.
3. Determine the dead (stained blue) and viable (unstained) cells under a light microscope.

YO-PRO-1 STINING

1. Add YO-PRO-1 to the protoplasts in WI solution to a final concentration of 0.5 *μM*.
2. Determine the dead cells (intense green fluorescence and nuclear fragmentation in the nuclei) under a fluorescent microscope.

Reporter Gene Assays

LUCIFERSE ACTIITY ASSAY

1. Take out the samples from –80°C freezer and add 100 μL of cell lysis buffer when they are still frozen.
2. Vortex vigorously for 2 s to lyse the protoplasts and keep the lysate on ice.
3. Spin down cell debris at 8000 to 10,000g for 1 min at 4°C.
4. Use 5 to 50 μL of cell extract to measure luciferase activity by using luciferase assay substrate with a luminometer.

GUS ACTIITY ASSAY

1. Add 10 μL of cell extract prepared from into 90 μL of 1 mM MUG in 10 mM Tris-HCl, pH 8.0, and 2 mM MgCl2, and mix well.
2. Incubate at 37°C for 30 to 90 min.
3. Add 0.9 mL of 0.2 M Na_2CO_3 to stop the reaction.
4. Measure the fluorescence of MU using a fluorometer.

Reerse Transcription Polymerase Chain Reaction Assay

1. Isolate total RNA by using TRIzol Reagent (Invitrogen) according to the hand book. Add 0.4 mL of TRIzol for 8 x 10^4 protoplasts.
2. Mix 1 μg of total RNA, 0.1 μL of oligo(dT) (500 ng/μL) and RNase-free H2O in a final volume of 14 μL.
3. Heat the mix at 65 to 70°C for 5 min and chill on ice.
4. Briefly spin down the samples.
5. Add 6 μL of cDNA synthesis cocktail (4 μL of 5X first-strand buffer, 1 μL of 2.5 mM dNTP, 0.4 μL of 0.1 M DTT, 0.4 μL of RNase inhibitor, and 0.2 μL of reverse transcriptase).
6. Incubate at 42°C for 1 h.
7. Add 20L of H_2O.
8. Take 1 μL of the first-strand cDNA template for each polymerase chain reaction (PCR) using primers of the interested genes or control genes, such as genes encoding actin, ubiquitin, or tubulin.
9. Alternatively, take 0.1 to 0.2 μL of complementary DNA template for real-time PCR analysis.

Notes

1. Prepare 10 mL of solution to digest 10 to 20 leaves, which could yield approximately one million protoplasts.
2. The growth condition of plants is most critical for experimental reproducibility. Researchers in each laboratory may need to work out the best plant growth conditions. The well-expanded third and fourth pairs of leaves are recommended for the protoplast isolation.
3. The quality of DNA is very important for protoplast transfection. Poor-quality DNA may kill protoplasts and fail to produce any results. It is recommended to use CsCl gradients for Maxi-plasmid DNA isolation.
4. The ratio of effector and reporter DNA could vary from 2:1 to 4:1.
5. The experiments can be easily scaled up or down as long as the recommended DNA/protoplasts ratio is followed. Use 200 µL of (4×10^4) cells for most experiments, such as Western blot analysis and kinase activation. However, reporter enzyme assays only require 50 µL (1×10^4) cells.
6. To prevent sticking of protoplasts to the plastic, the plates could be coated with 5% calf serum for 1 second before use. You can also use 12- or 24-well tissue culture plates for small amount of cells.
7. The protoplasts could be treated with different PAMPs, such as bacterial flagellin and lipopolysaccharide, and fungal chitin.
8. The incubation conditions, such as light and temperature, depend on the purpose of experiments. For most experiments, protoplasts could be incubated at room temperature under low light (30–50 µ mol/m^{-2}/s).
9. The incubation time varies in different assays. The incubation time is 3 to 6 h for Western blot analysis and reporter enzyme assay and 1 to 6 h for reverse transcription (RT)-PCR analysis. The kinase activation could be detected within minutes after PAMP treatment.
10. Add DTT in cell lysis buffer right before use.
11. Dilute the cell extract with cell lysis buffer if the reading is over the linear range of the luminometer.
12. The RNA yield is 2 to 3 µg for 8×10^4 protoplasts, which is sufficient to analyze 40 to 50 genes by RT-PCR.
13. The number of PCR cycles depends on the abundance of the tested genes. It is usually 25 to 35 cycles.
14. It is necessary to carry out a control PCR using RNA as template without RT. If a PCR product is amplified from the control reaction, this means that there is genomic DNA contamination in your RNA samples. RNA samples could be treated with RNase-free DNase I (Invitrogen) to remove DNA before RT.
15. Try to design RT-PCR primers to cover an intron so that the size of PCR product from cDNA is smaller than that from genomic DNA, or to design one primer covering the sequences from two exons, so that the primer can only anneal to the cDNA but not genomic DNA.

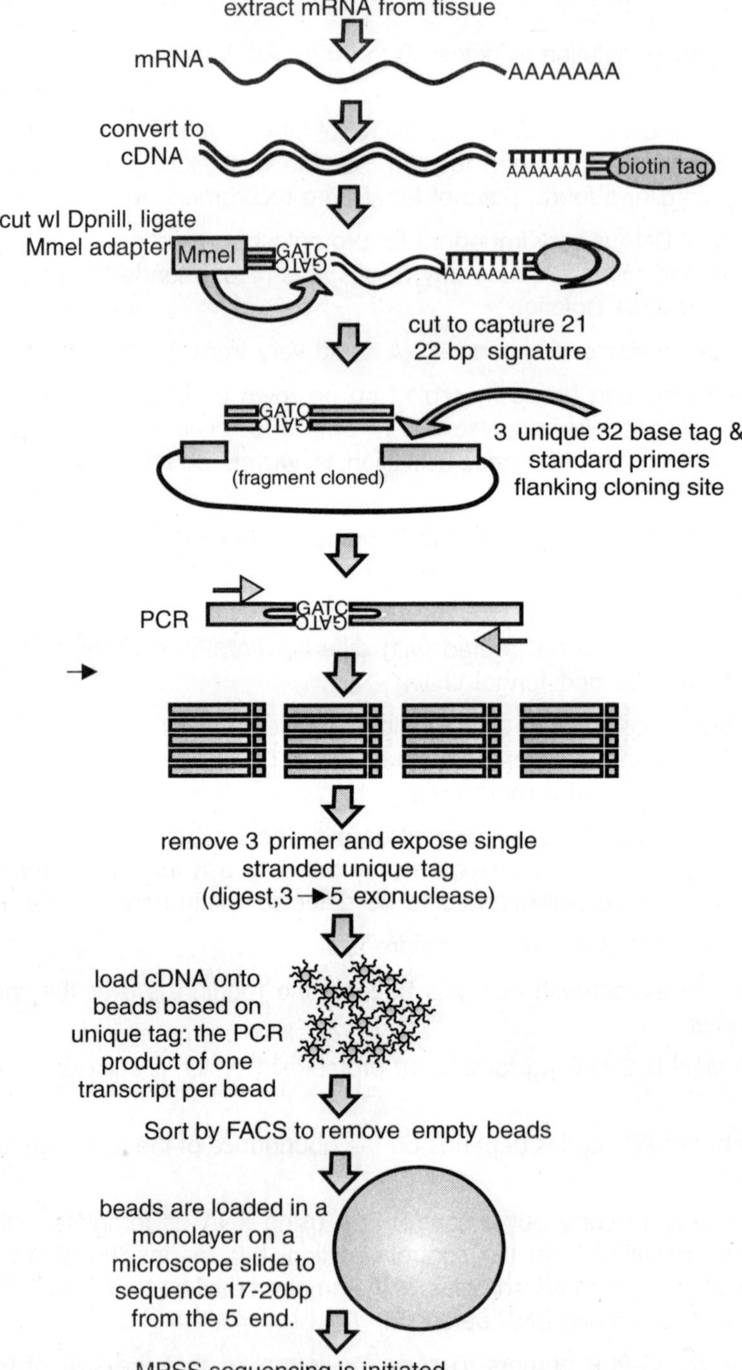

Figure 13.2: Overview of massively parallel signature sequencing (MPSS) library construction. Poly-A messenger RNA (mRNA) is converted into double-stranded complementary DNA (cDNA) and ultimately cloned onto microbeads for sequencing by MPSS. The 3'-most DpnII fragment of the cDNA is captured using a biotinylated oligo-dT primer.

USE OF MASSIVELY PARALLEL SIGNATURE SEQUENCING TO STUDY GENES EXPRESSED DURING THE PLANT DEFENSE RESPONSE

Massively parallel signature sequencing (MPSS) is a novel gene expression technology that has been used extensively in our laboratory for *transcriptional* analysis of *Arabidopsis* and rice. Like other gene expression technologies such as microarrays and serial analysis of gene expression, MPSS can be used to monitor the abundance of transcripts in plants faced by challenges, including biotic stress. The *characterization* of patterns and levels of *transcriptional* activity under such stress can be used to address specific *hypotheses* or can be used as the starting point for *quantitative* analyses of individual genes or gene families.

Although MPSS has its *limitations*, this technology offers certain advantages over other methods for whole-genome expression analyses. Other methods, such as *complementary* DNA (cDNA) microarrays, oligonucleotide microarrays, or SAGE may be less expensive and easier to perform in an individual laboratory, but MPSS provides *quantitative expression* information and allows the identification of novel transcripts.

However, MPSS data have several unique aspects and must be treated in a different way than SAGE data. Because of this, we have developed novel methods to link plant genomic data with tag- or signature-based *expression* information. Many of the analysis methods that we describe in this chapter could be used equally well to analyze SAGE data. MPSS has not been widely used to study plant–pathogen interactions, but the technology is well suited to monitoring transcriptional regulation in the host.

In the case of eukaryotic pathogens such as fungal, oomycete, or animal infection, it also may be possible to monitor in parallel the *transcriptional* events that take place in the pathogen. The limitation to *measurements* of eukaryotic and not prokaryotic transcription is a result of the use of polyadenylated RNA in MPSS.

Because of the depth of sequencing (>1 million transcripts sampled per library), eukaryotic pathogen transcripts may be detected in mixed tissues, although this detection may require high inoculation levels or an advanced stage in the infection. Because the MPSS signatures are generally 17 or 20 nucleotides in length, it is necessary to match these sequences to a *sequenced* genome or cDNA sequences. Whole- or partial-genome sequences of plants and many of their eukaryotic pathogens are increasingly available, and we believe that the approach of parallel measurements of host and pathogen genomes will be used more widely in the future.

In collaboration with the laboratory of Dr. Guo-liang Wang (The Ohio State University), we are currently evaluating host and pathogen transcription in MPSS libraries constructed from rice infected with the rice blast pathogen *Magnaporthe grisea*. Much of the genomic sequence of *Magnaporthe is* now available, and the rice genome is largely complete; therefore, rice-Magnaporthe represents one of the first plant–pathogen systems in which the host and pathogen can be simultaneously monitored by MPSS. To facilitate the use and interpretation of MPSS data, we have constructed a customized set of methods, tools, and databases.

Our database and a specialized web interface facilitates public access to gene expression data derived by MPSS. The MPSS data and *Arabidopsis* genomic sequence and annotation were used as the basis for the development of publicly available analysis and comparison tools. Our web site includes a genome viewer, a set of gene, signature and library analysis pages, an FTP site for retrieval of the data, and a signature extraction tool to allow specific sequence comparisons to the MPSS data.

Because the methods that we used for analyzing the MPSS data are critical for the interpretation of the results, we will describe these bioinformatics methods in this chapter.

MATERIALS

Isolation of Total RNA

1. TRIzol Reagent (Invitrogen/Life Technologies).
2. Diethyl pyrocarbonate (DEPC; Sigma).
3. DEPC-treated double-distilled H_2O (ddH_2O): incubate 0.05% DEPC at room temperature for at least 4 h and then autoclave for 45 min at 121°C. Alternatively, it can be purchased from Ambion.
4. 50-mL Polypropylene conical tubes.
5. Mortar and pestle, spatulas.
6. Chloroform.
7. Isopropanol.
8. Absolute ethanol.
9. 75% Ethanol (prepared with RNase-free water and stored at –20°C).
10. Liquid nitrogen.
11. RNase Zap (Invitrogen/Life Technologies).

Bioinformatics and Analysis of MPSS Data

1. Database server.
2. MySQL. Oracle will facilitate some analyses, but is not necessary.

METHODS

Isolation of Total RNA

1. Clean all equipment with soap and rinse with deionized water followed by 100% ethanol. Allow this to dry, and then treat with RNase Zap.
2. Plant tissues should be frozen before RNA extraction, preferably by rapid freezing in liquid nitrogen and storage at –80°C.
3. Chill the mortar by adding a small amount of liquid nitrogen and allowing this to boil away.
4. Homogenize 50 to 100 mg of tissue, transfer to a tube, and add 1 mL of TRIZOL reagent,

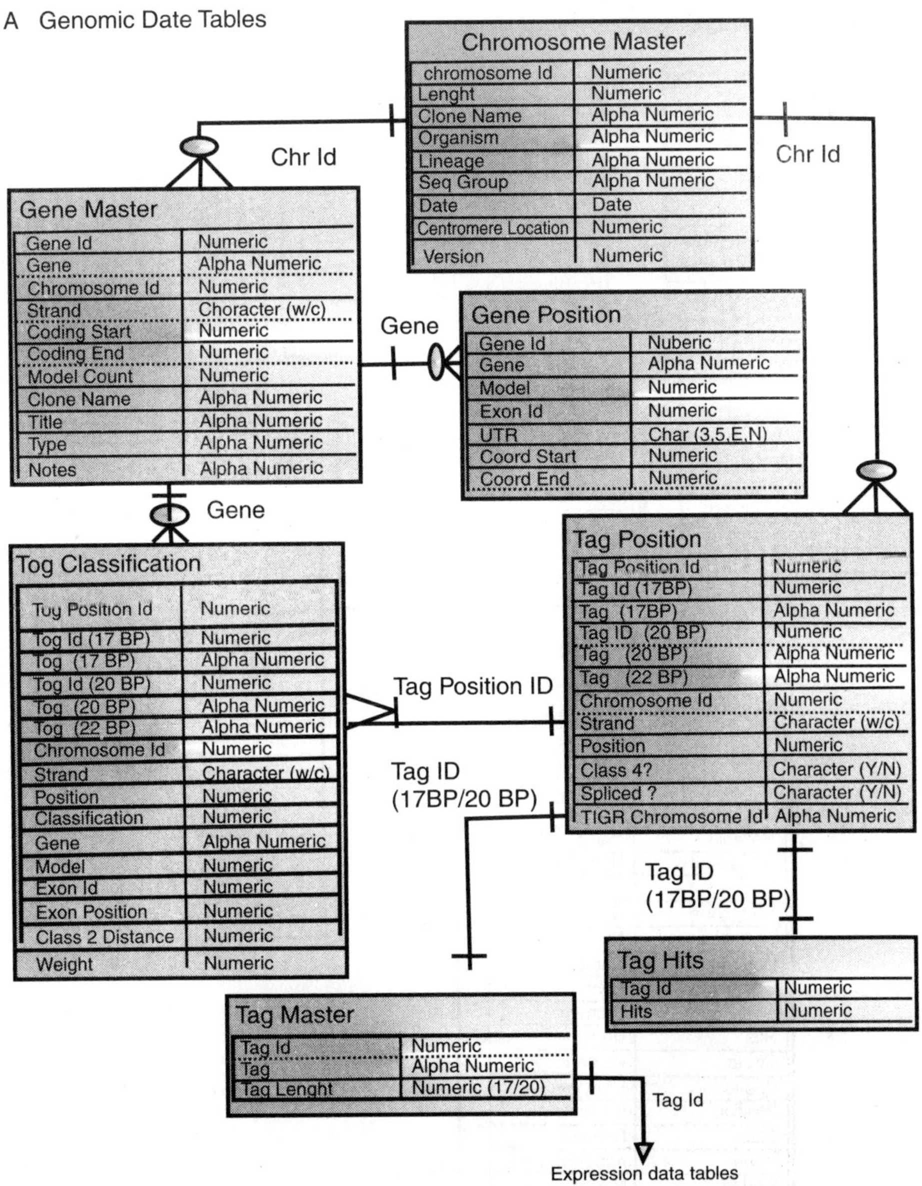

Figure 13.3: Schema for a massively parallel signature sequencing (MPSS) database. The database is designed with two major sets of tables, one that contains the genomic annotation and genomic signature information and a second that contains the MPSS expression data. These genomic and expression data are linked through the "tag_master" table. The tables and fields are shown for each of the two major sets of tables. The lines connecting tables indicate one-to-one (simple lines) or one-to-many (branched lines) relationships. (A) The genomic data tables contain information extracted from available genome annotation files. These files are based on coordinates and include gene and exon information, as well as the extracted genomic signature sequences, coordinates and classification data. (B) The expression data tables contain information about the raw sequence data, the sequencing runs, libraries, as well as the normalized data. The most important information about each expressed signature, including the normalized abundance value in each library and the genomic information is stored in the "summary" table. (Fig. Contd.)

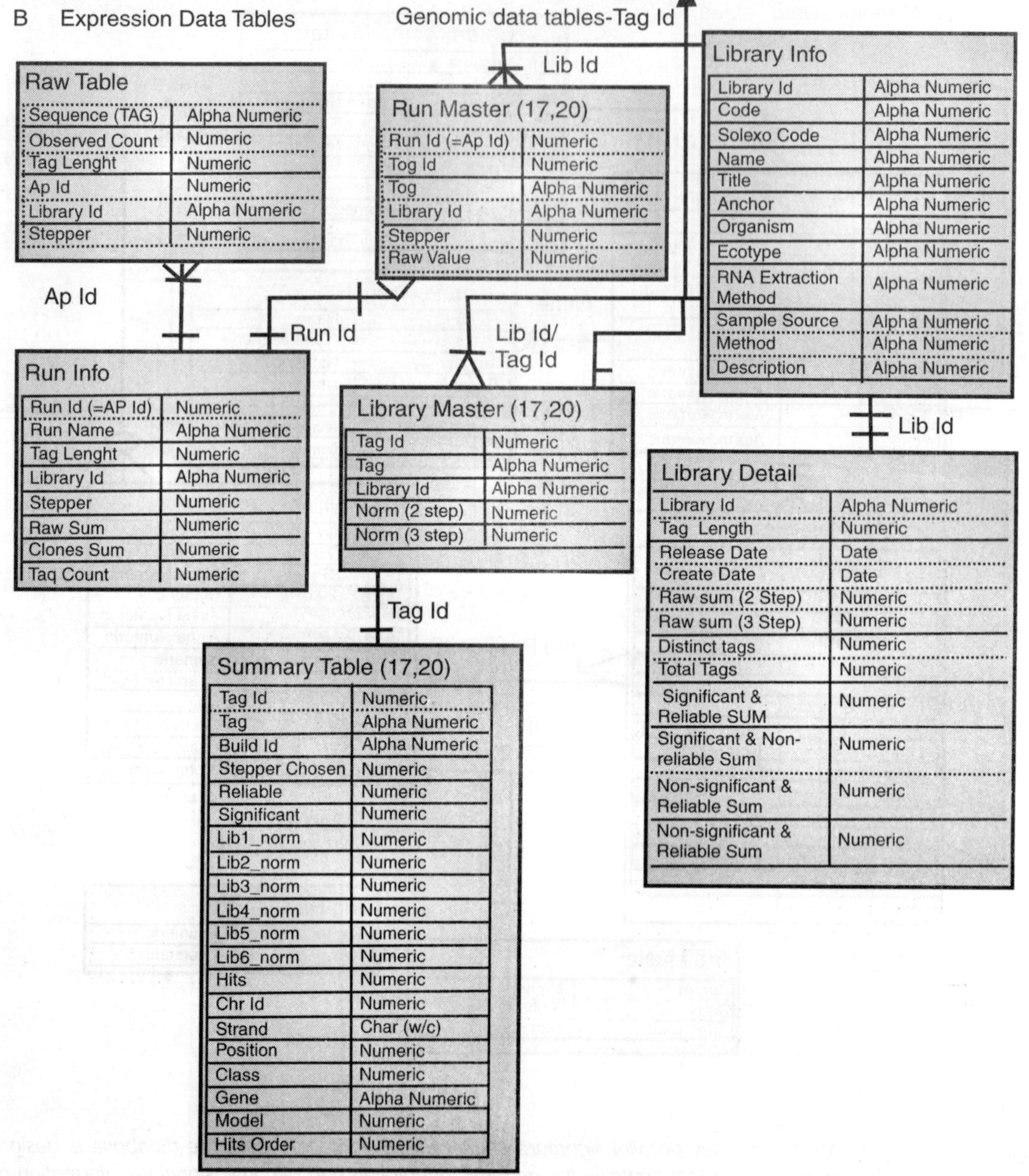

Figure 13.3: (Contd.)

with the sample not exceeding 10% of the volume of TRIZOL

5. Incubate the samples for 5 min at room temperature.
6. To remove insoluble material centrifuge at 12,000g for 10 min the supernatant contains the RNA. Transfer the cleared solution to a fresh tube.

7. Add 0.2 mL of chloroform for each milliliter of TRIZOL; cap tubes securely and shake vigorously for approx 15 s. Incubate at room temperature for approx 3 min. Centrifuge the samples at 12,000g for 15 min at 2 to 8°C. After centrifugation, the mixture separates into a lower phenol–chloroform phase, an interphase, and a colourless upper aqueous phase. RNA remains exclusively in the aqueous phase. The volume of the aqueous phase is about 60% of the volume of TRIZOL Reagent used for homogenization.
8. Add 0.5 mL of isopropanol and 0.5 mL of (0.8 *M* sodium citrate and 1.2 *M* NaCl) per milliliter of aqueous phase. Mix and leave at room temperature for 5 to 10 min.
9. Spin the precipitate at 12,000g for 20 min.
10. Remove the supernatant and wash with 75% EtOH, using 1 mL per milliliter of TRIZOL.
11. Spin sample at 12,000g for 15 min, then remove the supernatant; the pellet should appear clear and almost gelatinous.
12. Air dry and resuspend in the appropriate amount of water, using approx 0.5 mL to resuspend to a concentration of approx 1 mg/mL.
13. Typical yield is between 500 and 700 µg of RNA per gram of leaf tissue, with higher yields for tissues with more compact cells such as flowers.

RNA Quality Analysis and mRNA Purification

1. The total RNA should be analyzed by agarose gel electrophoresis (1% agarose, run at approx 100 V for 30 min) to assess the quality; good-quality RNA should show distinct bands representing ribosomal RNA and one transfer RNA band. If possible, a portion of this RNA may be used for assays of known markers using gel blot analysis, reverse transcription polymerase chain reaction (PCR), microarrays or other methods.
2. Because of the complexity of the methods and the equipment that is required, in practice, MPSS sequencing can only be performed at Solexa, Inc. Therefore, the next step is to send at least 20 µg of purified total RNA to Solexa for sequencing.
3. Solexa verifies the quality of total RNA using an Agilent Bioanalyzer, comparing the ratio of rRNAs and the distribution of RNA sizes.
4. Total RNA that passes initial quality assessments is treated at Solexa with DNase, and polyadenylated RNA is isolated using the Poly(A) Purist messenger RNA (mRNA) purification kit from Ambion. The mRNA is reassessed and quantified using the Agilent Bioanalyzer.

MPSS Library Construction and Sequencing

1. The first process in MPSS comprises library construction. Through this set of steps, Solexa clones a specific fragment from each mRNA molecule onto a single 5-µm bead; this is performed in parallel for millions of beads. The original process was described in Brenner et al and an overview is shown in Doublestranded cDNA is prepared with approx 100 ng of mRNA, using biotinylated oligo-dT for reverse transcription followed by second-strand cDNA synthesis.
2. The cDNA is next digested with the restriction enzyme *DpnII* (recognition sequence GATC) and the 3'-end fragments (cDNA fragments of *DpnII* to poly-A sites) are affinity-purified using Streptavidin beads.

3. The 5' cDNA termini are ligated to an adapter containing *a MmeI* (Type IIS) restriction enzyme site. Cleavage with this enzyme generates a DNA fragment containing the adapter and a 20- or 21-bp portion of the cDNA, including the GATC *DpnII* site.
4. The 3'-ends of these fragments are ligated to a second adapter, and these molecules are cloned directionally into a "signature cloning vector." This is a complex mixture of a single standard plasmid backbone that contains one copy out of 16.7 million different 32-bp "combitags". This step generates a library of tag-signature clones, which is titered to determine its complexity.
5. From an aliquot of the library containing approx 1.3×10^6 signatures, the tagsignature molecules are amplified by PCR using a fluorescently labeled oligonucleotide. The combitag is made single-stranded by T4 polymerase treatment, and the product is then hybridized to 5-μm "microbeads." Each bead is coated with a distinct combitag complementary to one of the combitags adjacent to the signatures. The hybridization step associates the PCR product of each specific cDNA-derived signature with a single bead (approx100,000 identical copies of each signature per bead).
6. Only microbeads with attached cDNA fragments will possess the fluorescent label from the PCR, and these are physically separated from "unloaded" beads using a MoFlo high-speed cell sorter (Dako-Cytomation, Inc., Fort Collins, CO). The purified microbeads are loaded and immobilized in a monolayer array in a microfluidic flow cell as described in Brenner et al.. This permits parallel sequencing of 20 nucleotides from each cDNA by MPSS.
7. The first step of sequencing is redigestion of cDNAs by *DpnII* and ligation of an adapter molecule to this site; the adapter includes *a BbvI* (Type IIS) restriction enzyme that cuts asymmetrically at positions 13 (5') and 9 (3') nucleotides away from the recognition site. The position of the *BbvI* site results in a four-base single-stranded overhang immediately adjacent to the *DpnII* site, and these four nucleotides are the first to be sequenced.
8. A set of "encoded adapters" is ligated to the four-base overhang; the 5'-end of these adapters contains all 256 combinations of four nucleotides, the 3'-end contains 1 of 16 10-nucleotide single-stranded decoding sequences, and there is an internal *BbvI* recognition site. Four types of encoded adapters are ligated to each bead, each of which contains a unique decoder sequence that identifies one of the four nucleotides of the cDNA signature.
9. The presence on each bead of the 4 of 16 decoding sequences is determined by 16 hybridization steps, each using a different fluorescently labeled decoder probe. This process determines the identity of four nucleotides in the cDNA signature.
10. The sequencing cycle is repeated up to four times by digestion with *BbvI*, removing the encoded adapter along with the four sequenced nucleotides, and exposing the adjacent four nucleotides for sequencing.

MPSS Image Acquisition and Base-Calling

1. A flow cell used for a single MPSS sequencing "run" contains approx 1.2 million beads, these beads are arranged in a two-dimensional array. The sequence information is

collected by recording and integrating the fluorescence signal on the beads with a chargecoupled device camera. The microscope objective *magnification* and the resolution of the camera define the number of beads visualized in a single image. Currently, each image covers 122nd of the useful surface of the flow cell and creates a 5 × 5 pixel area for each microbead. This results in approx 64,000 beads visualized in a single image, with 22 images or "tiles" for the entire flow cell. The images from each tile are analyzed and base-called independently.

2. The determination of the sequence of every bead occurs by *successively hybridizing* one of the 16 individual decoders and taking pictures corresponding to each tile by moving the modified microscope stage. This generates 16 × 22 (352) images for each four-nucleotide sequencing cycle. The sequencing protocol generates signatures 20 bases in length, *necessitating* 5 cycles × 352 images, for a total of 1760 images.
3. In addition to the fluorescent images, the sequencer takes a visible light (toplight) image of each of the 22 tiles at the beginning of the process to register the position of each bead in the different tiles. This is used to thread the bead's fluorescence across all the images collected across all the decoding steps for each cycle of four nucleotides of sequence information.
4. Once threading of the images is performed for each of the 16 images on a particular tile (for one sequencing cycle), there are three requirements for base-calling of an individual bead:
   a. A minimal fluorescence signal must be present.
   b. A minimal signal to local noise must be reached.
   c. The bead must have a minimal ratio of signal from one decoder to the next, to ensure that each bead has only one base at a particular nucleotide position.
5. Finally, all tiles from a single sequencing run are merged by summing the abundance count of identical sequences. This creates a final raw data set consisting of a list of distinct 17 or 20 base sequences with a raw abundance value representing the number of times that sequence was observed in the run.

Database Design and Implementation

We have designed a relational database for storage and handling of MPSS expression data and genomic sequence information. The schema for this database is shown in Figure elsewhere in this chapter, and the genomic data are stored *separately* from the MPSS expression data. For parallel analysis of both the host and pathogen genome, the database should contain separate information from each genome.

Genomic Data Tables

1. Genomic sequence and annotation data must be available; we typically use XML-formatted data from TIGR. Genomic tables indicated in Figure elsewhere in this chapter include coordinate, strand and chromosome data for each gene, exon, or other genomic feature. Because the database build process is procedure-oriented, we typically build the primary tables (*chromosome_master, gene_master, gene_position*) using Oracle and

export these tables to MySQL for data analysis and our web interface.

2. The gene_master table is a master table for all genes. The gene_position table is a master table for all the exons in a gene. So for each gene, the gene_position table contains its exons, introns, and untranslated regions. An intermediate table named generic_gene_master is created using the information in both gene_master and gene_position to make sure that there is one table with all the relevant *information* from both gene_master and gene_position. This avoids "select" commands that require multiple tables and maximizes the database performance.
3. The "potential" or "genomic" signatures are extracted from the chromosome sequences. This uses a specialized script written in C++ that identifies each *occurrence* of "GATC" and copies the GATC plus 13 or 16 of 3' nucleotides into the "tag_position" table along with information on the chromosome, position, and strand of the sequence. Signatures must be extracted from both strands of the *chromosome*. Because our MPSS expression data includes signatures of both 17 and 20 bases, genomic signatures of these lengths are *extracted* and stored.
4. The next step is the classification of the genomic signatures. This step is the most important and time-consuming step of process of building the database. Each *signature* in the tag_position table is classified based on comparisons to the genome annotation, requiring the gene_master, gene_position and generic_gene_master tables. The classes are as follows: Class 1, in an exon, same strand as ORF; Class 2, the longer of either an annotated 3'-untranslated region or 500 bp after the stop codon, same strand as ORF; Class 3, antisense of an exon; Class 4, matching the genome but not class 1, 2, 3, 5 or 6 (in an intergenic region, for example); Class 5, entirely within and on the same strand as an intron; Class 6, entirely within an intron, but on the anti-sense strand; Class 7, signature includes an exon/intron boundary and is spliced. Signatures that are identified by MPSS but do not match to the genome are listed as "Class 0." All *classified* genomic signatures are stored in the tag_class table. Class 7 signatures (those that span *annotated* splice sites) must be identified using a separate *signature extraction* script because they are derived from spliced transcripts and not the raw genomic sequence. The process and classes of the signatures are described in more detail in Meyers et al..
5. The tag_hits table contains distinct tags and the number of times it hits the genome (number of times it appears in the tag_position table). The tag_master is a list of all possible signature extracted from the genome, and after the *expression* data is added, the Class 0 signatures that do match the genome are added to this table.

Expression Data Tables

1. The expression data is received from Solexa in simple text files that contain, for each MPSS sequencing run, the signature sequenced by MPSS and a raw abundance level for each signature.
2. There are two primary sets of tables for the expression data. One set of tables includes the "run_master" table that contains a list of observed signatures and the abundance or *expression* level of those signatures found in each MPSS *sequencing* run. There are usually

four runs per library, and each run is sequenced in a particular "stepper". However, the expression data represented in these tables requires additional *processing* to merge the runs and the steppers, and to produce a final normalized value in "*transcripts* per million" (TPM) for each signature in the library. The "library_master" table stores intermediate data in which the runs, but not the steppers, have been merged.

3. The normalized MPSS expression data for all of the libraries is stored in a single, large table; this "summary" table includes the results of two filtering steps and the merged sequencing runs. The steps in the construction of this table are described in more detail elsewhere. In addition to the normalized expression data (in TPM), the "*summary*" table contains relational data that associate signatures with the genomic sequence and *annotation*.

   Much of these data are redundant with information stored in the other tables, and Figure elsewhere in this chapter, but genomic data for signatures duplicated in the genome is not stored in this table, nor are signatures found in the genome but not in the MPSS expression data.

   By creating a table that stores all of the required data, the disadvantages of data redundancy are outweighed by the improved functionality and enhanced performance of the database.

Graphic Interfaces for the Interpretation of MPSS Data

We developed web-based graphical interface and analysis tools specialized for MPSS data. The interface is written in PHP and requires the graphical library, GD. The interface accepts user inputs such as gene identifiers, query sequences, or chromosome position information to display the MPSS data matched to the genome. Although this is not an essential part of MPSS data analysis, the ability to *visualize* the data helps *immeasurably* in interpreting the results.

MPSS Data Analysis

Statistical Analysis Methods

MPSS provides an absolute, rather than relative, count of the abundance of a specific transcript in a specific sample. This is a "digital" *measurement* that is conducive to relatively simple statistical tests, whereas the sample size of more than one million signatures per sample provides a high level of confidence in statistical calculations.

The observed abundance of a given signature in MPSS data *demonstrates* a binomial *distribution*, and statistical models based on such a distribution are applicable to MPSS data. For example, the Z-test model described by Man et al. can be used to test if the level of a signature and the transcript from which it is derived is different in two samples. Under this model, if x_1 and x_2 represent the observed counts of a specific signature in samples 1 and 2, and n_1 and n_2 represent the total number of MPSS signatures sequenced from these samples, the proportions

$$p_1 = \frac{x_1}{n_1}$$

and

$$p_2 = \frac{X_2}{n_2}$$

each have a binomial distribution. Because n_1 and n_2 are large in MPSS (typically $>10^6$), the difference $(p_1 - p_2)$ follows an approximate normal distribution, defined as:

$$N\left((p_1 - p_2), \sqrt{pq\left(\frac{1}{n_1} + \frac{1}{n_2}\right)}\right)$$

where the unknown parameters p and q can be estimated as

$$\hat{p} = \frac{X_1 + X_2}{n_1 + n_2}$$

and,

$$q = 1 - p$$

respectively. In the following normally distributed statistical test, λ can be used with standard statistical tables to determine the Z score and the p value, providing an estimate of confidence in the difference between the signature abundance in the two samples:

$$\lambda = \frac{p_1 - p_2}{\sqrt{\hat{p}\hat{q}\left(\frac{1}{n_1} + \frac{1}{n_2}\right)}}$$

with $p < 0.001$, a twofold difference in expression can be detected for genes expressed at only 30 to 40 TPM, or smaller changes may be detected for genes expressed at a higher levels. Although the Z-test is suitable for most analyses, alternative *statistical* models have been described for use with MPSS data that are more stringent.

The Z-test allows MPSS users to quickly identify differentially *expressed* genes between two samples. However, like many other statistical tests for pairwise comparison, the Z-test has limited analytical power when applied to multiple samples.

Because of the digital nature of MPSS data, direct comparison of large numbers of samples is feasible. We have found that most commercial software packages that were originally designed for analyzing microarray data are equally useful in analyzing MPSS data sets.

The Use of Commercial Gene Expression Packages for MPSS Analysis

Numerous software packages are commercially available for the analysis of gene expression data. These tools can be applied to MPSS data by converting the MPSS data into the correct input file format; typically this requires summing the abundance of the signatures for each gene and entering the data based on gene identifier numbers.

These software packages include Spotfire, Resolver, Partek Pro, and GeneSpring. These *packages* facilitate analyses such as principal components analysis, hierarchical clustering, self-organizing maps, data filtering, pathway views, etc.

Gene Inventories and Analysis of Tissue Specificity

A basic application of expression analysis using MPSS is to generate "inventories" of expressed genes. This allows the user to catalog nearly every gene found in a specific tissue or treatment, and sort these based on absolute abundance level. As the database increases with the addition of more libraries, repeated identification of the same transcript across tissues will validate the expression of that gene.

Because sequence-based technologies for measuring gene expression, including estimated sequence tags, SAGE, and MPSS, require no previous knowledge of expressed transcripts, it is possible to discover novel transcripts that may play an important role in the biology of the sample that is being studied. These data can be used to annotate genomic sequence.

Another application of the data is the identification of regulatory sequences with specific expression characteristics. We have analyzed diverse *Arabidopsis* tissues to identify genes that show evidence of tissue specificity or low, moderate, or high levels of expression. The promoters of genes identified through such an approach may have useful experimental characteristics, and the genes may be good markers for the specific trait of interest.

Separating Expressed Signatures Derived From the Host or Pathogen

Sequence-based technologies such as MPSS are sensitive to sequence differences. With the genomic data for both a host and pathogen, it is possible to separate signatures derived from either organism. As described previously, the potential or genomic signatures can be extracted from both host and pathogen, and by comparing these two sets of sequences, it is possible to determine signatures that uniquely map to each genome or will map to both genomes; only signatures in the former category will be useful, but given that most host-pathogen interactions are across kingdoms, relatively few signatures are expected to be perfectly conserved across the two genomes. Most *signatures* matching both genomes will do so purely by chance and rarely because they are conserved. MPSS data derived from infected material with a eukaryotic host and pathogen will *simultaneously* measure host and pathogen gene expression.

Notes

1. Use polypropylene tubes (such as *standard microfuge tubes*) or glass, because some types of plastics will dissolve after treatment with phenol or chloroform.
2. The TRIZOL reagent is hazardous; be sure to wear gloves, eye protection, and a laboratory coat.
3. Homogenization can be performed using a mortar and pestle or, for larger volumes, a machine such as a Polytron can be used. Using a mortar and pestle, it may be easier to grind the frozen material without the TRIZOL, transfer the powdered tissue to a tube, then immediately add the TRIZOL; in a mortar prechilled with liquid nitrogen, the TRIZOL will solidify. For larger volumes, 10 mL of TRIZOL can be used per 1 g of tissue, with the later steps using 50-mL conical tubes centrifuged at 12,000g.
4. A chilled centrifuge at approx 4°C is preferable but not essential in each of the

centrifugation steps.

5. This step is listed as optional in some protocols; because many plant tissues contain high levels of *polysaccharides* or *extracellular* material, we routinely perform this step in our extractions to avoid problems in later stages.
6. Dispose of TRIZOL and chloroform *appropriately* because they are hazardous chemicals.
7. As described in Brenner et al., the 32 nucleotide combitags are produced by eight rounds of *combinatorial* synthesis using combinations of the four-nucleotide words CATT, TCAT, TACA, TTTC, CTAA, ACTA, ATCT, and AAAC. These tags are isothermal melting temperatures and the mixture is complex enough such that each cDNA is ligated to a unique combitag.
8. Samples are typically sequenced in two frames ("steppers") by the use of initiating adapters in which the *BbvI* site is offset by one or two bases.

USE OF MICROARRAY ANALYSIS TO DISSECT THE PLANT DEFENSE RESPONSE

Plants respond to pathogen attack by activation of a large number of inducible defense mechanisms. This response includes increased transcription of many genes. The fact that rapid activation of gene expression is correlated with resistance suggests that the identification of genes that undergo expression changes in response to pathogen attack and the elucidation of the signal transduction *mechanisms* that control their expression are essential for *understanding* how plants defend themselves from pathogens.

Consequently, gene-expression studies have long been a major component of research aimed at understanding the molecular basis of disease resistance. The development of genome-scale microarrays has provided a powerful new tool for gene expression studies. Microarrays consist of dense arrays of nucleic acid probes attached to a substrate such as a glass slide.

They can represent tens of thousands of genes, which is the entire genome for many organisms. The expression levels of essentially every gene in a genome can be monitored simultaneously by hybridizing fluorescently labeled RNA preparations to the arrays.

As discussed in this chapter, microarrays have been used for discovery of pathogen-induced genes, for studies of the nature of gene-for-gene resistance, in efforts to build genetic models of the signal *transduction* circuitry controlling defense gene expression, and for identification of promoter elements that may mediate coordinated expression of defense genes.

MATERIALS

Choice of Platform

Several different microarray platforms are in widespread use. Some arrays cannot be easily manufactured on site and therefore they must be purchased from commercial entities. Others are produced by spotting nucleic acids on glass slides, using equipment that is widely available.

Different platforms have different advantages and *disadvantages* that affect their suitability for specific microarray experiments. Affymetrix produces GeneChip® oligonucleotide arrays. These arrays consist of oligonucleotides synthesized directly on the arrays. Each *oligonucleotide* occupies a very small space, so one array can include more than 500,000 different *oligonucleotides.*

This feature allows each gene to be represented by multiple oligonucleotides. Typically, one gene is represented by 16 to 20 oligonucleotides that correspond perfectly to the target gene, and a corresponding set of *oligonucleotides* each containing a single base *mismatch.*

The expression level for each gene on the array is calculated by *combining* the signals from the perfect match oligonucleotides, correcting for nonspecific hybridization using the mismatch *oligonucleotides,* correcting for local background, and *normalizing* over the entire array. Thus, the gene expression level is provided as the expression level relative to the rest of the genes on the array.

For many species, GeneChip arrays have the capacity to represent the entire genome; therefore, they are very useful for experiments aimed at discovery of all genes that show certain expression patterns. Of course, such discovery is limited by the sensitivity of the arrays; genes expressed at very low levels or in only a few of the cells *composing* a sample cannot be detected.

GeneChip arrays have excellent technical reproducibility because of their consistency in manufacturing and the statistical power resulting from having multiple *measurements* for each gene. However, GeneChip arrays are only available for certain organisms. The design of the arrays cannot be altered by individual investigators because the initial costs of *production* of a new array are very great.

GeneChip experiments also are expensive relative to some other types of arrays; therefore, the costs of using them for large numbers of samples tend to become prohibitive. Arrays produced by *Nimblegen* are similar to GeneChip arrays in that the *oligonucleotides* are synthesized directly on the array at extremely high density and that each gene is represented by multiple oligonucleotides.

The major advantage of this platform is that a different *manufacturing* method allows the design of the arrays to be readily changed so that the use of custom-designed Nimblegen arrays by individual *investigators* may not be *prohibitively* expensive. To use these arrays, customers supply Nimblegen with RNA samples; the company conducts array production and hybridization and returns the data.

Using Nimblegen arrays is slightly more expensive than using GeneChips and, therefore, cost considerations tend to restrict use of this platform to *experiments* requiring relatively small numbers of samples. Arrays also may be produced by spotting nucleic acids onto glass slides. The *equipment* required for doing this is widely available at research *universities.* The density of such arrays is not as high as those of GeneChips or Nimblegen arrays, so each gene is usually represented by only one or two spots.

The spots may consist of long (60–70 mer) oligonucleotides, polymerase chain reaction (PCR) products, or complementary DNA (cDNA). Long oligonucleotides have the considerable advantage that they avoid the labor required for production of thousands of PCR products or cDNA clones.

Also, it is easier to distinguish members of gene families using long *oligonucleotides,* as they require shorter stretches of gene-specific sequence. However, the larger target sizes of PCR products and cDNA clones can result in increased *sensitivity* relative to long oligonucleotides. Spotted arrays generally show *substantial* array-to-array variation because of low uniformity in spotting. Consequently, these arrays are normally used with "two-colour" methods. Two samples are labeled with different dyes and hybridized to the same array. The signal is expressed as the ratio between the two signals. If the *investigator* is interested in comparing multiple samples to each other, the experiment should be designed accordingly.

One strategy is to use a single control sample labeled with one dye together with each of the experimental samples labeled with another dye. This control sample could be an RNA sample or a genomic DNA sample. In this case, the desired ratios can be derived computationally. Consider two arrays, one probed with sample A and reference R and the other probed with sample B and reference, R.

The ratio of A to B can be derived as (A/R) × (R/B). This method increases the error in the measur-ement of A/B relative to a measurement obtained by applying A and B to the same array. Nevertheless, it is useful for experiments that require comparisons among many samples in many combinations, as it reduces the numbers of arrays required.

There are several advantages to using spotted arrays. They offer great flexibility in design, allowing custom arrays to be produced as needed for specific applications. They are suitable for experiments requiring arrays representing a few hundred genes as well as for arrays representing tens of thousands of genes. They are relatively *inexpensive* to produce and use, and the required equipment for doing this is widely available.

The major disadvantage is that technical *reproducibility* is generally poorer than that of GeneChips, because of variations in spotting and reduced statistical power resulting from the use of only one or two spots to represent each gene. Data quality can be improved by using multiple technical replicates for each sample, which may be less expensive than using a single GeneChip array.

Also, for small arrays it is feasible to spot the entire array multiple times on each slide, thereby obtaining multiple measurements for the expression level of each gene and increasing statistical power.

METHODS

Statistical Considerations

Statistical analysis is important for many aspects of microarray data analysis. The primary data from a microarray experiment is an image showing variations in fluorescence intensity over the surface of the array. Conversion of this image into an expression measurement for each gene represented by the array is usually accomplished using a commercial software package that integrates the intensity of each pixel over the area occupied by each array element, corrects for local background, and in the case of arrays with multiple elements for each gene, combines the data from each element into a single measurement.

The data are then normalized to compensate for variations in labeling efficiency between samples. These operations do not generally require decision making by the investigator, so they are not discussed further here. A common question addressed by microarray experiments is, "Which genes are expressed at different levels in pathogen-infected tissue than in uninfected tissue?" This apparently simple question is actually quite difficult to answer with confidence. The difficulty arises from random variations that affect both the actual expression levels of genes and the measurements of expression levels, combined with the large number of genes tested. Consider a simple *experiment* with three biological replicates of mock-infected and infected plants.

Various statistical tests can be applied to select genes that are expressed at different levels in infected than in mock-infected plants at 95% confidence. However, if the array *represents* 10,000 genes, then 5% of these, or 500 genes, can be expected to pass the test even though they are not truly differentially expressed. Imposing more stringent statistical criteria reduces the number of such false-positives, but necessarily also increases the number of false-negatives.

Consequently, *investigators* must keep the limitations of the analyses in mind, and tailor statistical methods according to the goals of the work. For some purposes, it is better to reduce false-positives (for example, when the goal of the experiment is to identify a few reliable expression changes correlated with infection), whereas for others it is better to reduce false-negatives (for example, when the goal is to obtain an *expression* profile to compare with profiles obtained after infection by other pathogens).

Expression profiles also can be viewed as detailed descriptions of cell states. Similarities among expression profiles suggest similar host responses to different pathogens, genes that act at similar points in genetic regulatory networks, and groups of genes with similar biological functions.

Identification of similarities among expression profiles is a pattern recognition problem in a highdimensional space. Computational methods are required to identify the patterns and to display them in a form that is readily perceived by humans. Hierarchical clustering is one such method that arranges both genes and *experiments* according to *similarities* in expression patterns.

The software is freely available. Similarity relationships among genes and experiments are represented as tree diagrams that are familiar to biologists as they are similar to the tree diagrams used to represent phylogenetic relationships. Gene expression levels are represented using a simple two-colour scale (red/green is a common choice) that allows investigators to perceive patterns in the data easily.

The major advantages of hierarchical clustering are its simplicity and its easily visualized output. However, it suffers from some limitations. It is fundamentally a one-dimensional method. It only shows which profiles are most similar to each other overall; it cannot reveal that two profiles are both similar to a third, but in different ways.

It is also strongly affected by the sequence of the pair-wise clustering events that lead to the final tree. As a result, small changes in the compositions of the profiles can have dramatic effects on the structures of the trees. Some of these difficulties can be overcome by using other clustering methods such as selforganizing maps, K-means clustering, or principal component analysis. A method based on nonlinear dimensionality reduction, called local context finder, may prove useful for *multidimensional* pattern *recognition* in complex data sets.

Eamples of Microarray Analses of Plant Defense Responses

Relatively few microarray studies of plant–pathogen interactions have been published, but even these early studies demonstrate that microarray experiments can be used to address questions that are otherwise difficult to approach. A very basic question is, "Which plant genes undergo expression changes in response to pathogen attack?" The answer seems to be that an enormous number of genes undergo expression changes. From various experiments conducted using an Affymetrix chip representing 8000 *Arabidopsis* genes, 500 to 2000 genes showed expression changes, which is a large fraction of the genome. It is quite possible that some of these genes are not responding to the *pathogen* attack directly but rather are responding to insults to *homeostasis* that occur as a consequence of *pathogen* activity.

However, extensive gene expression changes are observed at very early times after infection, when it seems unlikely that the pathogen would have had much effect on host metabolism. A common motivation for searching for genes induced in response to infection is the idea that genes that show such induction are likely to be involved in resistance.

Of course, this is not necessarily true. Further work is needed to determine whether or not a pathogen-induced gene contributes to resistance. Fortunately, the development of powerful reverse genetics methods, including large transfer DNA insertion collections and RNAi technology, has made it feasible to conduct functional assays on quite large numbers of *candidate* genes.

Finding that knocking out a candidate gene by mutation or RNAi results in reduced resistance is very good evidence for a role for that gene in resistance. Expression profiling was used to investigate the molecular basis of induced systemic resistance (ISR), a jasmonate and ethylene-dependent, salicylateindependent resistance induced by *root-colonizing rhizobacteria.*

No significant changes in gene expression were associated with development of ISR in the absence of pathogen challenge. After challenge, ISR plants responded more rapidly to pathogen challenge than naive plants, suggesting that ISR is a consequence of potentiation of defense responses. Studies of expression profiles have led to new insights about gene-for-gene resistance.

Tao et al. found that the shapes of the profiles from plants undergoing gene-for-gene resistance in response to *Psm ES4326/avrRpt2* were very similar to those of plants responding to the virulent strain *Psm* ES4326. However, in the case of the avirulent strain, the amplitudes of the profiles were much greater.

The amplitude of the profile from plants responding to *Psm* ES4326 at 30 h after infection was similar to those of plants responding to avirulent strains at 6 h after infection. This indicates that many defense responses are common to basal resistance and gene-for-gene resistance, with the major difference between the two lying in the kinetics and/or intensity of defense activation, consistent with the suggestion made by Lamb et al. years ago.

R genes vary considerably with respect to genetic *requirements* for effective resistance. For example, *RPP4* requires PAD4, SA accumulation and SGT1b; *RPP7* requires SGT1b but is independent of PAD4 and SA, whereas *RPP8* is independent of SGT1b, PAD4, and SA. One interpretation of these results is that these three *R* genes trigger activation of different defense mechanisms that are under the control of different downstream signaling genes.

However, this does not appear to be the case, because the expression profiles of resistance reactions triggered by *RPP4*, *RPP7*, or *RPP8* are all very similar. Rather than triggering different defense responses, the three *R* genes seem to require different signaling elements to achieve very similar results, suggesting a convergence point in the signaling pathways triggered by each *R* gene product. In principle, expression profiles should be useful in efforts to model genetic

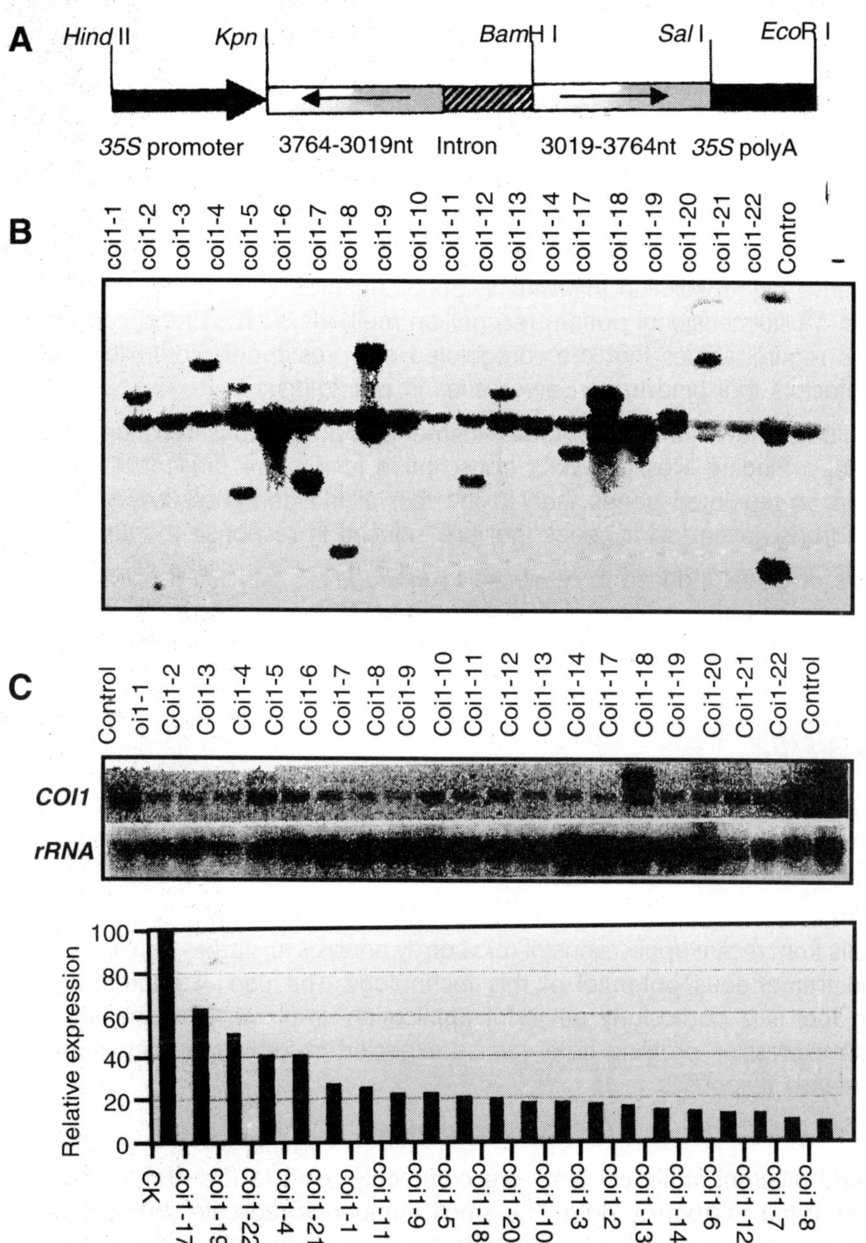

Figure 13.4: Suppression of rice COI1 gene by RNA inference (RNAi). (A) Schematic drawing of the COI1 RNAi construct. (B) Southern blot analysis of genomic DNA from the control and COI1 suppression lines after digestion with EcoRI and probed with the B fragment (3019–2764 nt of COI1). (C) Northern blot analysis of total RNAs from the control and COI1 suppression lines. A fragment of 825–2244 nt of COI1 was used as the probe and 25S rRNA was used as the loading control.

regulatory networks *controlling* gene expression. Wild-type and various regulatory mutant plants were subjected to *expression* profiling after infection with PsmES4326. From the data, it was possible to group mutations affecting SA and JA/ET signaling based on profile similarity.

The roles of four genes, *PAD1, PAD2, EDS3,* and *EDS8,* in signaling were predicted based on the expression profiles of the mutants. *PAD1* and *EDS8* were correctly predicted to affect JA signaling, and *EDS3* was correctly predicted to affect SA signaling. A prediction that *PAD2* affects SA signaling could not be verified.

These results are encouraging, but more sophisticated methods for pattern recognition are needed to derive high-resolution information about signaling network structure from expression profiling data. Multidimensional pattern recognition methods such as local context finder may be helpful in this regard. Genes that are coregulated are presumably controlled by similar sets of *transcription* factors that bind to conserved sites in promoters.

Several studies have defined promoter elements that are conserved among groups of co-regulated genes. Binding sites for WRKY transcription factors are present at substantially higher frequencies in SA-regulated genes than in the rest of the genome. They also are enriched in promoters of transcription factor genes that are induced in response to infection.

Promoters of genes induced in response to *RPP4, RPP7,* and *RPP8* activation were enriched for motifs associated with binding of WRKY-, TGA-, and ERF-type transcription factors. These studies provide useful clues about transcriptional *activation* of defense-related genes that can serve as a guide for further research.

CONCLUSION

Microarray analysis is a relatively new technology. Methods for array production and hybridization are still improving. The analysis of microarray data requires statistical analyses that are unfamiliar to many biologists. These methods must be tailored to fit the goals of particular experiments.

The results from recent applications of microarray analysis to studies of plant disease resistance illustrate the tremendous potential of this technology. The use of expression profiling as a phenotyping tool is a particularly powerful application. Improvements in methods for pattern *recognition* in expression profiling data can be expected to facilitate systems-level *investigations* into plant defense responses.

Notes

1. Several different statistical tests are commonly used to identify differentially expressed genes using microarray data. The small number of replicate data sets that are usually available presents challenges for tests such as t-tests, which are commonly used. The permutation testing method, called significance analysis of microarrays (SAM), is probably better for handling small numbers of replicates. SAM has the additional advantage of providing an estimate of the false discovery rate, which is helpful in deciding where to set the cut-off between genes judged to be differentially expressed and genes judged not to be. SAM software is freely available. Development of statistical methods for detecting differentially expressed genes in microarray data is an area of active research.

It is quite possible that there are methods better than SAM that the author is unfamiliar with.

2. A single microarray experiment yields expression level values for many genes from a particular sample. Collectively, these data constitute a snapshot of the expression pattern of the genome for that sample. This snapshot is called the expression profile of that sample.

3. Pattern-recognition *programs* represent expression profiles in *n-dimensional* space, where n is the number of genes represented in the profiles. Each profile is defined as a vector in the space, with the end point of the vector determined by the expression values of each gene in the profile. This is easy to imagine when each profile consists of data for only three genes. You can plot the vectors using three axes. When asked which vectors are most similar, you can easily answer based on which vectors are closest to each other in the *three-dimensional* space. The *mathematics* behind the problem do not change when the space has more dimensions, but humans cannot easily visualize the arrangements of the vectors.

USE OF RNA INTERFERENCE TO DISSECT DEFENSE-SIGNALING PATHAS IN RICE

RNA interference (RNAi) causes gene-specific silencing based on sequence homology-dependent *degradation* of cognate messenger RNA (mRNA). The *phenomenon*, also known as posttranscriptional gene silencing, was first discovered in petunia, in which overexpression of the *CHS* gene encoding a key enzyme (*chalcone synthase*) in *anthocyanin biosynthesis surprisingly* resulted in *downregulation* of anthocyanin levels (3,4). Recent studies show that RNAi is mediated by short-interfering RNAs or microRNAs, which result from the cleavage of a double-stranded RNA by an RNase III-related nuclease Dicer.

RNAi is universally present in plants, animals, and fungi and is now considered an important mechanism for *endogenous* gene regulation, development, and host defense. Furthermore, the gene-silencing method based on RNAi recently has emerged as a powerful tool for the functional discovery of eukaryotic genes and genetic engineering of host resistance against viral infection. In *comparison* with transfer DNA or transposon insertion, chemical or radiation treatment, and other *mutagenesis* approaches, there are a number of advantages for using RNAi to generate loss-of-function (knockout or knockdown) mutants, especially in a plant species with a large-size genome.

First, RNAi allows targeted and effective knockout or knockdown of specific genes at a high frequency without random and laborious screening of loss-of-function mutants from large mutant populations. Second, *simultaneous* suppression of *redundant* or homologous genes (e.g., multiple members of a same gene family) can be achieved with RNAi.

Third, inducible RNAi may provide an effective way for functional analysis of genes whose mutation will lead to embryonic or early developmental lethality. Furthermore, a large population of gene-silencing lines can be generated using high-throughput RNAi, which will complement other mutagenesis approaches for both forward and reverse genetics-based functional genomic studies. Stable gene-silencing lines can be generated via plant transformation using various RNAi constructs that may include sense, antisense, inverted repeat, or tandem inverted repeat of specific genes.

Chuang and Meyerowitz first reported the effective gene silencing in *Arabidopsis* using an RNAi construct composed of an inverted repeat of the gene of interest. Smith et al. proposed including an intron for more effective gene silencing based on the hypothesis that excision of the intron might improve alignment of the *complementary* sequences flanking the intron.

By incorporating a chemical-inducible promoter, Guo et al. constructed an inducible RNAi vector that should be advantageous to the study of lethal genes required for embryo or early development. For high efficient, large-scale gene silencing, Wesley et al. developed a high-throughput RNAi vector (pHELLSGATE) by *combining* with Invitrogen's Gateway recombination technology. With slight modifications, a similar vector (pANDA) was constructed for high-throughput RNAi in rice plants.

In addition, Brummell et al. developed an innovative method for high-throughput generation of specific RNAi contructs by one-step, simple cloning of any target gene fragment between *a 35S* promoter and an inverted repeat of a heterologous 3' untranslated region. As a result, a variety of RNAi vectors for different purposes are now available for generating stable gene silencing lines. During the past decade, significant progress has been made toward the understanding of defense signaling in *Arabidopsis*, a model dicot.

However, little is known about signal transduction and defense pathway interactions leading to host-defense response in rice, a model monocot and economically important food crop. Our laboratory has previously used RNAi to knockout/ knockdown rice *OsMAPK5* gene encoding a stress-responsive mitogen-activated protein kinase and successfully demonstrated the importance of *OsMAPK5* in rice biotic and abiotic signal transduction.

We also have generated by RNAi a series of transgenic rice lines deficient of or insensitive to major defense signal molecules such as jasmonic acid (JA) and ethylene. These transgenic RNAi lines may serve as powerful genetic tools for epistasis analysis and are important for *dissecting* defense signal pathways in rice. Here we describe a detailed procedure for *generating* JA-insensitive transgenic rice lines by RNAimediated *suppression* of a rice ortholog of *Arabidopsis COI1* gene that encodes a key component of JA signal transduction.

MATERIALS

Construction of RNi Vector

1. *Escherichia coli* strain DH5a.
2. *BamHI*, EcoRI, HindIII, *KpnI*, and SalI restriction enzymes, Taq DNA polymerase and T4 DNA Ligase.
3. 10 mM dNTPs.
4. 100-bp and 1-kb DNA ladders.
5. pCAMBIA 1300 vector.

6. QIAquick polymerase chain reaction (PCR) purification kit (Qiagen).
7. QIAprep® Spin miniprep kit (Qiagen).
8. Geneclean III kit (Q-Biogene).
9. Agarose.
10. Luria broth (LB) medium (1.0 L): 10 tryptone, 5 g of yeast extract, 10 g of NaCl, 15 g of agar.
11. Ampicillin.
12. Kanamycin.

Rice Transformation

1. Seeds of rice (*Oryza sativa* spp*a* japonic*a* cv. Nipponbare).
2. *Agrobacterium tumefaciens* strain EHA105 (rifampicin resistant).
3. Bio-Rad MicroPulser Electroporator and cuvet (gap size 2 mm).
4. YEP medium (1.0 L): 10 g of bacto-peptone, 10 g of yeast extract, 5 g of NaCl, 15 g of agar.
5. Chu (N6) basal salt (C1416), MS basal salt (M5524), and MS modified vitamin powder (M6896; Sigma).
6. Rifampicin: dissolve in methanol to make 25 mg/mL stock solution and store at –20°C.
7. Hygromycin B: 50 mg/mL; store at 4°C.
8. Cefotaxime: dissolve in distilled water to make 250 mg/mL stock solution and store at –20°C.
9. 1 M Acetosyringone stock: dissolve 196.2 mg of acetosyringone in 1 mL of dimethyl sulfoxide, store at –20°C.
10. Callus induction medium: 3.98 g of Chu (N6) basal salt, 0.1 g of MS modified vitamin powder, 2 mg of 2,4-D, 0.5 g casamino acids, 2.5 g of proline, 30 g of sucrose, and 2 g of Gelrite. Adjust pH to 5.7, add water to 1 L, and autoclave before use.
11. Suspension medium: 3.98 g of Chu (N6) basal salt, 0.1 g of MS-modified vitamin powder, 0.5 g of casamino acids, 30 g of sucrose, 10 g of glucose, and 100 *μM* acetosyringone (added after autoclave). Ajust pH to 5.2, add water to 1 L, autoclave before use.
12. Co-cultivation medium: 3.98 g of Chu (N6) basal salt, 0.1 g of MS modified vitamin powder, 1 g of casamino acids, 2 mg of 2,4-D, 30 g sucrose, 10 g of glucose, *100 μM* acetosyringone (added after autoclave), and 2 g of Gelrite. Adjust pH to 5.2, add water to 1 L, and autoclave before use.
13. Washing medium: 3.98 g of Chu (N6) basal salt, 0.1 g of MS-modified vitamin powder, 30 g of sucrose, and 250 mg of cefotaxime (added after autoclave). Adjust pH to 5.7, add water to 1 L, and autoclave before use.
14. Selection medium: same as the callus induction medium with addition of 250 mg/L cefotaxime and 50 mg/L hygromycin after autoclave.

15. Regeneration medium: 4.31 g of MS basal salt, 0.1 g of MS modified vitamin, 1 g of casamino acids, 2 mg of kinetin (or 2 mg benzyladenine), 0.1 mg of a-naphthaleneacetic acid, 30 g of sucrose, 30 g of sorbitol, 50 mg of hygromycin (added after autoclave), and 3 g of Gelrite. Adjust pH to 5.7, add water to 1 L, and autoclave before use.
16. Rooting medium: 4.31 g of MS basal salt, 0.1 g of MS-modified vitamin, 30 g of sucrose, 50 mg of hygromycin (added after autoclave), and 2 g of Gelrite. Adjust pH to 5.7, add water to 1 L, and autoclave before use.

Analysis of Transgenic Rice Plants

1. TRIzol reagent (Invitrogen).
2. 70% and 100% ethanol.
3. TE buffer: 10 mM Tris-HCl, 1 mM ethylene diamine tetraacetic acid, pH 8.0.
4. 3 M Sodium acetate, pH 5.2.
5. PerfectHyb™ plus hybridization buffer (Sigma).
6. 20X standard saline citrate (SSC): dissolve 175.3 g of NaCl and 88.2 g of sodium citrate in dH_2O, adjust pH to 7.0, and add water to 1 L.
7. 10% Sodium dodecyl sulfate.
8. Nylon membrane.
9. Random primed labeling kit.
10. Formaldehyde.
11. JA.
12. Methyl jasmonate (MeJA).

METHODS

Generation of Rice COI1 RNi Construct

1. Extract genomic DNA from young leaves of rice seedlings using CTAB method as previously described.
2. Design two pairs of rice COI1 gene-specific primers containing BamHI/KpnI and BamH/SalI sites, respectively. To generate intron-containing hairpin RNA, a 1-kb BamHI/KpnI fragment (a fragement, with a 258-bp COI1 intron) and a 0.7-kb BamH/SalI fragment (B fragment) were amplified from genomic DNA by PCR under the following program: 94°C for 2 min; 30 cycles of 94°C for 30 s, 54°C for 30 s, and 72°C for 1 min; finally 72°C for 10 min.
3. Check the PCR products on 1% agarose gel for specific amplification and verify by restriction enzyme digestion or DNA sequencing as needed.
4. urify the PCR fragments (A and B) with QIAquick® purification kit.
5. After being digested with *BamHI* and *KpnI*, the A fragement (approx 1.0 kb, with intron)

is purified with the Geneclean III kit and ligated to the BamHI/KpnI sites of pCAMBIA1300S, which is modified from pCAMBIA1300 and contains a double 35S promoter and a terminator (14).

6. Transform the ligation product into DH5a-competent cells by heat shock (42°C, 1 min) treatment. Plate the bacteria on LB medium with kanamycin (50 mg/L) and incubate at 37°C overnight.
7. Pick up 10 bacterial colonies for plasmid DNA extraction using QIAprep® Spin miniprep kit and identify the pCAMBIA1300S recombinant containing the COI1 A fragment by PCR or *BamHI/KpnI* digestion.
8. Digest the B fragment (approx 0.74 kb) with *BamHI* and *SalI.* After purification with the Geneclean III kit, the B fragment was ligated to the *BamHI/SalI* sites of the aforementioned recombinant plasmid. As a result, the final RNAi construct contains two complementary COI1 fragments flanking the COI1 intron, which will allow the formation of inverted repeats or intron-spliced hairpin in rice plants.

Preparation of Agrobacterium Suspension

1. Transform the COI1 RNAi construct into *Agrobacterium tumefaciens* strain EHA105 using the Bio-Rad MicroPulser electroporator according to the manufacturer's instruction.
2. After electroporation, immediately add 1 mL of YEP or LB liquid medium to the agrobacterial cells and incubate for 2 h at 28°C on a shaker.
3. Plate 50 to 100 μL of agrobacterial suspension onto YEP solid medium containing kanamycin (50 mg/L) and rifampicin (60 mg/L). Incubate the plates at 28°C for 2 d.
4. Pick up several agrobacterial colonies and identify true transformants carrying the COI1 RNAi construct by PCR and/or restriction digestion. Store the agrobacterial transformant in glycerol stock at –70°C as needed.
5. Streak the agrobacterial transformant on YEP agar medium containing kanamycin (50 mg/L) and rifampicin (60 mg/L) and incubate the plate at 28°C for 2 d.
6. Inoculate one to two loops of agrobacterial cells into 20 mL of YEP liquid medium containing kanamycin (50 mg/L), rifampicin (60 mg/L), and 100 μM of acetosyringone, incubate overnight at 28°C on a shaker (150 rpm).
7. ollect overnight agrobacterial cultures (OD600 = 1–2) in sterile centrifuge tubes by centrifugation (<3000g for 10 min).
8. Resuspend agrobacterial cells in 30 mL of suspension medium to a density of about OD_{600} = 0.05.

Agrobacterium-ediated Transformation of RNi Construct

1. Dehusk 100 immature or mature seeds of rice and surface sterilize with 70% ethanol for 1 min and then with 50% Clorox® (2.6% sodium hypochlorite) for 30 min with gently shaking.
2. Rinse seeds in sterile distilled water three times to remove residual Clorox.

3. Place seeds on the callus induction medium in 10-cm Petri dishes (10 seeds per plate), seal the plates with parafilm and incubate them under continuous light at 30°C.
4. After 2 wk, separate the calli derived from the scutella with scalpel and transfer them onto fresh callus induction medium and incubate for an additional 2 wk.
5. Select embryogenic calli and soak them in 30 mL of agrobacterial suspension (OD600 = 0.05) for 30 min with gentle shaking at room temperature.
6. Decant agrobacterial suspension and blot rice calli on sterile filter papers or Kimwipe tissues to remove excess bacteria.
7. Transfer the inoculated calli onto the cocultivation medium and incubate at 22°C in darkness for 2 d.
8. Collect the cocultivated calli in a 50-mL sterile tube; Wash the calli by gentle swirling for 6 times with sterile 6 × 30 mL dH_2O (1–2 min each time), followed by two-time washes (30 min each) with 2 × 30 mL washing medium.
9. Blot the calli on sterile tissue paper to remove excess washing medium.
10. Transfer the calli onto the selection medium and culture under continuous light at 30°C for 3 wk.
11. Transfer hygromycin-resistant calli to the regeneration medium and culture under continuous light at 30°C.
12. Once shoots are regenerated from calli, transfer them to the rooting medium in test tubes or plastic containers for regeneration of intact rice plantlets.
13. After 2 to 4 wk, rice plantlets are ready for transplanting to soil in pots.

Molecular Characterization of RNi Transgenic Lines

1. Perform Southern and Northern blot analyses to verify the introduction of COI1 RNAi construct into rice transgenic lines and to determine the suppression of endogenous COI1 gene expression, respectively.
2. Extract genomic DNA from leaves of control and transgenic rice seedlings using the CTAB method.
3. Digest 10 μg of genomic DNA with *EcoRI* in a 30-μL reaction at 37°C overnight.
4. Separate the digested DNA on a 0.8% agarose gel and transfer DNA onto a nylon membrane according to the standard Southern blot protocol.
5. Extract total RNA from leaves of control and transgenic rice seedlings with the TRIzol reagent by following the manufacturer's instruction.
6. Separate 15 μg of total RNA on a 1.2% agarose gel containing formaldehyde and transfer RNA onto a nylon membrane according to the standard Northern blot protocol.
7. Prepare the B fragment used in the COI1 RNAi construct as a probe for Southern hybridization and the PCR fragment corresponding to 825-1244 nt of rice COI1 gene as a probe for Northern hybridization.

8. Radiolabel the aforementioned probes with [α-^{32}P] dCTP using the random priming method.
9. Hybridize Southern and northern blots in PerfectHyb™ plus buffer at 62°C overnight with the radiolabeled probes, respectively.
10. After washing the membranes (2X SSC for 10 min at 62°C twice and then 1X SSC plus 0.5% sodium dodecyl sulfate at 62°C for 20 min twice), the Southern and Northern blots are autoradiographed and/or analyzed with a phosphoimager for relative levels of the COI1 gene expression in control plant and RNAi transgenic lines.

JA Sensitiity Test of COI1 Suppression Lines

1. Collect rice seeds from the control plant and RNAi lines with significant suppression of endogenous COI1 gene.
2. Place surface-sterilized seeds on half-strength MS medium containing 20 μM MeJA, and incubate at 25°C under the 14-h light /10-h dark condition for 9 d.
3. Measure both shoot and root lengths of control and RNAi transgenic seedlings. In comparison with the control, the COI1 suppression lines exhibit less inhibition of shoot growth by MeJA and thus are insensitive to jasmonate.

Effect of COI1 Suppression on J-Responsie Gene Epression

1. To determine the role of COI1 in mediating JA signaling, the expression of JA-responsive genes (e.g., *OsVSP* encoding rice vegetative storage protein, and *OsMPK7* encoding a JA-inducible mitogen-activated protein kinase) are examined in response to JA treatment.
2. Spray the leaves of 2-wk-old control and COI1 RNAi transgenic seedlings with 0.1 mM JA solution.
3. Sample water- and JA-treated young leaves at different time points (0, 1, 3, 6, 12, and 24 h after treatment), freeze them in liquid nitrogen immediately and store at –70°C until use.
4. Extract total RNA from leaf samples and prepare Northern blots as described previously.
5. Hybridize Northern blots with radiolabelled, JA-responsive gene probes (e.g., *OsVSP* and *OsMPK7*).
6. After washing, Northern blots are autoradiographed and analyzed with a phosphoimager for relative expression of JA-responsive genes in control and COI1 suppression lines following JA treatment. Reduced expression of *OsVSP* and *OsMPK7* are observed in the COI1 suppression lines in response to JA treatment, suggesting a positive role of the COI1 in mediating JA-responsive gene expression.

Disease Resistance Ealuation of RNi Transgenic Lines

1. Rice RNAi transgenic lines may be evaluated for altered disease resistance and susceptibility using different pathogens, such as *Magnapothe grisea* (rice blast) and *Xanthomonas oryzae* pv. *oryzae* (rice bacterial blight).
2. Preliminary tests can be conducted with first-generation transgenic lines by spot

inoculation of *M. grisea* on detached leaves. Further evaluation of disease resistance should be conducted with heterozygous seeds from the first generation transgenic lines and preferably homozygous seeds identified from the second generation transgenic lines.

3. For the blast infection, 2-wk-old seedlings are spray-inoculated with *M. grisea* at a concentration of 250,000 canidial spores/mL. After incubation in a dew chamber (22°C) for 24 h, rice plants are moved to a growth chamber and maintained at 28°C with a 14-h light/10-h dark cycle.
4. Disease rating as well as measurement of lesion size and number are conducted at 6 d after inoculation. The relative growth of *M. grisea* in control and RNAi transgenic lines can also be determined using a real-time PCR assay or Northern blot/phosphoimaging analysis.

Notes

1. Two pairs of specific primers were designed based on the sequence of rice COI1 gene (accession number BAB84399). The A fragment, corresponding to 27613764 nt (with a 258 bp COI1 intron), was amplified with the first pair of primers (COI1-BamHI-F1, 5'-CCT GGA TCC AGT TAA GTT CCC ACC CAG ATT ATG C; and COI1-KpnI-R, 5'-CCA GGT ACC GGC TAT CCA CAC AGG GTT CTC C). The B fragment, corresponding to 3019-3764 nt, was amplified with the second pair of primers (COI1-BamHI-F2, 5'-CGA GGA TCC GTG AGG AAC GTG ATA GGA GAT AGA GG; and COI1-SalI-R, 5'-CGT GTC GAC GGC TAT CCA CAC AGG GTT CTT CTC C).
2. Gene-specific sequences (e.g., 3' region) are usually selected for specific gene silencing. The inverted repeat should be at least 100-bp long for effective RNAi. Typically, complementary flanking sequences are 250- to 500-bp long and separated by a spacer or intron sequence of 200 to 300 bp. In this case, a 258-bp intron of rice COI1 gene was conveniently included in the RNAi construct because it was reported to improve the effectiveness of RNAi.
3. Besides the traditional cloning approach, RNAi construct can be made by high throughput cloning using Gateway recombination technology and inverted repeat of a heterologous 3'-untranslated sequence.
4. Alternatively, a freeze–thawed method can be used to introduce the RNAi construct into *Agrobacterium* cells. Briefly, *Agrobacterium* competent cells are added with 1 μg plasmid DNA and quickly frozen in liquid nitrogen. The microcentrifuge tubes containing agrobacterial cells were then taken out and immediately put in 37°C water bath for 5 min. After addition of 1 mL of YEP liquid medium, incubate the bacterial cells for 2 h at 28°C on a shaker before plating.
5. It is important to keep in moisture after transplanting. Transgenic plantlets should be covered with plastic cones or bags for 2 to 3 d to prevent moisture evaporation and facilitate root growth.
6. To detect the endogenous gene expression without the interference of RNAi transgene, the probe used for Northern hybridization must be different from the gene sequence region used to make the RNAi construct. If the 3' region of a gene is used to make RNAi

construct, a DNA sequence from the 5' region should be used as a probe to detect the suppression of endogenous gene expression in Northern analysis.

7. MeJA is much less expensive than JA and is adequate for the jasmonate sensitivity test. Although the growth of rice seedlings can be inhibited by MeJA at as low as 1 μM concentration, 20 μM appears to be an appropriate concentration for examining jasmonate insensitivity in rice.
8. Because the seeds from the primary transgenic plants are heterozygous and contain segregants that lose the RNAi transgene, they need to be further analyzed by PCR for the presence or absence of the RNAi transgene after the JA sensitivity test. Based on PCR results, MeJA sensitivity data can be corrected for the genetic segregation. Therefore, it is better to use homozygous seeds from the secondgeneration transgenic plants for JA sensitivity tests.
9. To obtain homozygous seeds, rice seeds from the second-generation plants should be harvested individually and tested for homozygosity by PCR. In addition, transgene segregation (the presence or absence of RNAi construct) may be detected based on hygromycin sensitivity. Rice seeds and leaf segments can be placed in Petri dishes containing 50 mg/L hybromycin solution and tested for inhibition of seed germination or browning of leaf segments, respectively.

14

Chapter

ELECTROPHORETIC METHODS FOR PLANT PROTEOMICS

Since the late 1970s, numerous studies have been performed seeking *changes in genome expression* (note that the term *proteome* was first coined in 1994) in different plant tissues, organs, subcellular compartments, and organelles to analyze the influence of internal or external stimuli, such as response to treatment with hormones, chemicals, or biotic/abiotic stresses (e.g., heat, cold, drought, or ozone) on protein *expression* patterns, or to monitor different *developmental* and growth stages (e.g., seed germination, maturation, or aging) in plants. Apart from the study of proteome variations within a given genome, proteome variations between different genomes were also widely analyzed.

For example, phylogenetic *relationships* and genetic distances were estimated within and between populations to generate genetic maps and to establish dendrograms based on *parsimony* analysis of qualitative and *quantitative* variations in protein patterns. Other applications included the assessment of genetic diversity/variability for the *differentiation* and characterization of closely related lines, cultivars, hybrids, or mutants.

A third field of research was identification of quality-related molecular markers (e.g., for baking, malting, and brewing quality), sensitivity/resistance to plant pathogens (such as fungi and viruses), and localization of quantitative trait loci (QTLs) for plant breeding purposes.

In more recent surveys, the scope has been extended to the study of pathogenic as well as symbiotic plant-microbe interactions (e.g., between legumes and nitrogen-fixing bacteria), identification and characterization of plant allergens in foods (e.g., wheat flour allergens), or assessment of substantial equivalence in genetically modified plants.

For a more comprehensive overview, the reader is referred to on these subjects. With the exception of a few studies *utilizing* liquid *chromatography/tandem* mass *spectrometry* (LC-MS/MS) in plant proteome analysis, the *overwhelming* number of the aforementioned surveys was based on *two-dimensional* gel *electrophoresis* (2-DE) technology.

2-DE is—and will probably remain for the foreseeable future—the only analytical technique that can be routinely applied for parallel quantitative expression profiling of large sets of complex protein mixtures, such as total cell or tissue extracts.

Combined with protein identification by mass spectrometry (MS), 2-DE is currently the workhorse for proteomics. 2-DE couples isoelectric focusing (IEF) and sodium dodecyl sulfate-polyacrylamide gel electrophoresis (SDS-PAGE) to resolve denatured proteins according to two independent parameters, i.e., isoelectric point (pI) in the first dimension and molecular mass (M_r) in the second.

Depending on the gel size and pH gradient used, 2-DE can resolve more than 5000 proteins simultaneously (approx 2000 proteins routinely) and can *visualize* and quantify <1 ng of protein per spot, given that a higly sensitive protein detection method has been applied.

Equally important, it delivers a map of intact proteins, which reflects changes in protein *expression* level, isoforms, or posttransla- tional *modifications*, which is in contrast to LC-MS/MS-based methods, which perform analysis on peptides, in which M_r and pI information is lost, and in which stable isotope labeling is required for quantitative analysis.

Initial proteome studies were performed by using the 2-DE technology originally described by O'Farrell in 1975, based on carrier ampholytegenerated pH gradients in the first dimension. The limitations of this 2-DE technology with respect to reproducibility, resolution, separation of very acidic and/or very basic proteins, and sample loading capacity have been largely overcome by the introduction of immobilized pH gradients (IPGs) for the first dimension of 2-DE.

IPGs are based on the use of the bifunctional ImmobilineR reagents, a series of 10 chemically well-defined acrylamide derivatives with the general structure CH_2=CH-CO-NH-R, where R contains either a carboxyl or an amino group. These form a series of buffers with different *pK* values between *pK* 1 and 13. Since the reactive end is copolymerized with the acrylamide matrix, extremely stable pH gradients are generated, allowing true steady-state IEF with increased reproducibility.

Narrow pH range IPGs not only provide increased resolution (ApI = 0.001) but also permit detection of lower abundance proteins, whereas alkaline proteins up to pH 12 have been separated under truly steady-state conditions using IPG technology.

Other technical improvements, such as more powerful reagents for solubilization of hydrophobic proteins, devices for *semiautomatic* running of multiple gels in parallel, highly sensitive protein detection procedures based on fluorescent dye technologies for improved *reproducibility*, more accurate quantitation, and simplified spot pattern comparison, as well as highly sophisticated computer software for the analysis of complex 2-D patterns, have also contributed to the widespread use of 2-DE in proteome analysis.

Generally speaking, proteome analysis is technically far more *challenging* than genome analysis owing to the highly diverse *physicochemical* properties and abundance of proteins. For instance, it has been estimated that >20,000 genes in higher eukaryotic organisms such as plants translate to 50,000 to 100,000 proteins, owing to alternative protein splicing, proteolytic cleavages, phosphorylation, glycosylation, and more than a hundred other possible posttranslational modifications. Fortunately, not all these protein variants are expressed in a given tissue at a given time, but nevertheless their number is likely to be in the range of at least 10,000 to 20,000. Moreover, *plant proteome analysis* faces several specific challenges. In particular, sample preparation is difficult because of the rigidity of plant cell walls and because of the accumulation of large quantities of interfering compounds in the central vacuole such as phenolics, pigments, and hydrolytic enzymes, which upon tissue disruption can lead to protein degradation and/or precipitation.

In addition, owing to the usually low protein content in green plant tissues (typically 2%, compared with approximately *20%* in *mammalian* tissues or microbial cells), methods for enrichment of proteins are usually required. Despite the aforementioned obstacles, 2-DE-MS has been successfully applied for proteome analysis on the whole plant tissue and subcellular levels.

The major steps of the classical 2-DE-MS workflow include: (1) sample preparation/ prefractionation and protein solubilization; (2) protein separation by2-DE; (3) protein detection and *quantitation*; (4) computer-assisted analysis of 2-DE patterns; (5) protein identification and characterization by MS; and (6)2-D protein database construction. The major emphasis of the following sections is on 2-DE technology with IPGs (method 1); the other methods will not be discussed here.

The reader is referred to the corresponding chapters of this book and reviews in and covering these aspects. Briefly, the first dimension (IEF) of 2-DE with immobilized pH gradients (IPG-Dalt), according to Gorg et al. (updated in 2000 and 2004) is performed in individual, 3-mm-wide and up to 24-cm-long IPG gel strips cast on GelBond PAGfilm (laboratory-made or commercial Immobiline Dry-Strips). Samples can be applied onto the IPG strips either by cup loading or by in-gel rehydration.

After completion of IEF, the IPG strips are equilibrated with SDS buffer in the presence of urea, glycerol, dithiothreitol (DTT), and iodoacetamide and applied onto horizontal or vertical SDS gels in the second dimension. 2-DE has been considerably simplified by the use of semiautomated devices such as the *IPGphor* in the first dimension and multiple SDS-PAGE apparatuses for running up to 20 different samples in parallel. After electrophoresis, the separated proteins are visualized by staining with silver nitrate, organic dyes, autoradiography (or phosphor imaging) of radiolabeled samples, or—preferably— by labeling or staining with fluorescent dye molecules.

MATERIALS

Equipment

The following were all procured from GE Healthcare/Amersham Biosciences:

1. Isoelectric focusing device (Multiphor II). 2. IPG DryStrip reswelling tray.
3. IPG DryStrip kit.
4. IPGphor.
5. IPGphor Strip holders.
6. IPGphor *Cup* loading Strip holders.

7. Multiple vertical SDS electrophoresis apparatus (Ettan DALT). 8. SDS gel casting *box.*
9. Cassette rack.
10. Glass plates.
11. Thermostatic circulator (Multitemp III).
12. Power supply (Multidrive XL).
13. Laboratory shaker.

Solutions

1. Urea lysis solution: 9.5 M urea, 2% (w/v) 3-[(3-cholamidopropyl)dimethylammonio]1-propane sulfonate (CHAPS), 2% (v/v) Pharmalyte pH 3 to 10, 1% (w/v) DTT.
   a. To prepare 50 mL of lysis solution, dissolve 30.0 g of urea (Merck, Darmstadt, Germany) in deionized water and make up to 50 mL.
   b. Add 0.5 g of Serdolite MB-1 mixed ion exchange resin (Serva, Heidelberg, Germany; stir for 10 min, and filter.
   c. Add 1.0 g of CHAPS (GE Healthcare, Freiburg, Germany), 0.5 g DTT (SigmaAldrich, Taufkirchen, Germany), 1.0 mL Pharmalyte pH 3.0 to 10 (GE Healthcare), and—immediately before use—50 mg of Pefabloc proteinase inhibitor (Merck) to 48 mL of the urea solution.
   d. For solubilization of the more hydrophobic proteins, use thiourea/urea lysis solution (2 M thiourea [Sigma-Aldrich], 5–7 M urea, 2–4% [w/v] CHAPS, and/or sulfobetaine detergents [e.g., SB 3-10), 1% DTT, 2% [v/v] carrier ampholytes) in combination with IPG strip rehydration solution, consisting of a mixture of urea/thiourea (6 M urea, 2 M thiourea, 1% [w/v] CHAPS, 15 mM DTT, 0.5% [v/v] Pharmalyte, pH 3.0–10.0).
2. IPG DryStrip rehydration buffer: 8 M urea, 0.5% (w/v) CHAPS, 15 mM DTT, 0.5% (v/v) Pharmalyte, pH 3.0 to 10.0.
   a. To prepare 50 mL of the solution, dissolve 25.0 g of urea (Merck) in deionized water and complete to 50 mL.
   b. Add 0.5 g of Serdolite MB-1 (Serva), stir for 10 min, and filter.
   c. To 48 mL of this solution add 0.25 g of CHAPS (GE Healthcare), 0.25 mL Pharmalyte, pH 3.0 to 10.0 (40% w/v; GE Healthcare), and 100 mg of DTT (Sigma-Aldrich) and complete to 50 mL with deionized water.
3. IPG strip equilibration buffer: 6 M urea, 30% (w/v) glycerol, 2% (w/v) SDS in 0.05 M Tris-HCl buffer, pH 8.6.
   a. To make 500 mL, add 180 g of urea (Merck), 150 g of glycerol (Merck), 10 g of SDS (Serva), 16.7 mL of SDS gel buffer, and a few grains of bromophenol blue (Serva).
   b. Dissolve in deionized water and fill up to 500 mL. The buffer can be stored at room temperature up to 2 weeks.
4. SDS gel buffer: 1.5 M Tris-HCl, pH 8.6 and 0.4% (w/v) SDS.

a. To make 500 mL, dissolve 90.85 g of Trizma base (Sigma-Aldrich) and 2.0 g of SDS (Serva) in about 400 mL of deionized water.

b. Adjust to pH 8.6 with 4 N HCl (Merck) and fill up to 500 mL with deionized water.

c. Add 50 mg of sodium azide (Merck) and filter. The buffer can be stored at 4°C up to 2 wk.

5. Electrode buffer stock solution: to make 5 L of electrode buffer stock solution, dissolve 58.0 g of Trizma base (Sigma-Aldrich), 299.6 g of glycine (SigmaAldrich), and 19.9 g of SDS (Serva) in deionized water and complete to 5.0 L.

6. Acrylamide/bisacrylamide solution (30.8% T, 2.6% C): 30% (w/v) acrylamide and 0.8% (w/v) methylenebisacrylamide in deionized water.

a. To make 500 mL, dissolve 150.0 g of acrylamide (GE Healthcare) and 4.0 g of methylenebisacrylamide (GE Healthcare) in deionized water and fill up to 500 mL.

b. Add 1 to 2 g of Serdolit MB-1 (Serva), stir for 10 min, and filter.

c. The solution can be stored up to 2 wk in a refrigerator.

7. Ammonium persulfate solution: 10% (w/v) of ammonium persulfate in deionized water. To prepare 10 mL of the solution, dissolve 1.0 g of ammonium persulfate (GE Healthcare) in 10 mL of deionized water. This solution should be prepared freshly just before use.

8. Displacing solution: 50% (v/v) glycerol in deionized water and 0.01% (w/v) bromophenol blue. To make 500 mL, mix 250 mL of glycerol (100%; Merck) with 250 mL of deionized water, add 50 mg of bromophenol blue (Serva), and stir for a few minutes.

9. Overlay buffer: buffer-saturated 2-butanol. To make 30 mL, mix 20 mL of SDS gel buffer with 30 mL of 2-butanol (Merck) , wait for a few minutes until the two phases have separated, and remove the butanol layer with a pipet.

10. Agarose solution: Suspend 0.5% (w/v) agarose (GE Healthcare) in electrode buffer and melt it in a boiling water bath or in a microwave oven.

METHODS

First Dimension: Isoelectric Focusing in IPG Strips (IPG-IEF)

The first dimension of 2-DE with IPGs (IPG-Dalt), isoelectric focusing (IEF), is performed in individual 3-mm-wide IPG gel strips cast on a supporting GelBond PAG film™ plastic sheet. A multitude of 7-, 11-, 18-, and/or 24-cmlong ready-made Immobiline DryStrips of almost any desired pH range has been made available, e.g., wide pH ranges of IPG 3 to 10 and IPG 3 to 11, medium pH ranges of IPG 4 to 7 or IPG 6 to 9, and narrow pH ranges of IPG 4 to 5 or IPG 4.5 to 5.5 (e.g., GE Healthcare, Bio-Rad, Sigma-Aldrich, Serva). Alternatively, laboratory-made IPG DryStrips can be used.

For details on IPG gel casting, the interested reader is referred to previously published protocols. IPG DryStrips have to be rehydrated before IEF and are then applied onto the cooling plate of a horizontal isoelectric focusing apparatus. More recently, IPG-IEF has been simplified

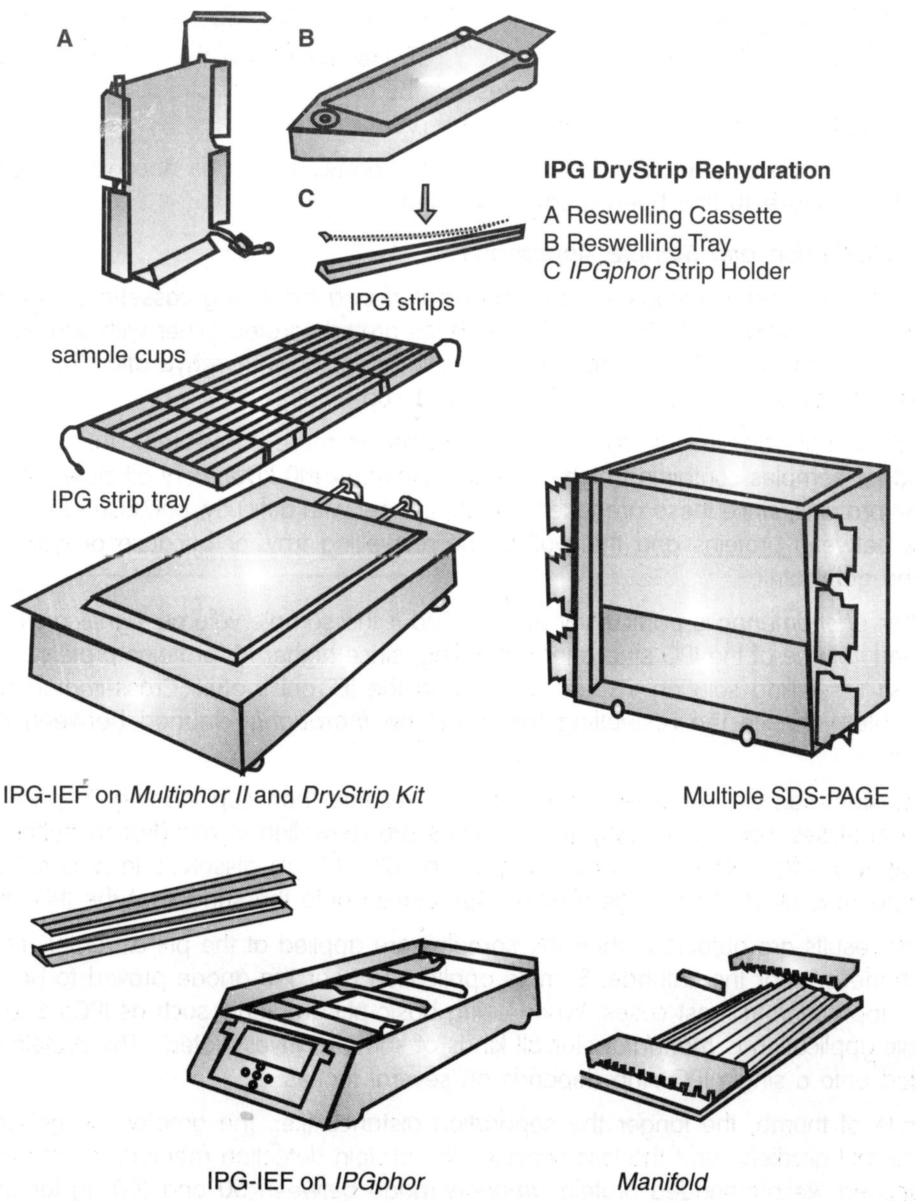

Figure 14.2: Procedure of 2-DE with IPGs (IPG-Dalt) based on the protocol of Gorg et al.

and accelerated by using an integrated system, the *IPGphor.* This instrument features strip holders that provide rehydration of individual IPG DryStrips with or without sample, optional separate sample cup loading, and subsequent IEF with high (8000 V) voltage, without handling the IPG strip after it is placed in a ceramic strip holder. With IPGs above pH 9, horizontal streaking owing to DTT depletion can occur at the basic end.

To avoid streaking, the cysteines should be stabilized as mixed disulfides by using hydroxyethyl-disulfide (HED) reagent (DeStreak™, GE Healthcare) in the IPG strip rehydration solution instead of a reductant such as DTT. Besides the elimination of streaking, the use of HED results in a simplified spot pattern and improved reproducibility.

Moreover, with the IPGphor, IEF of strongly alkaline proteins such as ribosomal and nuclear proteins with pIs above 10 has been greatly simplified.

IPG Strip Rehydration and Sample Application

Prior to IEF, the dried IPG strips must be rehydrated in a reswelling cassette or a reswelling tray to their original thickness of 0.5 mm. IPG DryStrips are rehydrated either with sample already dissolved in rehydration buffer (sample in-gel rehydration) or with rehydration buffer without sample, followed by sample application by so-called cup-loading.

Although sample application by *in-gel rehydration* is more convenient, this procedure is discouraged for samples containing high molecular weight (<100 kDa), very alkaline, and/or very hydrophobic proteins, since these are taken up into the gel with difficulty, because of hydrophobic interactions between proteins and the wall of the reswelling tray, or because of size-exclusion effects of the gel matrix.

The latter *phenomenon* is particularly pronounced if the sample volume *significantly* exceeds the calculated volume of the IPG strip after reswelling, since higher M_r proteins preferably remain in the excess reswelling solution instead of entering the IEF gel matrix. Cross-contamination is another problem, hence the reswelling tray must be *thoroughly* cleaned between different *experiments.*

In conclusion, sample in-gel rehydration is less reliable than cup loading, in particular for quantitative analyses. For *cup loading,* IPG DryStrips are reswollen in rehydration buffer, without sample, however. After IPG strip reydration, samples (20–100 µL) dissolved in lysis solution are put into disposable plastic or silicone rubber cups placed onto the surface of the IPG strip.

The best results are obtained when the samples are applied at the pH extremes, i.e., either near the anode or near the cathode. Sample application near the anode proved to be superior to cathodic application in most cases. When using basic pH gradients such as IPGs 6 to 12 or 9 to 12, anodic application is *mandatory* for all kinds of samples investigated . The protein amount to be loaded onto a single IPG strip depends on several factors.

As a rule of thumb, the longer the separation distance (i.e., the greater the gel size), the narrower the pH gradient, and the less sensitive the protein detection method, the more protein must be applied. Recommended protein *amounts* range between 50 and 100 µg for analytical (silver-stained) 20 x 20 cm^2 2-DE gels, and up to 1 mg (or even more) for micropreparative gels. In case of very narrow pH range IPGs, we strongly recommended application of prefractionated samples only .

Rehydration of IPG DryStrips Using the IPG DryStrip Reswelling Tray

1. For *sample in-gel rehydration,* directly solubilize a cell lysate or tissue sample (5–10 mg protein/mL) in an appropriate quantity of IPG DryStrip *rehydration* buffer. For 180-mm-long and 3-mm-wide IPG DryStrips, pipet 350 tL of this solution into the grooves of the

IPG DryStrip reswelling tray. For longer or shorter IPG strips, the rehydration volume has to be calculated *accordingly* (e.g., 450 tL for 240-mm-long IPG DryStrips).

2. Remove the protective covers from the surface of the IPG DryStrips and apply the IPG strips, gel side down, into the grooves without *trapping* air bubbles. Then cover the IPG strip, which must still be moveable and not stick to the tray, with IPG Drystrip cover fluid (which prevents drying out during *reswelling*), and rehydrate the IPG strips overnight at approximately 20°C. Higher temperatures (>37°C) hold the risk of protein *carbamylation*, whereas lower temperatures (< 15°C) should be avoided to prevent urea crystallization on the IPG gel.
3. For *cup loading*, the IPG dry strips are rehydrated overnight in rehydration buffer in the reswelling tray as described above in step 2 but without sample, however.

Saple In-Gel Rehydration of Ipg Drystrips Using Ipghor Strip Holders

1. Solubilize proteins in sample solubilization buffer (i.e., urea or urea/thiourea lysis solution) and dilute the extract with IPG DryStrip rehydration buffer.
2. Apply the required number of IPGphor strip holders onto the cooling plate/electrode contact area of the IPGphor.
3. Pipette 350 tL of sample-containing rehydration solution (for 180-mm-long IPG strips) into the strip holder base.
4. Peel off the protective cover sheets from the IPG strip and slowly lower the IPG strip (gel side down) onto the rehydration solution. Avoid trapping air bubbles. The IPG strip must still be moveable and not stick to the tray. Cover the IPG strips with 1 to 2 mL of IPG DryStrip cover fluid and apply the plastic cover. Pressure blocks on the *underside* of the cover ensure that the IPG strip keeps in good contact with the electrodes as the gel swells. Apply low voltage (30–50 V) during rehydration for *improved* entry of high M_r proteins.

IPG-IEF on a Flat-Bed Apparatus (Multiphor II Unit)

The rehydrated IPG strips may be directly applied onto the cooling plate of the IEF apparatus if running time does not exceed 12 h (which is usually the case for wide and medium pH range IPGs such as IPG 3–10 or 4–7), if the pH gradient does not exceed pH 10.0, and (3) if only small sample volumes (20 μL) are applied by cup loading . The application of higher sample volumes (up to 100 μL) is facilitated when the sample cups of the Immobiline

DryStrip Kit are used. When IEF is performed using the DryStrip Kit, the IPG gel strips can be covered by a layer of silicone oil or DryStrip cover fluid. This is mandatory for IEF in very basic (pH > 10.0), or for micropreparative runs in narrow pH gradients (pH range < 1 unit), whereas in case of broad pH *gradients* not exceeding pH 10.0 (e.g., IPG 4–7 or 3–10), the DryStrip Kit can be used without oil overlay.

1. Place the cooling plate into the Multiphor II electrophoresis unit. Pipet 3 to 4 mL of kerosene or IPG DryStrip cover fluid onto the cooling plate, and position the Immobiline DryStrip tray on the cooling plate. Avoid trapping air bubbles between the tray and the cooling plate.

Table 14.1 Isoelectric Focusing (IEF) Running Conditions Using the Multiphor II IEF Unita

IPG strip length	180 mm
Temperature	20°C
Current max.	0.05 mA per IPG strip
Power max.	0.2 W per IPG strip
Voltage max.	3500 V
1. Analytical IEF	
Initial IEF:	
Cup loading (20–50 μL)	
150 V, 1–3 h[b]	
300 V, 1–3 h[b]	
600 V, 1 h	
In-gel rehydration (350 μL)	
150 V, 1–3 h[b]	
300 V, 1–3 h[b]	
IEF to the steady state at 3500 V:	
1–1.5 pH units	
e.g., IPG 5–6	24 h
e.g., IPG 4–5.5	20 h
3 pH units	
IPG 4–7	12 h
IPG 6–9	12 h
4 pH units	
IPG 4–8	10 h
IPG 6–10	10 h
5–6 pH units	
IPG 4–9	8 h
IPG 6–12	8
h 7–8 pH units	
IPG 3–10	6 h
IPG 3–11	6 h

8–9 pH units

IPG 3–12	6 h
IPG 4–12 NL8 h	

2. Extended separation distances (240 mm)

IEF to the steady-state at 3500 V:

IPG 3–12	8 h
IPG 4–12NL	12 h
IPG 5–6	40 h

3. Micropreparative IEF

Initial IEF:

Cup loading (100 μL)

50 V, 12–16 h

300 V, 1 h

In-gel rehydration (350 μL)

50 V, 12–16 h 300 V, 1 h

IEF to the steady state at 3500 V:

Focusing time of analytical IEF plus approximately 50%

2. Connect the electrode leads on the tray to the Multiphor II unit.
3. Pour about 10 mL of IPG DryStrip cover fluid or silicone oil into the tray, and place the corrugated *Immobiline* strip aligner into the tray on top of the oil.
4. After the IPG strips have been rehydrated, use clean foreceps to remove the reswollen IPG strips from the reswelling tray. Rinse them with deionized water and blot for a few seconds between two sheets of moist filter paper to remove excess liquid in order to prevent urea crystallization on the surface of the gel during IEF. Transfer the rehydrated IPG gel strips (gel side up and acidic end toward the anode) into adjacent grooves of the aligner in the tray. Align the IPG strips such that the anodic gel edges are lined up.
5. Cut two IEF electrode strips (GE Healthcare) or paper strips prepared from 2-mmthick filter paper (e.g., MN 440, Macherey & Nagel, Germany) to a length corresponding to the width of all IPG gel strips lying in the tray. Soak the electrode strips with deionized water, remove excessive moisture by blotting with filter paper, and place the moistened IEF electrode strips on top of the aligned strips near the cathode and anode.
6. Position the electrodes and press them gently down on top of the IEF electrode strips.
7. If samples have already been applied by in-gel rehydration, cover the IPG strips with approx 80 mL of DryStrip cover fluid, and continue with **step** 12. In case of sample application by cup loading, continue with **step 8.**

8. Place the sample cups on the sample cup bar, but avoid touching the gel surface with the cups. Moreover, make sure that there is a distance of a few millimeters between the sample cups and the anode (or cathode, in case of cathodic sample application).
9. Move the sample cups into position, one sample cup above each IPG strip, and gently press down the sample cups. The sample cups should form a good seal with the IPG strips but not damage their surface!
10. Once the sample cups are properly *positioned*, pour about 80 mL of DryStrip cover fluid into the tray so that the IPG gel strips are completely covered. If the oil leainto the sample cups, suck the oil out, readjust the sample cups, and check for leakage again. Fill up each sample cup with a few drops of DryStrip cover fluid. In case of IEF using wide pH gradients in the range between pH 3 and 10, the oil step can be omitted.
11. Pipet the samples into the cups by underlaying. *Watch again for leakage.*
12. Close the lid of the electrofocusing chamber and start the run according to the parameters given in. For improved sample entry, voltage should be lim-ited to 150 to 300 to 600 V for the first few hours. Then continue with *maximum* settings of 3500 V to the steady state. Current is limited to 0.05 mA/ IPG strip. Optimum focusing *temperature* is 20°C .
13. When the IEF run is completed, remove the electrodes, sample cup bar, and IEF electrode strips from the tray. Use clean forceps and remove the IPG gel strips from the tray. Those IPG gel strips that are not used immediately for a seconddimension run and/or are kept for further reference are stored between two sheets of plastic film at –70°C up to several months.

IPG-IEF Using the IPGphor Unit

IPG-IEF for 2D *electrophoresis* can be simplified by the use of an integrated instrument, the IPGphor. (GE Healthcare; a similar device has recently been developed by Bio-Rad). The IPGphor includes a Peltier element for precise temperature control (between 19.5°C and 20.5°C) and a programmable power supply. The central part of this *instrument* is the so-called strip holders of different lengths (7, 11, 13, 18, or 24 cm) made from an *aluminium* oxide ceramic, in which IPG strip *rehydration* with sample solution and IEF are performed without further handling after the strip is placed into the strip holder.

Table 14.2: IPGphor Running Conditions. (for Sample In-Gel Rehydration and for Cup Loading)

Gel length	180 mm
Temperature	20°C
Current max	0.05 mA per IPG strip
Voltage max	8000 V
1. Analytical IEF	
Reswellingb:	

(Table Contd.)

30 V, 12–16 h[b]

Initial IEF:

200 V, 1 h

500 V, 1 h

1000 V, 1 h

IEF to the steady state:

Gradient from 1000 to 8000 V within 30 min

8000 V to the steady state, depending on the pH used:

1–1.5 pH units	
e.g., IPG 5–6	8 h
e.g., IPG 4–5.5	8 h
3 pH units	
IPG 4–7	4 **h**
4 pH units	
IPG 4–8	4 **h**
5–6 pH units	
IPG 4–9	4 **h**
7 pH units	
IPG 3–10 L	3 h
IPG 3–10 NL	3 **h**
8–9 pH units	
IPG 3–12	3 h
IPG 4–12	3 h

2. Micropreparative IEF Reswellingb:

30 V, 12–16 h*

IEF to the steady state:

Focusing time of analytical IEF + additional 50% (approx)

The IPGphor is *programmable* and can store up to ten different programs. A delayed start is also possible, which allows the user to load the strip holders with sample dissolved in rehydration buffer in the afternoon and *automatically* start IEF during the night so that IEF is finished the next morning.

When protein separation is performed in alkaline pH ranges (>pH 10.0), much better separations are obtained by applying sample via cup loading on separately rehydrated IPG strips than by sample in-gel rehydration. Sample cup loading is accomplished with special cup loading ("universal") IPGphor strip holders, or with a multiple cup loading strip holder ("Manifold"), which allow(s) the application of quantities up to 100 μL.

Table 14.3: IPGphor Running Conditions for Very Alkaline Immobilized pH Gradients (IPGs) (Sample Cup Loading).

Gel length	180 mm
Temperature	20°C
Current max	0.07 mA per IPG strip
Voltage max	8000 V
IPGs 6-12, 9-12, 10-12	
Sample application	anodic
Initial IEF:	
150 V, 1 h	
300 V, 1h	
600 V, 1h	
IEF to the steady-state:	
Gradient from 600 to 8000 V within 30 min	
8000 V to the steady state	
Total volt hours: 32,000 Vh	

The strip holder platform regulates temperature and serves as the electrical connector for the strip holders. Besides easier handling, a second advantage of the IPGphor is shorter focusing time, since IEF can be performed at rather high voltage (up to 8000 V). Typical running conditions for IEF using the IPGphor are given in.

As indicated earlier, low voltage (30–50 V) should be applied during the rehydration step for *improved* sample entry of high M_r proteins into the *polyacrylamide* gel, which otherwise can be a problem with sample in-gel *rehydration* (*15,28*). Then voltage is increased stepwise up to 8000 V. If IPG strips with *separation* distances <_11 cm are used, voltage should be limited to 5000 V.

For optimum results for samples with high salt *concentrations*, or when narrow pH intervals are used, it is beneficial to insert moist filter paper pads (size: 4 × 4 mm^2) between the electrodes and the IPG strip prior to raising the voltage to 8000 V. After termination of IEF, the IPG strips are stored as described above in.

IEF OF IN-GEL REHYDRATED SAMPLES

1. Apply the required number of strip holders onto the cooling plate/electrode contact area of the IPGphor. Pipet the appropriate amount (e.g., 350 μL in case of a 180-mm-long IPG strip) of sample-containing IPG DryStrip *rehydration* buffer into the strip holders, lower

Table 14.4: IPG Strip Equilibration Protocol		
Reagent	***Effect***	
50 mM Tris-HCl, pH 8.8		
+2% SDS		
+6 M urea	Improved protein transfer from IPG strip to SDS gel	
+30% glycerol		
+1% DTT		
+4% iodoacetamide	Removes point streaking	
	Alkylation of SH groups	
	DTT	***Iodoacetamide***
1. 15 min (10 min)	+	–
2. 15 min (10 min)	–	+

the IPG dry strips gel side down into the rehydration buffer, and overlay with IPG DryStrip cover fluid as described in.

2. Program the IPGphor (desired *rehydration* time, volt hours, voltage gradient).
3. After the IPG gel strips have been rehydrated (which requires 6 h at least, but typically overnight), IEF starts according to the *programmed* parameters listed in.
4. After completion of IEF, store those IPG gel strips that are not used immediately for a second-dimension run between two sheets of plastic film at –70°C.

IEF USING CUP LOADING STRIP HOLDERS

1. Rehydrate IPG DryStrips with rehydration buffer, without sample, however, in a reswelling tray. After the IPG strips have been rehydrated, use clean foreceps to remove the reswollen IPG strips from the *reswelling* tray or *reswelling* cassette. Rinse them with deionized water and blot for a few seconds between two sheets of moist filter paper to remove excess liquid in order to prevent urea *crystallization* on the surface of the gel during IEF, as described in .
2. Apply the required number of the cup loading IPGphor strip holders onto the cooling plate/electrode contact area of the IPGphor instrument, and make sure that the pointed (anodic) ends contact the anodic *electrode* area. Instead of individual strip holders, the Manifold device may be used.
3. Apply the rehydrated IPG gel strips into the cup loading strip holders (or into the Manifold), gel side upward and pointed (acidic) ends facing toward the anode. Make sure that the cathodic end of the IPG strip is *approximately* 1.5 cm from the end of the channel and in electrical contact with the electrode rails via the electrode clips.

4. Moisten two filter paper electrode pads (size: 4 × 10 mm^2) with deionzed water, remove excess liquid by blotting with a filter paper, and apply the moistended filter paper pads on the surface IPG gel at the anodic and cathodic ends of the IPG strip between the IPG gel and the electrodes. If necessary (e.g., when the sample contains high amounts of salt), these filter papers should be replaced by fresh ones after several hours.
5. Position the movable electrodes above the electrode filter paper pads. Clip the electrodes firmly onto the electrode paper pads.
6. Position the movable sample cup either near the anode or cathode, and gently press the sample cup onto the surface of the IPG gel strip. The sample cup should form a good seal with the IPG strip but not damage its surface.
7. To confirm that the sample cup does not leak, pipet 100 µL of IPG cover fluid into the cup. If a leak is detected, remove the fluid and use a tissue paper to remove the cover fluid and reposition the sample cup. Check again for leakage. Remove the cover fluid before loading the sample.
8. Overlay each IPG strip with 2 to 4 mL of IPG strip cover fluid (the use of silicone oil or kerosene instead of IPG strip cover fluid is discouraged). In case the cover fluid leaks into the sample cup, rearrange the cup and use tissue paper to remove the cover fluid from the cup. Check again for leakage, and pipet the sample (20– 100 µL) in the sample cup.
9. Program the instrument (desired volt hours, voltage gradient, temperature, and so on) and run IEF according to the recommended settings in. Omit the low-voltage rehydration step recommended for sample in-gel rehydration in.
10. After IEF is complete, proceed with equilibration and second-dimension IEF (SDS-PAGE), or store the IPG strips, for up to several months, between two plastic sheets at –70°C.

IPG Strip Equilibration

Prior to the second-dimension separation (SDS-PAGE), it is essential that the IPG strips be *equilibrated* to allow the separated proteins to interact fully with SDS. Because the focused proteins bind more strongly to the fixed charged groups of the IPG gel matrix than to carrier *ampholyte* gels, relatively long equilibration times (10–15 min), as well as urea and glycerol to reduce *electroendosmotic* effects, are required to improve protein transfer from the first to the second *dimension*.

The by far most popular protocol is to incubate the IPG strips for 10 to 15 min in the buffer originally described by Gorg et al. (*13*) (50 mM Tris-HCl [pH 8.8], containing 2% [w/v] SDS, 1% [w/v] DTT, 6 M urea, and 30% [w/v] glycerol). This is followed by a further 10- to 15-min equilibration in the same solution containing 4% (w/v) iodoacetamide instead of DTT.

The latter step is used to alkylate any free DTT, as otherwise it migrates through the second-dimension SDS-PAGE gel, resulting in an artifact known as point streaking that can be observed after silver staining. More importantly, iodoacetamide alkylates sulfhydryl groups and prevents their reoxidation.

This two-step reduction/alkylation procedure is highly recommended, since it *considerably* simplifies *downstream* sample *preparation* (protein in-gel digestion) for spot *identification* by mass spectrometry. After equilibration, the IPG strips are applied onto the surface of the second-dimension horizontal or vertical SDS-PAGE gels.

1. Dissolve 100 mg of DTT (Sigma-Aldrich) in 10 mL of equilibration buffer to make equilibration buffer I.
   a. Make 10 mL per IPG strip.
   b. Place the focused IPG strips into individual test tubes (250 mm long; 20 mm internal diameter), and add 10 mL of equilibration buffer I to each tube.
   c. Seal the tubes with Parafilm, rock them for 15 min on a shaker, and pour off the equilibration buffer. Shorter equilibration times (10 min) can be applied, at the risk, however, that some proteins may not migrate out of the IPG gel strip during sample entry into the SDS-PAGE. In this case it is advisable to check, by staining the IPG strip after removal from the SDS gel, whether all proteins have left the IPG strip.
2. Dissolve 0.4 g of iodoacetamide (Sigma-Aldrich) in 10 mL of equilibration buffer to make equilibration buffer II.
   a. Make 10 mL per IPG strip.
   b. Add this buffer and 50 µL of bromophenol blue (Serva) solution as tracking dye for SDS-PAGE to each tube, and equilibrate for another 15 min with gentle agitation.
3. Pour off equilibration buffer II, and proceed to SDS-PAGE. If SDS-PAGE is performed on a horizontal electrophoresis unit (e.g., Multiphor II), briefly rinse the IPG gel strip with deionized water, and place it on a piece of filter paper at one edge for a few minutes to drain off excess equilibration buffer. If SDS-PAGE is performed on a vertical electrophoresis unit (e.g., Ettan Dalt), briefly rinse the equilibrated IPG strip with electrode buffer.

Second Dimension: Multiple Vertical SDS-PAGE

SDS-PAGE can be performed on horizontal or vertical systems. Horizontal setups are ideally suited for ready-made gels (e.g., ExcelGel SDS; Amersham Biosciences/GE Healthcare), whereas vertical systems are preferred for multiple runs in parallel, in particular for large-scale proteome analysis, which usually requires simultaneous electrophoresis of batches of second-dimension SDS-PAGE gels for higher throughput and maximal reproducibility.

SDS Gel Casting

1. The gel casting cassettes (200 × 250 mm^2) are made in the shape of books consisting of two 3-mm-thick glass plates, connected by a hinge strip, and two 1.0-mm-thick spacers in between them. Stack 14 cassettes vertically into the gel casting box of the Ettan Dalt II apparatus with the hinge strips to the right, interspersed with plastic sheets (e.g., 0.05-mm-thick polyester sheets).
2. Place the front plate of the casting box in place and screw on the nuts (hand tight).
3. Connect a polyethylene tube (i.d. 5 mm) to a funnel held in a ring-stand at a level of about 30 cm above the top of the casting box. The other end of the tube is placed in the grommet in the casting box side chamber.

Table 14.5: Recipes for Casting Vertical SDS Gels (7.5, 10, 12.5, or 15% T).

	7.5% 2.6% C	T 10% T 2.6% C	12.5% T 2.6% C	15% T 2.6% C
Acrylamide/bisacrylamide (30.8% T, 2.6% C)	244 mL	325 mL	406 mL	487 mL
Gel buffer	250 mL	250 mL	250 mL	250 mL
Glycerol (100%)	50.0 g	50.0 g	50.0 g	50.0 g
Deionized water	461 mL	380 mL	299 mL	218 mL
TEMED (100%)	50 μL	50 μL	50 μL	50 μL
Ammonium persulfate (10%)	7.0 mL	7.0 mL	7.0 mL	7.0 mL
Final volume	1000 mL	1000 mL	1000 mL	1000 mL

4. Fill the side chamber with 100 mL of displacing solution.
5. Immediately before gel casting, add TEMED and ammonium persulfate solutions to the gel solution. To cast the gels, the gel solution (830 mL) is poured into the funnel. *Avoid introduction of any air bubbles into the tube.* Do not fill the cassettes with acrylamide solution completely, as some space at the top (approx 10 mm) is needed to fix the IPG strip to the SDS gel with hot agarose.
6. When pouring is complete, the tube is removed from the side chamber grommet so that the level of the displacing solution in the side chamber falls.
7. Very carefully pipet about 1 mL of overlay buffer onto the top of each gel to obtain a smooth, flat gel top surface.
8. Allow the gels to polymerize at approximately 20°C for at least 3 h, but preferably overnight for higher reproducibility.
9. After gel polymerization, remove the front of the casting box, and carefully unload the gel cassettes from the box, using a blade to separate the cassettes. Remove the polyester sheets that had been placed between the individual cassettes.
10. Wash each cassette with water to remove any acrylamide on the outer surface, and drain excess liquid off the top surface. Since only 12 gel cassettes fit into the electrophoresis unit, unsatisfactory gels should be discarded, in particular gels with uneven thickness, i.e., usually those at the outer edges of the gel casting cassette.
11. Gels that are not needed immediately can be wrapped in plastic wrap and stored in a refrigerator (4°C) for up to 2 d.

Multiple SDS-PAGE

Using the Ettan Dalt II Vertical Electrophoresis Unit

1. Add 1875 mL of electrode buffer stock solution and 5625 mL of deionized H2O to the

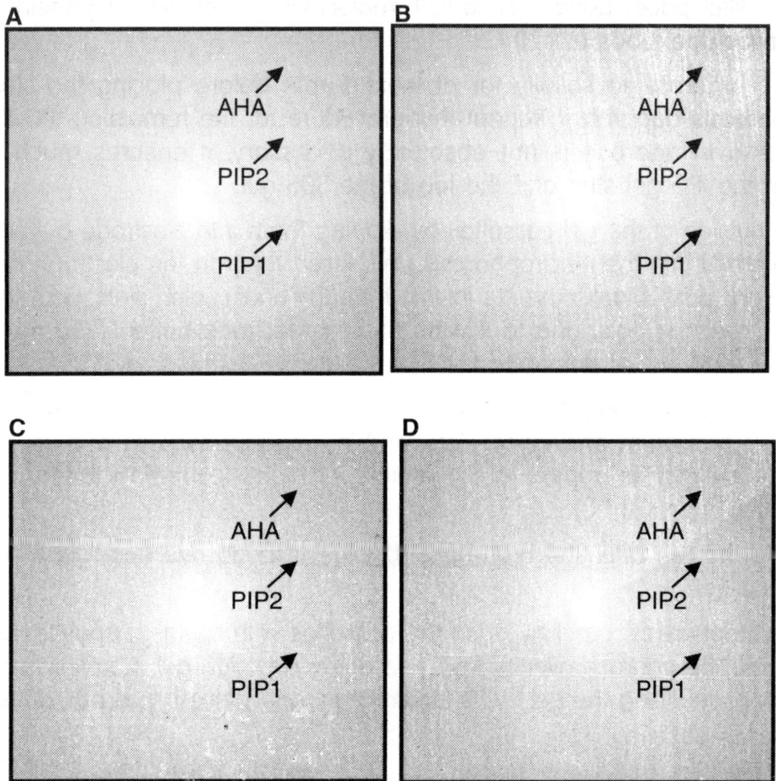

Figure 14.2: 2D electrophoretic separation of A. thaliana *leaf plasma membrane proteins: 60 µg of leaf plasma membrane proteins were loaded on the 2D gels. The first dimension was a 3 to10 linear pH gradient, and the second dimension a 10% acrylamide gel. H+ ATPase (AHA), aquaporin monomer (PIP1), and aquaporin dimer (PIP2) are indicated by arrows. The proteins were extracted and focused in a solution containing 7 M urea, 2 M thiourea, 20 mM DTT, 0.4% carrier ampholytes, and (A) 4% CHAPS, (B) 2% dodecyl maltoside, (C) 2% C7BzO, and (D) 2% ASB14.*

lower electrophoresis buffer tank of the Ettan DALT II unit. Mix, and turn on cooling (25°C).

2. Support the DALT gel cassettes (containing the SDS gels) in a vertical position on the cassette rack to facilitate the application of the IPG gel strips.
3. Briefly rinse the equlibrated IPG strip with electrode buffer (diluted 1:1 with H2O) and place the IPG strip on top of the DALT gel cassette.
   a. Use a thin spatula or a ruler to push against the plastic backing of the IPG strip and slide it into the gap between the two glass plates.
   b. Add 2 mL of hot (75°C) agarose solution, and continue to slide the strip down onto the surface of the SDS gel until good contact is achieved. Avoid trapping air bubbles between the IPG strip and the SDS gel surface.
   c. For co-electrophoresis of molecular weight marker proteins, soak a filter paper pad (2 H 4 mm²) with 5 µL of SDS marker proteins dissolved in electrophore sis buffer, let it dry, and apply it to the left or right of the IPG strip.

d. Dried filter paper pads soaked with molecular weight marker proteins can be stored in microfuge tubes at –70°C.

4. Allow the agarose to solidify for at least 5 min before placing the slab gel into the electrophoresis apparatus. Repeat this procedure for the remaining IPG strips. Although embedding in agarose is not absolutely necessary, it ensures much better contact between the IPG gel strip and the top of the SDS gel.
5. Wet the outside of the gel cassettes by dipping them into electrode buffer to make them fit more easily into the electrophoresis unit. Insert them in the electrophoresis apparatus. If necessary, push blank cassette inserts into any unoccupied slots. Seat the upper buffer chamber over the gels, and fill it with 2.5 L of electrode buffer (1250 mL of buffer stock solution + 1250 mL of deionized H2O).
6. Place the safety lid on the electrophoresis unit and start SDS-PAGE with 5 mA per SDS gel (100 V maximum setting) for approx 2 h. Continue with 15 mA per SDS gel (200 V maximum setting) for approx 16 h overnight, or higher current for faster runs (30 mA per SDS gel for approx 8 h).
7. Terminate the run after the bromophenol blue tracking dye has migrated off the lower end of the gel.
8. After electrophoresis, carefully open the cassettes with a plastic spatula. Use the spatula to separate the agarose overlay from the polyacrylamide gel. Carefully peel the gel from the glass plate, lifting the gel by its lower edge, and place it in a box of fixing or staining solution.

Notes

1. Lysis solution and IPG strip rehydration buffer should always be prepared freshly. Alternatively, small aliquots (1 mL) can be stored at –70°C. Lysis and rehydration solution thawed once should not be refrozen again.
2. Never heat urea solutions above 37°C, to avoid carbamylation of protein amino groups. It is important that the urea solution be deionized with an ion exchanger prior to adding the other chemicals, because urea in aqueous solution exists in equilibrium with ammonium cyanate, which can react with the NH_3+ of protein side chains (e.g,. lysine) and introduce charge artifacts (= additional spots) on the 2-D gel. In addition, always include carrier ampholytes in the lysis solution and rehydration buffer as cyanate scavenger.
3. If sample volumes exceeding 100 µL are to be applied onto IPG strips by cup loading, pipet in 100 µL, and run IEF with limited voltage (max. 300 V) until the sample has migrated out of the cup. Then apply another 100 µL and repeat the procedure until the whole sample has been applied.
4. Theoretically, no further intervention is required after the start of IPG-IEF until IEF has been completed. In practice, however, superior results are obtained if the electrode filter pads between the IPG strip and the electrodes are replaced by new ones after the sample has entered the IPG strip. This is particularly important for samples that contain high

amounts of salts and/or protein, because salt contaminants have quickly moved through the gel and have now collected in the electrode papers, or when very alkaline IPGs (e.g., IPG 6–12 or IPG 9–12) are used. In these cases, the filter paper pads should be replaced every 2 h. For IEF with very alkaline, narrow range IPGs, such as IPG 10 to 12, this procedure should be repeated once an hour.

5. When basic IPG gradients exceeding pH 10.0 are used for the first dimension (e.g., IPG 6–12 or IPG 9–12), horizontal streaking can often be observed at the basic end of 2-D protein profiles. This problem may be resolved substituting DTT in the rehydration buffer of the IPG strip by a disulfide such as hydroxyethyl disulfide (23) and by application of high voltages (8000 V) for short running times.

DETERGENTS AND CHAOTROPES FOR PROTEIN SOLUBILIZATION BEFORE TWO-DIMENSIONAL ELECTROPHORESIS

The solubilization of proteins for 2D electrophoresis-based proteomics is a *difficult* task. For example, proteins must reach their pI, which is also the minimum for solubility, and still stay soluble at the pI. Moreover, as protein *mobility* decreases when proteins get close to their pI, isoelectric focusing (IEF) is performed under strong electric fields (200 V/cm is not *uncommon* compared with the 10–20 V/cm used for sodium dodecyl *sulfate-polyacrylamide* gel electrophoresis [SDS-PAGE]).

This means, in turn, that salts and ionic compounds in general are almost precluded in IEF. Moreover, any solubilizing agent used prior to IEF must not change the original pI of the proteins. Consequently, this precludes the use of strong ionic detergents such as SDS. However, low amounts (up to 0.03% w/v) of ionic detergents can be used, provided that con-ditions favouring the *exchange* of SDS for other, nonionic *detergents* are used in IEF.

This ensures removal of bound SDS from the proteins but also means that the benefits of SDS are lost for the IEF dimension. However, the use of SDS has been often *recommended* as a way to ensure a complete initial solubilization before IEF. Apart from these problems arising from the nature of proteins, other problems are frequently *encountered* in many biological samples arising from the nonproteinaceous compounds that can be present in the sample.

A canonical example is that of nucleic acids, which completely blur the 2D electrophoresis pattern when present at too high a concentration. Nucleic acids act as mobile ion exchangers at the low ionic strength required by IEF, thereby creating severe artifacts. Other classes of compounds (lipids, salts, and so on) can be encountered in many samples and create their artifacts.

This is especially true for plantderived samples, owing to the ability of plant tissues to synthetize a host of compounds with varied structures. For example, phenolics and chlorophyll can be very abundant in some plant tissues and can *completely* ruin protein *separation* in proteomics techniques. There are thus different problems *depending* on the starting material.

When one starts from whole tissues, the main problem is generally the *interference* arising from nonproteinaceous compounds. These aspects are described in other chapters in this book. This chapter will therefore focus on the other aspect of the problem, i.e., the intrinsic solubilization of proteins, and will mainly deal with the chaotropes and detergents used for initial *solubilization* and for IEF. As mentioned earlier, the *constraints* present in IEF limit the choice to *chemicals* showing no net electric charge in solution over the pH range used for IEF, i.e., to nonionic or *zwitterionic compounds.*

This narrows the choice for chaotropes to the amide and urea families, as guanidines and amidines are charged below pH 12. Among the possible chaotropes, urea has been used for quite a long time. More recently, the addition of thiourea to urea as an additional chaotrope has shown interesting features for protein solubilization but also to limit protease action.

The role of chaotropes in the solubilization process is to break the noncovalent interactions between the various molecules present in the sample (e.g., hydrogen bonds, dipole-dipole interactions, and hydrophobic interactions) and to unfold the proteins.

Although ionic bonds are not directly affected by nonionic chaotropes such as urea and thiourea, the influence of these chaotropes on the dielectric constant of water also alters the strength of the ionic bonds. On the detergent side, it is quite clear that the uncharged detergents are clearly much less efficient than the ionic ones.

Ionic detergents make a charged "coat" on the protein molecules, so that the protein-detergent complexes repel each other via ionic interactions, thereby preventing protein aggregation. Unfortunately, ionic detergents cannot be used for IEF-based 2D separation. However, they can be used in methods using differential zone electrophoresis, and a short example will be given in this chapter. Among the wide choice of *commercially* available *uncharged* detergents, two subfamilies can be distinguished. Nonionic detergents have no charges on the molecule, whereas zwitterionic detergents have an equal number of negative and positive charges on their molecules.

Depending on the *pK* of the ionizable groups on the molecules, some detergents can be ionic in a certain pH range (in which at least one group is titrated) and zwitterionic in another pH range, whereas other detergents can be zwitterionic on the complete pH range. As an example of the two cases, classical betaines (bearing a quaternary ammonium and a carboxylic group) are positively charged at low pH, when the carboxylic group is not fully deprotonated.

When the carboxylic group is fully deprotonated, i.e., more than 2 pH units above the *pK*, they behave as a zwitterionic detergent. In contrast, sulfobetaines, bearing a quaternary ammonium and a sulfonic group, are zwitterionic over the 0 to 14 pH range, as both groups are ionized in this range. As a matter of fact, only detergents completely zwitterionic over the pH range of interest can be used for IEF. The two "historical" detergents used for 2D electrophoresis are Triton X100 (or NP-40), a nonionic detergent, and the zwitterionic detergent 3[3-cholaminopropyl diethylammonio]-1-propane sulfate (CHAPS).

Both have been used extensively in *combination* with urea and have not proved very efficient for the solubilization of sparingly soluble proteins, e.g., membrane proteins. However, recent work has shown that either specially designed *zwitterionic detergents*, or carefully selected nonionic detergents can *solubilize* membrane proteins. It is interesting to note that Triton X-100 is not very efficient when used with urea alone and much more efficient in urea-thiourea.

This observation extends to other detergents of the oligo ethylene glycol family, such as the Brij® detergents, i.e., linear alkyl oligo ethylene glycol compounds. However, the most efficient nonionic detergents belong to the glycoside family (e.g., octyl glucoside, dodecylmaltoside), and the latter seems to be efficient for urea alone as well as for urea-thiourea. An example of the variations in protein solubilization induced by the choice of detergent can be seen in.

The multiple variables playing a role in the solubilization process have also been investigated in. However, it should not be concluded from the preceding discussion that dodecyl maltoside is the absolute best choice for protein *solubilization* for 2D electrophoresis. Although the optimal detergent will depend on the sample, a kind of shortlist exists, based on previous work.

Thus the best *candidates* for protein *solubilization*, at least as a first screen, can be chosen from dodecyl maltoside, ASB14, C7BzO, Brij56, and C13E10, CHAPS being a good choice for soluble proteins.

MATERIALS

Biological Material

Arabidopsis thaliana membrane preparations are obtained according to Santoni.

Equipment

1. A tabletop ultracentrifuge, used for membrane preparation and to remove unsolubilized proteins.
2. Immobilized pH gradient (immobilized pH gradient [IPG], linear and nonlinear pH gradient from 3 to 10, 18 cm length; Amersham Pharmacia Biotech).
3. IPGphor apparatus: for isoelectrofocalization of proteins (Amersham Pharmacia Biotech).
4. Tube gels electrophoresis setup (Bio-Rad), for first-dimension gel electrophoresis. 5. Protean II: for SDS-PAGE electrophoresis (Bio-Rad).

Products and Stock Solutions

1. Dodecyl maltoside, Triton X-100, and CHAPS are best used from 20% (w/v) stock solutions in water. These solutions should be stored at 4°C and show limited conservation (a few weeks).
2. C13E10, Brij 56, and ASB14 are best used from 20% (w/v) stock solutions in ethanol/water (50/50 v/v). These solutions are stable for months at room temperature.
3. Cationic detergents (dodecyl trimethylammonium bromide [DTAB], cetyl trimethyl ammonium bromide [CTAB], benzalkonium chloride) are used as a 20% (w/v) stock solution in water. These solutions are stable at room temperature but are very sensitive to temperature. They sometimes need to be warmed at 37 to 40°C to redissolve the detergent prior to use.
4. Urea stock solution for IEF. It is difficult to go beyond 9 *M* urea at room temperature, which is the concentration used when urea is the sole chaotrope. This means that urea is added as a solid.
5. Urea-thiourea stock solution: the final chaotrope concentrations are 7 *M* urea and 2 *M* thiourea. This means that a 1.25X concentrated solution can be prepared, which is then simpler to use than reweighing small amounts of solid urea and thiourea for each sample.

For 10 mL of this concentrated solution, weigh 5.25 g of urea and 1.9 g of thiourea.

Some detergents (e.g., CHAPS or Triton X-100) which are fully compatible with urea, can be added at this stage. Other detergents, which show a more limited urea compatibility (e.g., ASB14), must be added only when the solution is diluted to the final strength. A total volume of 4.2 mL of liquid must be added to the urea and thiourea to make up 10 mL. This solution is stable for months if stored frozen at –20°C.

6. Urea solution for zone electrophoresis: as urea is used at 4 *M* final concentration, it is quite convenient to prepare an 8 *M* stock solution. For 10 mL, dissolve 4.8 g of urea in 6.4 mL of water. This solution is stable to 2 to 3 d at 4°C.
7. Acidic solubilization buffer for zone electrophoresis: 1 *M* potassium dihydrogen phosphate + 1 μL/mL 85% phosphoric acid.
8 Tri-butylphosphine (TBP) is a liquid (4 *M* when pure). A 40-fold dilution in dimethyl formamide is made just prior to use. This solution is further diluted 50-fold in the sample solution.
9. Tris-carboxyethyl phosphine (TCEP) is a solid. A 1 *M* stock solution in water is made, which is stable for months at –20°C.

METHODS

Solubilization in Urea for IEF

Solubilization from a Solid Sample (e.g., Tissue or Cell Pellet)

In this case, the sample volume can often be neglected in the final solubilization volume. A sample solution containing urea, the selected detergent at 2 to 4% (w/v) concentration, carrier ampholytes (0.4% w/v for IPG, 2% w/v for carrier ampholyte [CA]-IEF), and a reducing agent (50 mM DTT or 5 mM TBP or 5 mM TCEP).

This solution is added to the solid sample, resulting in a liquid extract.Protein extraction is helped by sonication in a water bath sonicator for approx 30 min. Unsolubilized material is best removed by ultracentrifugation for 30 min at 200,000g at room temperature.

Solubilization from a Suspension or Solution

In this case, the volume of the sample must generally be taken into account. It is thus necessary to calculate the final solution. As a rule of thumb, the sample volume can represent up to 35% of the final *extraction* volume. Solid urea, water, and stock solutions of the *detergent, ampholytes,* and reducer are used in addition to the liquid sample to build the extraction solution.

Solubilization in Urea-Thiourea for IEF

With the spreading of IPG using sample application by in-gel rehydration, rather large sample volumes can be used. This is especially true when home-made strips are used, as these can be made wider than commercial IPG strips and can thus accommodate a larger volume (up to 1 mL). It is thus often possible to use a dilution approach with the concentrated chaotrope solution and the solid sample resuspended in a minimal volume of water or the liquid sample.

If the detergent can be predissolved in the concentrated chaotrope solution, which can also contain the reducer, then the sample volume can represent up to 20% of the total extraction volume. If the detergent must be added only at the end, with a urea concentration not exceeding 8 M, then it is more convenient to use 1 vol of sample, 1 vol of detergent stock solution, and to add 8 vol of concentrated chaotrope solution. If this approach leads to too high a volume, two alternate approaches can be considered:

1. Introduce in a sample tube a volume of a stock detergent solution equal to the sample volume. Evaporate the solvent in a SpeedVac. Add the sample and 4 vol of concentrated chaotrope solution.
2. Consider that the sample volume *will* make 40% of the final sample volume. Weigh the corresponding amounts of urea, thiourea, and solid detergent. Dissolve with the sample in a bath sonicator.

In all cases, an extraction time of 30 to 60 min at room temperature is optimal before centrifugation (200,000g, 30 min, room temperature) to remove unsolubilized material.

Solubilization for Zone Electrophoresis

Solubilization with SDS is not considered in this section, which will deal only with solubilization in urea and cationic detergents prior to off-diagonal electrophoresis. The calculations for making the extraction solution are simple, since the sample represents 1/4 of the final extraction volume. To the liquid sample are added (in this order, and expressed as a fraction of the initial sample volume):

0.4 vol of 20% (w/v) cationic detergent stock solution. 0.2 vol of reducer.

0.4 vol of acidic phosphate buffer. 2 vol of 8 *M* urea

The solution is extracted for 30 min in a bath sonicator. Centrifugation at 10,000g for 15 min at room temperature may be needed to remove precipitated material.

Notes

1. The partial specific volume of urea is a useful number to know to make concentrated urea solution. One gram of urea occupies 0.75 mL in solution. In the same order for most detergents, 1 g occupies 1 mL. This is also true for thiourea. Urea must also never be warmed above 37°C to limit protein carbamylation. As an example, 1 mL of aqueous extract is added to 900 mg urea. This results in 1.675 mL of a 9 *M* urea solution. Otherwise, 1 g urea is added to 1 mL aqueous extract, resulting in 1.75 mL of a 9.5 *M* urea solution.
2. As a matter of fact, most of these extraction solution contain less than 50% water. This means that dissolution of the solids is rather difficult to perform, especially because high temperatures cannot be used (see previous note). The use of a water bath sonicator (marketed for cleaning objects and glassware) is of great help for this difficult solubilization.
3. Many detergents are not fully compatible with urea. Depending on the detergent structure, on the urea concentration, and on the temperature, insoluble detergenturea complexes can form. Detergents with linear alkyl chains are especially prone to this problem (e.g., ASB 14, Brij 56), which completely prevents the use of commercial linear sulfobetaines, which do not stand more than 4 *M* urea.

4. Many detergents strongly interfere with some popular protein assay methods, whereas others are plagued by reducing compounds (see the corresponding chapter in this book). It is therefore recommended to take this dimension into account when performing protein solubilization. In some cases, one can end with a solubilization cocktail that is incompatible with any protein assay method. In this case, it is often advisable to determine the protein concentration in the initial sample, prior to extraction, especially if the sample is a suspension. This means in turn that it will not be possible to assess the efficiency of the solubilization process.

TO-DIMENSIONAL DIFFERENTIAL IN-GEL ELECTROPHORESIS (DIGE) OF LEAF AND ROOTS OF *LCOPERSICON ESCULENTUM*

One mainstay technology of proteomics is high-resolution two-dimensional gel electrophoresis (2-DE) used in conjunction with protein identification by mass spectrometry. The most common implementation of this technique is to couple a first dimension of protein separation by charge (isoelectric focusing) with a second dimension of protein separation by apparent size (sodium dodecyl sulfate-polyacrylamide gel electrophoresis [SDS-PAGE]). A 2D gel system can be loaded with amounts of up to milligrams of protein and can separate and visualize thousands of protein spots. Although this technique has been very widely applied, several technical limitations still exist.

The protein expression patterns on a given 2D gel can usually not be exactly replicated, because of subtle changes in *experimental* conditions, principally *involving* protein solubility. This makes it difficult to pinpoint subtle changes in protein expression level between gels and even more difficult to *quantitate* such changes accurately.

A comparison of protein expression profiles from two different samples run on parallel gels can be performed using various software programs, but these analyses typically require a significant amount of image *manipulation* in order to align protein spots exactly. These and other difficulties, such as the limited useful *dynamic* range of many protein staining techniques, limit both the speed and accuracy of quantitation of protein spots in 2D gel electrophoresis. Two-dimensional differential gel *electrophoresis technology* (2D-DIGE) adds an accurate *quantitative* component to *comparative* 2D gel analysis.

It can also be used to compare protein abundance changes across multiple samples simultaneously with concurrent statistical *measurements* of confidence. Protein samples to be compared are first labeled with high-sensitivity cyanine fluorescent dyes and then mixed together and run on the same 2D gel. This removes any gel-to-gel variability, thus allowing for much higher accuracy of relative quantitation. Protein abundance changes can be quantified using Cy dyes over approximately 4 orders of magnitude.

The 2-DE gel pattern is visualized by fluorescent image analysis using *sequential* acquisitions at excitation wavelengths appropriate for the dyes used. The protein/dye ratio is kept deliberately high (>20:1), to try and limit the labeling of proteins to one molecule of dye per protein. The charge and mass of the *fluorescence* dyes used are *carefully matched* to reduce protein *migration* shifts during *electrophoresis* that are caused by the dye group.

A numerical comparison of the images generated by scanning of the DIGE gel at two wavelengths allows for the relative *quantitation* of protein, or proteins, present in the spot. The use of a third dye *permits* the inclusion of a pooled internal *standard*, which can then be used to standardize quantitation across a series of comparative gels of different samples. There are still, however, some technical limitations *inherent* in the DIGE approach.

The fact that only a small portion of the protein present in the gel is actually labeled means that if the fluorescent spot is excised from the gel for further analysis, the majority of the protein can be left behind in the gel, with obvious adverse effects. This is *especially* true of very small proteins: even a single dye molecule can make an observable difference in protein migration.

To avoid this problem, most users poststain the gel after *fluorescent* image analysis with Sypro ruby or silver (as we describe in this report). This in turn can lead to some problems with the alignment of fluorescent and poststaining images to make sure the correct spots are excised, especially in very crowded gel regions, but it is preferable to the alternative.

An alternative DIGE strategy using different reagents that label proteins to saturation, avoiding these problems, has been reported in an evaluation study but is not yet widely available. The significance of quantitative analysis results from DIGE is another area of some contention.

The system is currently available only from Amersham Biosciences (now part of GE Healthcare) and is thus very expensive owing to lack of competition. It is possible to circumvent this problem by synthesizing and using the *N-hydroxysuccinimide* esters of commercially available cyanine dyes. The *accompanying* deCyder software (also from Amersham Biosciences) distinguishes protein expression changes that are statistically significant.

The comparison of *fluorescent* spot *intensities* using the software module is relatively more objective and accurate than conventional approaches involving manual adjustment of the brightness and contrast of two side-byside images.

It is essential to try and match protein loading of the two samples for comparison as closely as possible, to avoid skewing the results. Most users also report at least the results of duplicate experiments, since there is still some inherent variation, probably owing to small protein solubility differences rather than electrophoretic parameters.

Attempts have been made to overcome these problems using both *integration* of DNA expression results with proteomics data and an alternative normalization algorithm. More accurate *quantitation*, especially among sets of samples run on different gels, can be achieved by using all three cyanine dyes in a more complicated protocol involving the use of a pooled internal standard.

The utility of 2D-DIGE has already been *demonstrated* for a number of different biological applications, including, for example, analysis of protein changes in esophageal, gastric, colon, and breast cancers, leukemia cells, mouse liver toxicity, brain tissue, and *E. coli* and yeast grown under stress conditions.

To the best of our knowledge, however, there is as yet no published report showing the use of 2D-DIGE analysis on plant protein preparations. This is most likely because 2D gels of plant tissue protein extract preparations are somewhat *problematic* in themselves, owing to the high levels of other organic compounds found in such preparations.

These include complex carbohydrates, DNA, organic acids, polyphenols, and other structural

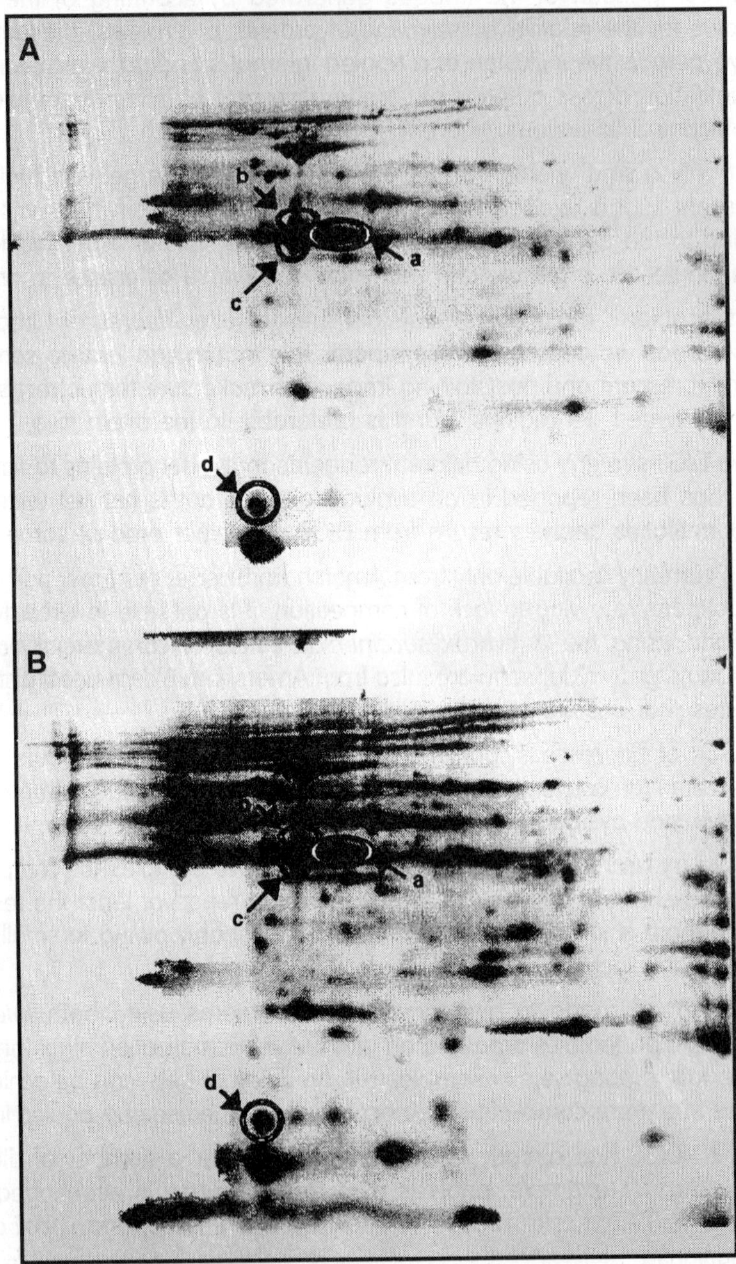

Figure. 14.3: (A) Single-channel fluorescence scanned image of the same gel as in Fig. 1, showing only those spots fluorescent at the Cy3 wavelength. The circled spots labeled a to d correspond to those labeled in the histogram in Fig. 3 and represent relatively abundant spots that are overexpressed in the Cy3-labeled sample by greater than 1.5-fold. (B) Single-channel fluorescence-scanned image of the same gel as in (A), showing only those spots fluorescent at the Cy5 wavelength, with the same spots circled.

components of mature adult plants. We have adapted the 2D-DIGE technology for use in our laboratory, in studies of protein expression in tomato plants subjected to heat, light, and salt stress conditions. We first prepare protein extracts from leaf or root tissue using the trichloroacetic acid (TCA)/acetone precipitation procedure used in many previous studies.

The TCA/acetone powder is then washed briefly to remove excess acid and extracted in 2D gel sample rehydration buffer containing thiourea. Following quantitation of each of a pair of samples and *adjustment* to equal concentrations, the protein extracts are labeled with cyanine dyes and then mixed together prior to first-dimension isoelectric focusing.

Using this approach we have been able to quantify *accurately* a large number of protein *expression* changes induced in the leaves and roots of tomato plants by the imposition of different abiotic stresses. We have also further characterized these protein *expression* changes by identifying proteins contained in differentially expressed 2D gel spots using nano liquid chromatography (nanoLC)-tandem mass *spectrometry* (nanoLC-MS/MS) and database searching.

The following detailed protocol describes the methods used for our *experiments* on tomato leaf and root tissue and includes protein extraction, CyDye labeling, 2-DE, fluorescent image analysis, and poststaining.

Results are also presented from a representative experiment showing quantitation of differential protein expression in tomato roots grown under salt stress conditions.

MATERIALS

Instrumentation

1. Bio-Rad PROTEAN IEF Cell.
2. Bio-Rad 24-cm Isoelectric Focusing Tray with Lid.
3. Amersham Biosciences Ettan DALTsix Electrophoresis Unit.
4. Amersham Biosciences Ettan DALT casting chamber.
5. Amersham Biosciences Electrophoresis Power Supply EPS 3501 XL.
6. Amersham Biosciences Typhoon multiwavelength fluorescent scanner.
7. Thermolyne Maxi Mix II vortexer.
8. Eppendorf Centrifuge 5415 D.
9. Beckman Coulter Avanti Centrifuge J-20.
10. Sigma Aldrich MicroCentrifuge SD.
11. Fisher Scientific Isotemp Incubator.
12. Ceramic mortar and pestle.
13. Low-fluorescence glass plates for casting 24 × 20-cm polyacrylamide gels (The Gel Company, San Francisco, CA).

Consumables

1. Amersham Biosciences Immobiline DryStrip pH 3.0–10.0 NL, 24 cm.

Chemicals, Buffers, and Solutions

1. Milli-Q water (18 MSZ resistance).
2. Leaf resuspension buffer: 10% TCA (EMD Chemical, Gibbstown, NJ), 0.07% (v/v) mercaptoethanol (EMD Chemical) in acetone (EMD Chemical).
3. EDTA wash solution: 0.07% (v/v) mercaptoethanol, 2 mM EDTA in acetone.
4. 100 mM Tris-HCl (Sigma, St. Louis, MO), pH 8.5.
5. Thiourea sample buffer: *2 M* thiourea (Bio-Rad, Hercules, CA), *7 M* urea (BioRad), 2% (w/v) dithiothreitol (DTT) (Bio-Rad), 4% (w/v) 3[3-cholaminopropyl diethylammonio]-1-propane sulfonate (CHAPS; Amersham Biosciences, Piscataway, NJ).
6. N,N-dimethylformamide (DMF; DriSolv, EMD Chemical).
7. CyDye DIGE Fluor Cy2, Cy3, Cy5 (Amersham Biosciences).
8. 10 mM Lysine (Sigma).
9. Immobilized pH gradient (IPG) rehydration buffer: 0.012% Amersham Biosciences DeStreak Reagent, *2 M* thiourea, *7 M* urea, 2% (w/v) DTT, 4% (w/v) CHAPS, 2% (v/v) Amersham Biosciences IPG buffer, pH 3–10 pH NL, trace Bromophenol Blue.

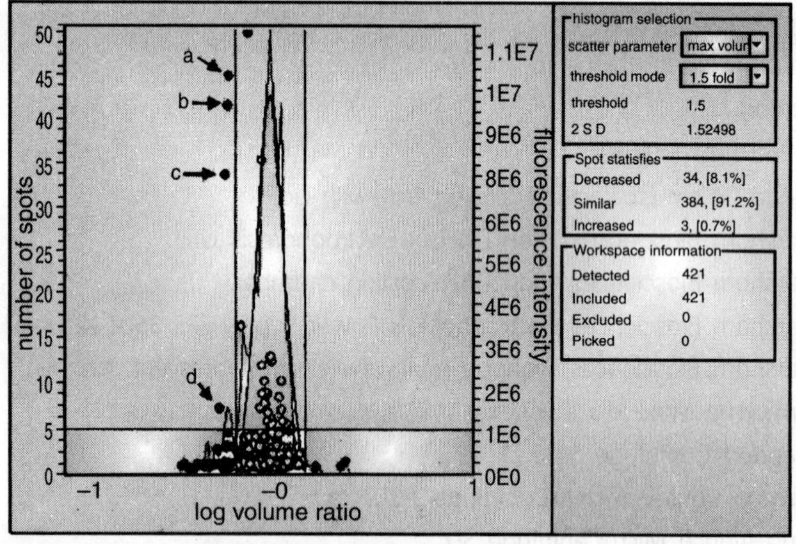

Figure 14.4: Histogram of protein expression values for spots shown on gel in Figs. 1 and 2. The x-axis indicates log volume ratio of spot fluorescence between the two acquired images, the left y-axis indicates spot frequency on the histogram curves, and the right y-axis indicates absolute fluorescent intensity for the individual spots. The blue curve indicates a model spot frequency histogram, which is based on the actual histogram data shown in the red curve. The four spots labeled a to d, which are outside the normal distribution curve on the x-axis and above the background threshold level on the y-axis indicated by the dashed line, represent relatively abundant spots that are overexpressed in the Cy3-labeled sample by greater than 1.5-fold. These correspond to the same spots labeled and circled in.

10. Mineral oil (Bio-Rad).
11. Ethanol (EMD Chemical).
12. Glacial acetic acid (J. T. Baker, Phillipsburg, NJ).
13. Bind-Silane and Repel-Silane (Amersham Biosciences).
14. Bind-Silane working solution: 8 mL ethanol, 200 μL glacial acetic acid, 1.8 mL Milli-Q H2O, 10 μL Bind-Silane.
15. 40% Acrylamide stock solution (40% acrylamide/bisacrylamide 37.5:1, 2.6% C; EMD Chemical).
16. 1.5 M Tris-HCl, pH 8.8.
17. 10% (w/v) SDS (Bio-Rad) solution.
18. 10% (w/v) Ammonium persulfate (J.T. Baker) solution.
19. TEMED (N,N,N,N-tetramethylethylenediamine; Bio-Rad).
20. Reduction reequilibration buffer: 50 mM Tris-HCl, 6 M urea, 30% (v/v) 87% glycerol (EMD Chemical), 2% (w/v) SDS, 0.5% (w/v) DTT.
21. Alkylation reequilibration buffer: 50 mM Tris-HCl, 6 M urea, 30% (v/v) 87% glycerol, 2% (w/v) SDS, 4.5% (w/v) iodoacetamide (Bio-Rad).
22. SDS electrophoresis running buffer: 25 mM Tris-HCl, 192 mM glycine, 0.2% (w/v) SDS.
23. 0.5% (w/v) Agarose sealing solution: 0.5% (w/v) agarose (Bio-Rad), 25 mM TrisHCl, 192 mM glycine (Bio-Rad), 0.2% (w/v) SDS, trace Bromophenol Blue. Heat up in a microwave immediately prior to use to liquefy the agarose.

METHODS

Leaf and Root Tissue Haresting and Protein Precipitation

1. Cut green nonsenescent leaves from the middle of the plant, and immediately place in a Ziploc plastic bag and freeze. If there is no freezer readily available, place on ice until leaves can be stored at –20°C.
2. For root tissue, dig up plants and cut the roots off at least 1 in. below the start of the green stem. Shake the roots and dip rapidly into three successive large beakers of water to remove soil; then immediately place in a Ziploc plastic bag and freeze, or temporarily store on ice.
3. Weigh out 2.5 g of frozen leaf or root tissue with stems and other extraneous materials removed. Cut the tissue into small pieces with chilled scissors into a chilled ceramic mortar and pestle. Grind to a fine powder in the presence of liquid nitrogen. More than one application of liquid nitrogen may be necessary to keep the tissue frozen.
4. To begin protein extraction, transfer the frozen leaf or root tissue powder to a 40-mL centrifuge tube, and resuspend the powder in 25 mL leaf resuspension buffer. Shake and mix thoroughly to ensure that all the powder has been resuspended. Let stand at –

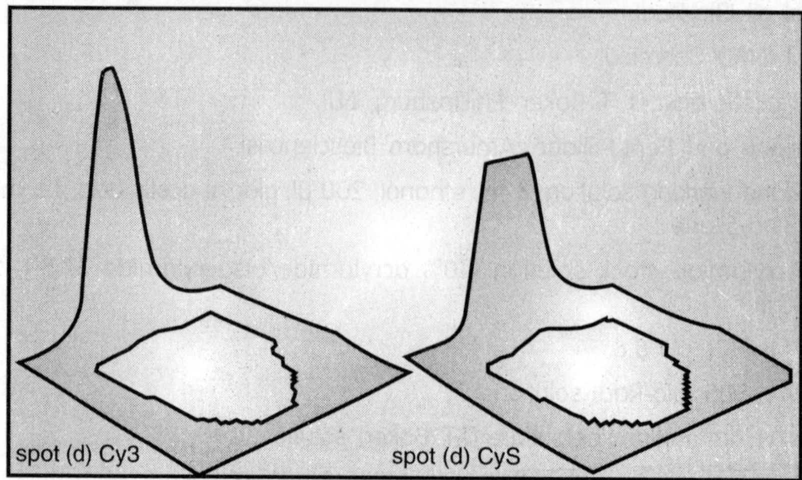

Figure 14.5: Three-dimensional representation of fluorescent imaging of spot d indicated in Figs. 2 and 3, showing quantitation of differential expression.

20°C for 45 min and shake again. Centrifuge the sample for 15 min at 35,000g.

5. Remove the supernatant with a glass pipette, trying not to disturb the pellet. Wash the pellet with the same volume of EDTA wash solution. Shake to disperse the pellet, let sit for 3–4 min on ice, and centrifuge again for 15 minutes at 35,000g. Repeat washings at least twice, or more if needed until leaf tissue protein pellet is no longer green.
6. Lyophilize the pellet. The resulting dry powder should be pale brown; it contains protein as well as other cell wall and fibrous materials.

Preparing Protein Sample from TC/cetone Poder

1. Weigh out 15 mg of the TCA/acetone powder from two plant tissue samples you wish to compare, and place in 1.5-mL microcentrifuge tubes with screw caps.
2. Pipet 1mL of 100 mM Tris-HCl (pH 8.5) into each tube, vortex for 1 min, and spin at 18,000g for 10 min. Remove and discard supernatant.
3. Pipet 1 mL of thiourea sample buffer into each tube. To extract protein from the powder, vortex for 5 min, place in a sonicating water bath for 10 min, and then place on a rotating shaker for 30 min.
4. Spin the pellet down at 18,000g for 10 min. The supernatant now contains protein extracted from the leaf tissue, usually in the range 0.5 to 1 μg/μL.

Preparing CDes and Fluorescent Protein Labeling

1. Obtain fresh DMF to reconstitute the CyDyes. DMF is degraded by oxygen, so always use fresh when preparing CyDyes.
2. To create a working solution of CyDye, add 1.5 μL of DMF to every 1 μL of CyDye. One microliter of working solution is intended to label 50 μg of protein. Owing to the light sensitivity of the CyDyes, keep them in the dark by wrapping the tubes with aluminum foil.

3. Pipet out 100 μL of each sample, containing 50 μg of protein, into 0.6-mL microcentrifuge tubes. (The samples should still be separate at this time.) Pipet 1 μL of the working solution of Cy3 into one sample and 1 μL of the work ing solution of Cy5 into the other sample.
4. Vortex for 10 s and spin briefly in a countertop centrifuge. Repeat twice.
5. Incubate on ice for 30 min, making sure to keep the tubes away from light.
6. To quench the labeling reaction, add 1 μL of 10 mM lysine to each sample. Incu bate on ice (covered) for 10 more min.
7. Combine 250 μL of the IPG rehydration buffer with the 100 μL fluorescently labeled aliquot of both samples for a final volume of 450 μL in a solution ready for isoelectric focusing.

First-Dimension Protein Separation: Isoelectric Focusing

1. Place the Bio-Rad focusing tray in the Bio-Rad focusing unit. Pipet the sample into the tray, aiming for the middle of the lane, and avoiding bubbles.
2. Peel away the plastic cover from the 24-cm 3 to 10 IPG dry strip using tweezers. Place it into the tray, gel side down, ensuring that the positive side is lined up with positive end of the tray and the negative side at the negative end of the tray. By carefully raising and lowering one end of the strip, make sure the entire length of the strip is covering the sample and there are no bubbles remaining.
3. Cover the strip with mineral oil to prevent it from drying out when voltage is applied.
4. A minimum of 67,000 volt-hours is usually needed for complete focusing.
    a. A recommended program is as follows: 50 V active rehydration for 12 h; 500 V for 1 h; 1000 V for 1 h; 8000 V for 9 h, 100 V for 5 h.
    b. The first step is the rehydration step and should not be less than 12 h. The CyDyes are light sensitive, so the focusing tray should be covered with some thing that will block out the light, or the whole apparatus should be used in a darkened room.
    c. The last step in the program is the removal window. The strip can be removed at any point during this time. The low voltage ensures that the strip remains completely focused until removed from the tray.

Preparing SDS-PGE Second-Dimension Gels

1. Low-fluorescence glass plates must be used to minimize background interference.
    a. For ease of handling during scanning, the gels are bound to the spacer side glass plate using Bind-Silane, and the opposing plates are treated with RepelSilane to facilitate removal.
    b. Wipe all the nonspacer side low-fluorescence glass plates with Milli-Q H2O and paper towels, and wipe again with ethanol and Kimwipes.
    c. Apply Bind-Silane working solution to the nonspacer side plates in 0.5-mL increments, spreading evenly over plate with a crew wipe, until each plate has received a total of 1 mL of working solution
2. Wipe spacer-side plates with ethanol first, followed by water, using a Kimwipe. Apply

Repel-Silane to the spacer side opposing plates in 0.5-mL increments, spreading evenly over plate with a crew wipe, until each plate has received a total of 1 mL. Let dry for 5–10 min, remove excess Repel-Silane with a crew wipe, wipe the plates with ethanol and milli-Q water, and remove excess liquid.

3. Allow plates to dry in a dust-free environment for a minimum of 3 h, noting that Bind-Silane- and Repel-Silane-treated plates should be stored in physically separate areas. Use nitrogen gas to spray off any dust from plates prior to assembly in gel casting apparatus.
4. Apply paper dot markers to the treated side of each Bind-Silane-treated (nonspacer) plate. One marker should be placed at the midpoint of each of the shorter edges of the plate approximately 1.5 cm from the edge of the plate, and markers should be placed at the bottom of the plate about ¬Q cm from the edge according to a numbering system that will allow you to distinguish the gels from each other during subsequent handling steps.
5. Prepare 12.5% gels using reagents as follows. This makes 900 mL, which is enough for 12 gels: 281 mL acrylamide 40% stock; 225 mL 1.5 M Tris-HCl, pH 8.8; 376 mL Milli-Q H2O; 9 mL 10% SDS solution; 9 mL 10% ammonium persulfate solution; 125 µL TEMED. Add acrylamide, Milli-Q H2O, Tris-HCl, and SDS to a gel beaker, mix thoroughly, filter the solution (0.2 µM), and place in a reservoir attached to the gel casting chamber by a peristaltic pump.
6. Turn stirrers on. Add ammonium persulfate and TEMED and quickly (within 15 s) turn on peristaltic pump.
   a. When almost all the solution has been pumped in, turn the mixer on its side to get as much of the solution in as possible.
   b. Before air reaches the tubing, add displacement solution to the hose side chamber.
   c. Continue pumping slowly until 12.5% solution is just below the top of the no spacer plates.
7. Turn off the pump and add enough water-saturated butanol to the top and back of the gels to cover them completely. (Butanol is on top of the partitioned solution.) Allow gels to set for 1 h, then pour off butanol, and cover with Milli-Q H2O.
8. Ideally gels should be left to sit overnight at room temperature, or they can be run after they have completely solidified. Disassemble the casting apparatus and rinse acrylamide from outside of plates. Use gels immediately or wrap in plastic wrap with water and refrigerate for later use (typically 1-wk storage maximum).

Second-Dimension Protein Separation: SDS-PGE

1. Remove the isoelectric focusing strip from the tray, and gently wipe the front and back with a Kimwipe to remove excess mineral oil. Take care not to press the Kimwipe into the gel. The strip can be frozen at this point at –80°C if desired.
2. Place the strip in a reequilibration tray, gel side up.
   a. Wash in 2 mL of freshly prepared reduction reequilibration buffer for 30 min on a shaker.

b. Decant the solution and add 2 mL of freshly prepared alkylation reequilibration buffer.

c. Place on the shaker for another 30 min. Decant the alkylation buffer.

d. Add 2 mL of SDS electrophoresis running buffer to the strip, and let sit on shaker for an additional 5 min.

3. Carefully place the strip between the plates of a previously prepared 24-cm gel. Placing the plastic side of the strip against the back plate will make it easier to slide the gel strip down onto the gel so it lays flat. There should be no bubbles between the edge of the strip and the surface of the gel. Seal the strip in place using the 0.5% (w/v) agarose sealing solution. Pipet the solution on top of the gel and allow it to solidify at room temperature.

4. Place the gel cassette rack inside the Ettan DALTsix Electrophoresis Unit, which is connected to a refrigerated recirculating water cooler.

   a. Place the gel in the outermost slot. When running multiple gels, place them evenly on both sides, all facing the same direction.

   b. Blank cassette inserts should be placed in the tank until each slot is filled.

   c. Fill the tank running buffer until it has reached the maximum fill line.

   d. Connect the lid to the power supply, and start electrophoresis.

   e. If the unit is running overnight, set the power supply to 2 W/gel. In the morn ing, or when running the gel during the day, the wattage can be increased but should not exceed 8 W/gel.

Scanning Fluorescent Image

1. Turn on the Typhoon scanner, and wait approx 30 min for the *instrument* to warm up before you start the first scan. *Scanning* before the *instrument* is warmed up can affect the accuracy of a scan.

2. Clean the Typhoon platen before and after using the Typhoon; use ethanol and crew wipes. It is easy to scratch the glass surface, so do not use paper towels.

3. Remove glass cassettes from the gel tank and wipe outside surfaces clean. Place *fluorescently* labeled gel, still inside the glass cassette, on the scanner platen in the correct *orientation.*

4. Scan image at the *appropriate* emissions filters and wavelengths for each label used: Cy 3 with 580 BP emission filter, green laser 532 nm; Cy 5 with 670 BP emission filter, red laser 633 nm; and Cy2 with 520 BP emission filter, blue laser, 488 nm.

   Initial scan parameters are press sample, depth +3 mm, 600 V photomultiplier tube setting and 500 gm pixel size, before repeating at 50 to 100 gm pixel size for a high-resolution scan.

5. Images are acquired using the Typhoon scanner control software, saved in Data Set format with an associated folder *containing* gel images, which can be opened in ImageQuant, and then resaved as either Data Set format for further processing in DeCyder (Amersham

Biosciences), or Tiff format for further processing by other image analysis packages, such as Progenesis (Nonlinear Dynamics).

Image Analsis

The acquired images from the different *wavelengths* scanned are then compared and analyzed using DeCyder. The programs are too *complicated* to describe in detail here, but the basic steps to follow in DeCyder are as follows.

1. Ensure that images are cropped to match each other exactly in size, then use the "*Process* gel *image*" *command* to measure and *normalize* spot volume intensity ratios for each spot, and generate a histogram showing number of unchanged, increased, and decreased spots between the two acquired images.
2. Use the exclude filter, or manually remove any obvious nonprotein background spots or streaks. Adjust histogram threshold parameters as required to display spots, which are over- or *underexpressed* by the amount specified, such as greater than twofold difference.
3. Manually inspect histogram and images, to determine which spots indicated as being significantly changed in expression are present in sufficient quantity to enable *identification* by mass spectrometric *techniques*. In most analyses, many of the protein features that are indicated as being significantly differentially *expressed* are present in such low amounts that the quantitation is not reliable and subsequent identification is not feasible.

   The spots of greatest interest are usually those which are displayed as being of both moderate to high abundance with a significant degree of differential expression.
4. For more accurate quantitation of differential expression, especially in the case of subtle differences in expression level, the entire process should be repeated in triplicate and the results analyzed for statistical significance.

Post-Staining for Further Processing

Following image analysis, the spacer side plate of the glass cassette is removed. Gels are then stained with silver while still immobilized on the remaining glass plate. This provides a visible stain that can then be used to manually cut out protein spots for *identification* by mass spectrometric techniques.

It also ensures that the majority of protein present is actually stained, in contrast to CyDye labeling, which only labels a small proportion of the protein present. The majority of spots visible with CyDye staining are usually visible with silver staining, although we have noticed that this does vary between different samples. Care must be taken to ensure that the right spot is cut out, by matching the silver-stained image to the CyDye labeled image as closely as possible.

This is not usually a problem, but at lower molecular weights the mass of the CyDye label can cause *significant* shifts, and also some discrepancies are caused by the fact that some proteins simply stain better with one technique than another. Results Eample: Eamining Protein Epression Differences Caused by Salt Stress in the Roots of Tomato Plants

Salt Stress Experimental Design

1. Tomato plants (*Lycopersicon esculentum*, Betterboy variety, purchased from Mesquite Valley

growers, Tucson, Arizona) were grown from 4-inch seedlings in a temperature-controlled greenhouse.

2. Twenty seedlings were grown in Hoglan's compost with the addition of MiracleGro fertilizer at the initial planting.
3. All 20 plants were watered daily with normal greenhouse water for 20 d, and then 10 plants were labeled as salt-stress plants and given 25 mM NaCl for 1 wk, which was ramped up to a 100 mM NaCl watering solution over the course of 1 mo.
4. The ten control plants were watered with water throughout the course of the experiment.

Harvesting Leaf and Root Tissue

1. At the 7-wk stage, after two consecutive weeks of 100 mM NaCl watering, 3 to 5 g of healthy leaf tissue were clipped from both control and salt-stress plants two times a week until leaves were necrotic.
2. As soon as leaves had become necrotic, each plant was uprooted, and roots were washed clean of soil and divided into proximal (nearest stem) and distal (furthest from stem) sections of half the total root length; they were immediately frozen in liquid nitrogen.

Extracting Protein from Proximal Roots

1. Harvested proximal root sections were ground to a fine powder using a mortar and pestle.
2. Protein precipitates were prepared using TCA/acetone.
3. Since root tissue contains less extractable protein than leaf, for the gel shown below, it was necessary to prepare three aliquots of protein extract, as described elsewhere in this chapter, then perform an additional TCA/acetone precipitation step and combine the pellets from all three into a single tube prior to CyDye labeling.
4. The samples were combined, the gel was run and analyzed using the protocols described in and the resulting images are shown below in.

Image Analysis of DIGE Gel Comparing Normal and Salt-Stressed Proximal Root Sections

The data produced in the gel from were analyzed using the DeCyder Differential In-Gel Analysis (DIA) module as outlined in. The initial output from the DeCyder program is a false-coloured representation, as shown in. Proteins present solely in the Cy5-labeled control material appear as red spots; those present solely in the Cy3-labeled material from salt-stressed plants appear as green spots. Proteins present in both samples at similar levels appear as yellow.

This provides an easily readable initial scan for obvious protein expression differences, as clearly red and green spots can be readily detected by visual inspection. In the example shown here, the majority of spots are yellow, indicating that there are few gross *differences* in protein *expression* between the two samples. The separated gel images of the two *fluorescent* wavelengths scanned are presented in Figure elsewhere in this chapter. These images can be used for in-depth analysis of protein *expression differences.*

The two images shown here look similar to each other, but there appears to be more protein present in. There is some difference apparent in overall protein loading (or labeling efficiency), which serves to highlight one of the advantages of this approach.

The software can be used to normalize the *quantitative* comparison of the two gel images, so that all the spots are considered, and only those with an expression ratio that is significantly more or less than the mean value are indicated as being differentially expressed. Quantitative analysis of the two images produced the histogram shown in. As shown in the spot detection window, after *normalization* and processing, the software detects 421 protein spots, of which 384 are unchanged, 34 are of higher intensity in the Cy3-labeled sample, and 3 are of higher intensity in the Cy5-labeled material.

The x-axis is in log volume ratio of spot fluorescence between the two acquired images, the left y-axis indicates spot frequency on the histogram curves, and the right y-axis indicates absolute *fluorescent* intensity for the individual spots, indicative of protein amount. The blue curve indicates a model spot frequency histogram, which is based on the actual histogram data shown in the red curve. The two vertical lines indicate the x-axis points at which a 1.5-fold protein expression difference is reached in each direction, with spots between the two lines considered to be unchanged in expression between the two samples.

The colour scheme is the same as in yellow indicates unchanged proteins, green indicates higher expression in the Cy3-labeled material, and red indicates higher expression in the Cy5-labeled material. In this type of analysis, many spots are often detected that are significantly altered in expression but are present at such low levels that it is not feasible to identify them. In the analysis shown in Fig. 3, we imposed an absolute fluorescence threshold value of 10^6 fluorescence units, as indicated by a dashed line, to exclude these.

There are four spots, labeled a to d, that are outside the normal *distribution* curve on the x-axis and above the specified 10^6 fluorescence unit background level on the y-axis. These spots therefore represent proteins that are overexpressed in the Cy3-labeled sample by greater than 1.5-fold and present at a level compatible with mass spectrometric identification.

There are numerous other spots on the histogram that represent proteins with significantly different *expression* levels but that are below our specified abundance threshold. The four *highlighted* protein features (a-d) thus represent proteins that are upregulated in expression by greater than 1.5-fold in response to salt stress.

The first three of the indicated spots, labeled A, b, and c, highlight one of the disadvantages of this approach, as they all appear to be part of a single poorly resolved protein feature, which has been artificially divided into spots by the analysis software. One spot, however, labeled d in , is a well-resolved feature with a fluorescence volume ratio of 1.79, shown in. This gel spot is a good candidate for excision and identification by mass spectrometric techniques, in order to identify the corresponding protein.

Notes

1. The EDTA wash solution contains both mercaptoethanol and EDTA to help minimize proteolytic breakdown occurring during the protein extraction. The mercaptoethanol also helps prevent formation of disulfide bonds.
2. This washing step is essential to ensure that the pH of the extracted protein mixture is not too low owing to residual TCA. A certain indicator that residual acid is present is to extract another aliquot of the TCA/acetone plant tissue powder using standard SDS-PAGE sample buffer; if the Bromophenol Blue turns bright yellow, the pH is acidic.

3. The CyDye labeling reagents are commercial products of Amersham Biosciences, known as Cy2, Cy3, and Cy5. They are all cyanine derivatives, which are closely matched, but not identical, in terms of charge and molecular weight, but they have different fluorescent profiles.
4. The approach described here relies on minimal protein labeling, hence the use of a 1:20 ratio off dye/protein. The majority of the protein remains unlabeled. One acknowledged problem with this approach is that the slight differences in size of the CyDyes can cause a difference in migration of the labeled and unlabeled protein, especially for low-molecular-weight proteins. This problem is addressed by the use of poststaining techniques to stain the remainder of the unlabeled protein and excising spots for further processing based on poststaining locations.
5. The protein concentration in each sample should be quantified as accurately as possible using standard quantitative protein assay techniques on a separate protein extract aliquot, and volumes should be adjusted so that equal amounts of protein are labeled with each fluorescent dye. In the event that the extracted protein concentration is too low, multiple aliquots can be extracted, precipitated with TCA/ acetone, and combined prior to labeling.
6. For a simple pairwise comparison experiment, two samples can be labeled with Cy3 and Cy5 and their relative fluorescent intensities compared directly. For comparing relative expression differences across replicates or across different members of a sample set, it is also necessary to include an aliquot of a pooled internal standard labeled with Cy2 to use in relative *quantitation* statistics.
7. When the aim of the *experiment* is to identify *differentially* expressed proteins by mass spectrometric techniques, it is usually necessary to spike the fluorescently labeled protein mixture with an equal volume of each of the unlabeled samples, for a final amount of 500 µg of total prôtein loaded in the gel. When doing this, care must be taken to ensure that the concentration of IPG buffer in the final mixture is between 0.5 and 1.0%.
8. For pairwise comparison gels, the image analysis is performed using the DeCyder DIA (differential in-gel analysis) module. If the experiment has been performed in at least triplicate and includes a Cy2-labeled pooled internal standard, the images can be analyzed in the DeCyder BVA (biological variation analysis) module. This analysis allows for the statistical evaluation, such as a Student's t-test, to determine the significance of any observed protein expression changes.
9. Fluorescent stains such as Sypro Ruby or Deep Purple can also be used for post staining, but this requires the use of a *fluorescent*-compatible spot-cutting robotic excision system.

VISIBLE AND FLUORESCENT STAINING OF TO-DIMENSIONAL GELS

Protein separation by electrophoresis is largely used in proteomic approaches because of high resolution, availability of powerful image analysis software for gel comparison, and compatibility with subsequent protein *characterization* by mass spectrometry (MS). For these various aspects, the selection of the protein staining procedure is of major importance.

Des Available for Gel Staining

Classically, *Coomassie blue* was the most widely used dye. However, it suffers from a low sensitivity in protein detection (a few tens of nanograms), which can be improved, however, by using the colloidal version (CCB). The binding behaviour is attributed to van der Waals forces and hydrophobic interactions. Such noncovalent binding of the dye allowed an excellent compatibility with matrix-assisted laser desorption ionization-time of flight (MALDI-TOF) MS.

CCB has been considered for long time as a convenient dye for largescale proteomic analysis. The other classical protein stain, *silver nitrate* (*SN*), displays an excellent sensitivity (approx1 ng) but shows staining saturation and could interfere with protein analysis by MS. Two categories of silver staining were used to visualize proteins in gel: the acidic silver nitrate and the alkaline silver diamine procedure, which differ in binding specificity, sensitivity, cost, and safety risk.

Proteins bind silver by salt formation on an acidic group of glutamic and aspartic acid residues or by complex formation through nucleophilic groups of histidyl, cysteyl, methionyl, or lysyl residues. Several modified protocols have been proposed since the first *descriptions* 25 yr ago. Owing to the complex chemistry involved, many *modifications* have been proposed to decrease background and increase *sensitivity*.

More recently, efforts were oriented toward the *compatibility* of silver stains with MS. Because of the oxidative attacks of silver ions on the proteins and the use of various sensitizing pretreatments of gels, irreversible modifications of amino acids have limited peptide mass fingerprint analysis or other MS analysis.

Most adaptations consisted of omitting crosslinking and sensitizing agents such as glutaraldehyde and formaldehyde and to adapt the destaining method before enzymatic digestion. In this chapter, we describe an acidic procedure that displays a good sensitivity and a clear background and is compatible with MS.

In the last few years, different *fluorescent* dyes have been introduced and have proved to combine high sensitivity and *compatibility* with MS. These include both commercially available stains, such as the series of Sypros and *ruthenium* red-based dyes for which *synthesis* procedures have been published . *Sypro Ruby* (SR) apparently includes the favourable features of CCB. SR is a luminescent ruthenium complex that interacts noncovalently with proteins.

As no irreversible modification of amino acids occurs during staining, satisfactory MS compatibility is expected . Additionally, SR was shown to have a broader linear dynamic range and a higher sensitivity than SN , suggesting a profitable use of this dye for large-scale proteomic analysis despite some propensity to saturation .

Nevertheless, the necessity for a fluorescent scanner, added to the cost of the dye itself, even if a ruthenium copy existed, has limited the use of SR up to now. Very recently, alternative molecules have been proposed. *Deep Purple* (DP) initially named "Lightning Fast," a sensitive fluorescent-based stain based on a natural compound extracted from the fungus *Epicoccum nigrum*, is now commercially availabie (Amersham Biosciences).

The fluorescent polyketide is able to bind with proteins and possibly to react on lysyl residues for fluorescence emission. It was described as more sensitive than SR and compatible with MALDI-TOF mass spectrometry. However, staining saturation was also demonstrated. Fluorescein derivatives constitute another recent fluorescent alternative. Three derivative with various lengths of hydrocarbon tails were examined as fluorescent dyes for gel staining . In this chapter, we present the C16 fluo*rescein* (C16-F) procedure, which exhibits good sensitivity and clear background, with no propensity to saturation.

Importance of Des Selection in a Proteomic Analsis

Large-scale proteome comparison from 2D gels is now widely used for various goals, such as the characterization of genetic diversity, the analysis of developmental processes, or the study of interactions with both abiotic and biotic environments. The selection of dyes in a proteomic analysis is of crucial importance, and time of manipulation, cost of chemical products, equipment required, sensitivity of dyes, and *compatibility* with MS constitute *determinant* parameters for the choice.

In a recent work, five procedures were compared to help in the selection of a protein stain convenient for large-scale comparison of 2D gel electrophoresis patterns by image analysis. It was shown that, for the same protein amount, the number of detected spots increased by a factor of 3 in a sequence CCB < C16-F < SN = DP < SR.

This was correlated with differences in sensitivity between dyes that led to the detection of additional spots belonging to classes of lower abundance. Analysis of the *distribution* of variation coefficients for spots from replicates also showed differences in staining *reproducibility* between dyes that in the order SR > C16-F > DP > SN > CCB. The present chapter details the gel staining protocols for the five dyes. To help readers in the selection of a procedure, results obtained with a same 100 μg protein sample from an *Arabidopsis* cell suspension are shown.

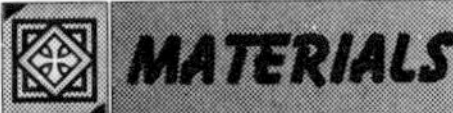

MATERIALS

Use only ultrapure water for all solutions and washing steps.

Protein Etraction and 2DE

1. Extraction solution: 90% cold (–20°C) acetone (v/v), 10% trichloroacetic acid (TCA) solution (v/v), 0.07% 2-mercaptoethanol (v/v).
2. Washing solution: 90% cold (–20°C) acetone (v/v), 0.07% 2-mercaptoethanol (v/v). 3. Solubilization solution: 9 M urea, 4% CHAPS (w/v), 0.05% Triton X-100 (v/v), 65 mM dithiothreitol (DTT).
4. First-dimension solution: 0.5% immobilized pH gradient (IPG) buffer 4 to 7 (v/v), 0.002% (w/v) bromophenol blue.
5. Reduction solution: 50 mM Tris-HCl, pH 8.8, 6 M urea, 30% (v/v) glycerol, 2% (w/v) SDS, 130 mM DTT.
6. Alkylation solution: 50 mM Tris-HCl, pH 8.8, 6 M urea, 30% (v/v) glycerol, 2% (w/v) SDS, 130 mM iodoacetamide.
7. Agarose solution: 0.6% (w/v) low-melt agarose in running buffer with traces of bromophenol blue.
8. Acrylamide gel solution: 11% acrylamide, 0.375 M Tris-HCl, pH 8.8, 0.1% SDS, 0.05%

ammonium persulfate, 0.0003% TEMED. 9. Running buffer: 25 mM Tris, 192 mM glycine, 0.1% SDS (*w/v*).

Gel Staining with CCB

1. Fixation solution: 50% (v/v) ethanol, 2% (v/v) phosphoric acid.
2. Incubation solution: 34% (v/v) methanol, 2% (v/v) phosphoric acid, 17% (w/v) ammonium sulfate. First mix ammonium sulfate, phosphoric acid, and water (180 mL of water for a 300 mL total volume), and then gently add methanol under agitation.
3. Staining solution: 34% (v/v) methanol, 2% (v/v) phosphoric acid, 17% (w/v) ammonium sulfate, 0.05% (w/v) Coomassie brillant blue G-250. First mix ammonium sulfate, phosphoric acid, Coomassie brillant blue, and water (180 mL of water for a 300 mL total volume), and then gently add methanol under agitation.

Gel Staining with SN

1. Fixation solution: 50% (v/v) methanol, 12% (v/v) acetic acid, 0.05% (v/v) formaldehyde.
2. Washing solution: 35% (v/v) ethanol.
3. Sensitization solution: 0.02% (w/v) Na2S2O3.
4. Staining solution: 0.2% (w/v) silver nitrate, 0.076% (v/v) formaldehyde.
5. Development solution: 6% (w/v) Na_2CO_3, 0.05% (v/v) formaldehyde, 0.0004% (w/v) Na2S2O3.
6. Stop solution: 50% (v/v) methanol, 12% (v/v) acetic acid. 7. Storage solution: 1% (v/v) acetic acid.

Gel Staining with SR

1. Fixation solution: 30% (v/v) ethanol, 10% (v/v) acetic acid.
2. Staining solution: Sypro Ruby ready to use commercial solution. 3. Washing solution: 10% (v/v) ethanol, 7% acetic acid.

Gel Staining with DP

1. Fixation solution: 10% (v/v) methanol, 7.5% (v/v) acetic acid.
2. Staining solution: 200X Deep Purple commercial solution diluted 200 times in water.
3. Washing solution 1: 0.1% (v/v) NH4OH.
4. Washing solution 2: 0.75% (v/v) acetic acid.

Gel Staining with C16-F

1. Fixation and staining solution: 30% (v/v) ethanol, 7.5% (v/v) acetic acid, 1 µM 5-hexadecanoylamino-fluorescein.
2. Washing solution: 7.5% acetic acid.

Equipment

1. 2DE: standard modern equipment for two-dimensional gel electrophoresis using IPG strips

in the first dimension (like the IPG-Phor and the Dalt systems [Amersham Bioscience, Buckinghamshire, UK] for the gels shown here).

2. Image acquisition: visible scanner or densitometer (like the GS-710 [Bio-Rad, Hercules, CA] for gels shown here), multiwavelength fluorescent scanner (like a FLA-5000 [Fuji Photo Film Company, Tokyo, Japan] for the gels shown here).

METHODS

Preparation of Total Soluble Protein Etracts (TC/cetone Protocol)

1. *Collect Arabidopsis thaliana* cells from a cell suspension.
2. Grind in liquid nitrogen.
3. Mix the fine powder with extraction solution and incubate at –20°C for at least 30 min.
4. Centrifuge insoluble material at 42,000g and wash pellets three times (*see* **Note 8)** with washing solution.
5. Allows pellets to dry in air.
6. Solubilize proteins by shacking pellets for 2 h in solubilization solution.
7. Estimate the protein amount according to the method of Bradford (34) using an aliquot sample and serum albumin as standard (in the presence of the same proportion of solubilization solution as introduced by aliquot samples).

To-Dimensional Gel Electrophoresis

1. Hydrate IPG strips (18 cm, pH 4.0–7.0) directly with 100 μg protein solution complemented with first-dimension solution.
2. Perform isoelectric focusing until 100 kV/h.
3. Before the second dimension, reduce and then alkylate proteins in reduction solution and then in alkylation solution for 15 min each.
4. Embed strips using agarose solution on the top of a 11% acrylamide gel.
5. Perform SDS-PAGE in running buffer, at 15 mA per gel, overnight and at 10°C (*see* **Note 10)**.

Staining and Imaging with CCB

Gels are stained according to the method of Neuhoff et al.. For each step, use 300 mL vol per 2D gel. Perform all steps under shaking.

1. Fix proteins by immerging gels in fixation solution for at least 2 h and up to overnight.
2. Wash gels three times for 30 min with water.
3. Transfer gels to incubation solution for 1 h and then to staining solution.
4. Allow staining to develop for 5 d under mild agitation.
5. Wash gels in water and acquire an image using a densitometer at 300 dpi.

Staining and Imaging with SN

Gels are stained according to the "Vorum" method of Mortz et al.. For each step, use 300 mL vol per 2D gel. Perform all steps under shaking.

1. Fix proteins by immerging gels in fixation solution for at least 2 h or overnight.
2. Wash gels three times for 20 min in washing solution.
3. Briefly incubate gels (2 min) in sensitization solution (*see* **Note 15).**
4. Wash gels three times for 5 min with water.
5. Incubate gels for 20 min in staining solution.
6. Quickly wash gels two times for 1 min in water.
7. Allow colour to develop for 5 to 10 min in development solution (*see* and **Note 16).**
8. Stop development for 5 min in the stop solution and transfer to the storage solution. 9. Acquire an image using a densitometer at 300 dpi.

Staining and Imaging with SR

For each step, use 300 mL vol per 2D gel. Perform all steps under shaking.

1. Fix proteins by immerging gels in fixation solution for 30 min.
2. Incubate gels for least 30 min (up to 1 d) in staining solution.
3. Wash gels for 30 min in washing solution.
4. Acquire an image using a FLA-5000 analyzer with 473-nm laser excitation and a long-pass filter, Y510. Select 100 gm resolution with a 16-bit gray scale level, and apply 700 V to the photomultiplier tube.

Staining and Imaging with DP (see Notes 11, 12, and 17)

For each step, use 300 mL vol per 2D gel. Perform all steps under shaking.

1. Fix proteins by immerging gels in fixation solution for 1 h.
2. Wash gels four times for 10 min each in water.
3. Incubate gels for 1 h in staining solution.
4. Wash gels two times for 10 min in washing solution 1 and then two times for 10 min in washing solution 2.
5. Acquire an image using an FLA-5000 analyzer with 532-nm laser excitation and a long-pass filter, O575. Select 100 gm resolution with a 16-bit gray scale level and apply 700 V to the photomultiplier tube.

Staining and Imaging with C16-F

Gels are stained according to Kang et al.. For each step, use 300 mL vol per 2D gel. Perform all steps under shaking.

1. Simultaneously fix and stain proteins by immerging gels in the fixation and staining solution for 1 h.

2. Wash gels at least two times for 5 min each in washing solution.
3. Acquire an image using an FLA-5000 analyzer with 473-nm laser excitation and a long-pass filter, Y510. Select 100 gm resolution with a 16-bit gray scale level and apply 700 V to the photomultiplier tube.

Comparison Beteen Des

When using 100 gg total soluble proteins from *Arabidopsis*, the staining procedures above allow for the routine detection of 250 spots with CCB, 450 spots with C16-F, 550 spots with DP, 600 spots with SN, and 800 spots with SR. The main features of the procedures are summarized in.

Protocols are compared according to the type of dye, number of steps, and duration of manipulation. The quality of images is estimated as a function of the background, spot saturation, and sensitivity of the dyes.

Notes

1. For gels to be stained using Deep Purple, high-quality SDS is recommended to avoid nonspecific stain and background (such as thal from USB, Cleveland, OH).
2. Different sources of CCB are available. In this work, Coomassie brillant blue G-250 (Ref 161-0406, Biorad, Hercules, CA, USA) was used.
3. Different sources of SN are available. In this work, silver nitrate (cat. no. S-0139, Sigma-Aldrich, St. Louis, MO) was used.
4. Sypro Ruby is distributed by different purchasers. In this work, SR (cat. no. 1703125, Bio-Rad) was used.
5. Deep Purple (cat. no. RPN6305, Amersham Biosciences, Buckinghamshire, UK). 6. 5-Hexadecanoylamino-fluorescein (cat. no. H-110, Molecular Probes, Eugene, OR).

Table 14.7: Comparison of Practical Features of the Staining Procedures

	CCB	*SN*	*SR*	*DP*	*C16-F*
Type of stain	Visible	Visible	Fluorescent	Fluorescent	Fluorescent
Steps to stain	4	8	3	5	2
Duration	5 d	4 h	3 h	3 h	3 h
Quality of background	+/–	++	+	+/–	
Absence of spot saturation	++	+/–	+/–	+/–	+
Sensitivity	+/–	+	++	++	+
Cost	++	++	–	–	++

++, very good/cheap; +, good; +/–, medium; –, bad/expensive.

7. Protein precipitation is usually sufficient after 30 min at –20°C. Prolonged storage at –

20°C (overnight) is recommended to improve precipitation, especially for small and/or hydrophilic molecules.

8. Pellets (without supernatant) can be stored at –20°C.
9. Solubilization can be improved by a prolonged agitation step (overnight) and/or by a short sonication step using a bath sonicator.
10. After the first 2 h, in which proteins are electroeluted from the strip into the second dimension gel at 15 mA per gel, the current can be increased to 30 mA to accelerate migration.
11. After migration (when the bromophenol blue front arrives at the bottom of gel), gels need to be quickly transferred to fixation solution to limit passive protein diffusion.
12. To limit manipulation of large gels (and to avoid gel breaking), it is recommended to change solution between consecutive steps by aspiration using a vacuum pump apparatus.
13. To prepare CCB incubation solution and staining solution, be careful to add methanol solution slowly (with Coomassie blue for staining solution) to the ammonium sulfate solution under agitation. When mixing methanol solution and ammonium sulfate solution, ammonium sulfate can precipitate, resulting in the formation of a compact block. To avoid such problems, always add methanol very carefully to the ammonium sulfate solution (dissolve ammonium sulfate in about 180 mL water for a 300 mL final volume), and never add ammonium sulfate solution to methanol solution. During the mixing step, a white ammonium sulfate precipitate can occur. To help ammonium sulfate resolubilization, just add a small volume of water.
14. To avoid the formation of Coomassie blue crystals, prepare staining solution 2 h before use. Coomassie blue needs to be first dissolved in methanol for 1 h. Methanol with Coomassie blue needs to be mixed with ammonium sulfate at least 1 h to obtain an homogeneous blue solution.
15. Sensitization is a quick but very important step allowing further silver fixation to proteins. This step is done in an alcohol-free solution; as the preceding steps involve a 35% alcohol solution, gels tend to float. To avoid such problems, increase agitation during this step, and use a two times greater volume.
16. Convenient development is obtained within 5 to 10 min. As the chemical reaction does not stop immediately when development solution is removed to the stop solution, an anticipation of the final staining is needed to limit overstaining of protein spots.
17. In the case of fluorescent dyes, all staining steps must be carried out in the dark. Aluminum foil can be used to envelop staining containers.
18. For the C16 fluorescein stain, a glass container must be used to avoid dye adsorption to plastic.

15 Chapter

FINGER PRINTING

It is noteworthy that protein identification by mass *spectrometry* (MS) was made possible by the development of "soft" *ionization techniques* developed in the late 1980's in Europe by Michael Karas and Franz Hillenkamp for matrix-assisted laser desorption ionization (MALDI) and in the United States by John Fenn for *electrospray ionization* (ESI).

Half of the 2002 Chemistry Nobel Prize for chemistry was Awarded to Fenn and Tanaka "for their development of soft desorption ionisation methods for mass *spectrometric* analyses of *biological macromolecules.*" Sufficient separation of the protein extract by 2D gels and compatible staining, make them amenable to protein *identification* by MS. However, MS of the entire protein does not lead to the direct identification of the protein.

To obtain a good *sensitivity*, enough mass accuracy, and access to the most of the data (i.e., to cover the sequence), the protein must be digested (cut) by an endoprotease to generate multiple protein-specific fragments.

If the *endoprotease* has sufficient *specificity* and if the protein is available in databases, a computed comparison of the experimental peptide masses (*generated* by the *endoproteic* digestion and detected by MALDI-time of flight [TOF] mass *spectrometry*) with the peptide masses of the in-silico digestion of all the proteins present in the database will allow the identification of protein candidates.

Concerning plant proteomics, the techniques used are similar to those of other proteome studies. However, particular attention must be paid to the specificity of the biological origin of the sample. Indeed, partly owing to their very large genomes (and often to polyploidy) compared with animal genomes, very few plant species genomes have been *sequenced* and annotated. *Arabidopsis thaliana* genome was the first to be completed, in December 2000. For nonsequenced plants, partial protein information may be available, but for the vast majority of plants (mainly trees and cereals, apart from rice), very few genomic information is available. Thus, for *nonsequenced* species, protein *identification* may be achieved by homology.

For unidentified proteins, MS/MS peptide *sequencing* will be the only possibility of identification (either by MALDI-TOF/TOF or by nano-liquid chromatography [LC]-ESI-MS/MS). Moreover, plant-specific databases cannot be used first, as most protein *contaminants* have a human or *mammalian* origin.

The protocol steps are as follows:

1. Spot excision from the gel and in-gel digestion with trypsin.
2. Digest deposition on the MALDI target and MALDI-TOF spectrum *acquisition* and annotation.
3. Database search and careful survey of search engine hits.

MATERIALS

Equipment

1. Clean vacuum centrifuge.
2. Clean oven (*up* to 56°C)
3. Modern (i.e., end of the 1990s as the limit) MALDI-TOF mass spectrometer equipped with delayed or pulsed ion extraction technology and an *electrostatic* mirror (reflectron [MALDI-reTOF]).

Reagents

In-Gel Digestion

1. Water (HPLC grade, or Milli-Q grade).
2. High-purity ammonium bicarbonate.
3. Acetonitrile (high-performance liquid *chromatography* [HPLC] grade).
4. Trifluoroacetic acid (TFA; HPLC grade).
5. n-Octyl-glycopyranoside (n-OGP).
6. Sequencing grade porcine trypsin. Although other good-quality sequencing-grade porcine trypsin sources are available, for homogeneity and convenience, the protocols and data presented below are optimized with Promega Porcine trypsin (sequencing grade, modified).
7. Ammonium bicarbonate buffer: 25 mM ammonium bicarbonate buffer, pH 7.8.
8. Acetonitrile/ammonium bicarbonate buffer: 50/50 (v/v) acetonitrile/25 MM ammonium bicarbonate buffer (pH 7.8).

9. Digestion buffer (prepare in ice): 0.0125 μg/μL sequencing grade trypsin in 25 mM ammonium bicarbonate buffer containing 5 mM n-OGP.

MAL DI-TOF Mass Spectrometry

1. a-Cyano-4-hydroxycinnamic acid powder, re-crystallized.
2. Water (HPLC grade, or milli-Q grade).
3. Acetonitrile (HPLC grade).
4. Acetone (HPLC grade).
5. Ethanol (HPLC grade).
6. TFA solution: 0.1% TFA (HPLC grade) solution in water.
7. Acetonitrile/TFA solution: 3:2 (v/v) acetonitrile/water acidified by 0.1% TFA.
8. Acetonitrile/TFA solution: 1:1 (v/v) acetonitrile/water acidified by 0.1% TFA.
9. Ethanol/acetone/TFA solution: 6:3:1 (v/v/v) ethanol/acetone/0.1 % TFA in water.

METHODS

Many gel staining protocols are suitable for subsequent MALDI-TOF peptide mass fingerprinting (PMF) analysis, although some give better results than others. Comparative performance and compatibility are discussed in other chapter of this book.

In-Gel Digestion

The in-gel digestion protocol is adapted from Jensen et al..

For gel handling and digestion steps, careful sample handling is compul sory, to minimize sample contamination with exogenous keratins (hair, skin, wool clothes, draughts, air-cooling, and so on).

Excision of Protein Spots from 2D Polyacrylamide Gels

1. Cut with a razor blade a 1 mL pipet tip 5 mm above the tip to make a punch (stamp-out diameter of 1-2 mm).
2. Stamp out (excise) the gel spot, and transfer it to a microcentrifuge tube.

Washing of Gel Pieces

1. Wash the spot with 100 μL of ammonium bicarbonate buffer (15 min vortex), and then discard the supernatant.
2. Wash the spot twice with 100 μL of acetonitrile/ammonium bicarbonate buffer (15 min vortex), and then discard supernatant.
3. Wash the spot with 100 μL HPLC grade acetonitrile to shrink the gel piece (15 min vortex), and then discard supernatant.
4. Dry gel fragments under vacuum on a centrifugal evaporator.

If needed, after these washing steps, the gel fragment can be stored at –20°C for few weeks before in-gel digestion. For 2D gels, reduction and alkylation of cystein residues prior to in-gel digestion are not necessary.

In-Gel Digestion

1. On ice, add 8 μL of cooled digestion buffer.
2. After 20 min, adjust with the minimum ammonium bicarbonate buffer volume to cover the gel.
3. Incubate at 37°C for 4 h.

Peptide Extraction, Concentration, and Desalting

Peptide Extraction

1. Extract the resulting tryptic peptide fragments by adding 20 μL of the TFA solution (sonicate for 15 min). Preserve the supernatant in a 500-μL microcentrifuge tube.
2. Extract with 20 tL of 3:2 acetonitrile/TFA solution in an ultrasonic bath for 15 min. Pool the supernatants.

Peptide Concentration and Desalting

1. Concentrate the pooled supernatant to a final volume of approx 10 μL in a vacuum centrifuge. This step removes acetonitrile (which otherwise will not allow peptide fixation on the hydrophobic chromatographic media) and reduces the volume to an acceptable one for next chromatographic step.
2. Firmly fix the ZipTip to a 10-μL adjustable pipet (back pressure may be high).
3. Wash the ZipTip five times with 10 μL of 3:2 acetonitrile/TFA solution. Discard the dispensed liquid.
4. Equilibrate the ZipTip five times with 10 μL of TFA solution. Discard the dispensed liquid.
5. Fix the peptides by slowly aspirating and dispensing 10 μL of the concentrated supernatant ten times without lifting the pipet tip.
6. Wash (desalt) the fixed peptides four times with 10 μL of TFA solution. Discard the dispensed liquid.
7. Elute the peptides in a 500-μL microcentrifuge vial with 2 μL of 1:1 acetonitrile/ TFA solution.

MALDI-TOF Peptide Mass Fingerprinting

1. In the MALDI processes, the sample is cocrystallized with an organic aromatic, usually an acidic compound (the matrix), whose main property is to absorb at the wavelength of the UV laser (generally 337 nm for nitrogen lasers or 355 nm for the tripled frequency of Nd:YAG lasers). During the desorption step, when the UV laser is firing, the aromatic moiety of the matrix (present in a very large excess of approx 10,000:1) absorbs the UV energy and the matrix sublimates, protecting the sample from decomposition or dissociation and driving it to a gaseous phase.
2. The ionization step is a proton exchange in the dense but dilating and cooling gas phase (the plume) between initially charged matrix ions or clusters and neutral peptides (ion-

molecule reactions).

3. Once desorbed and ionized, the sample is accelerated in the TOF tube toward the detector, its speed (i.e., TOF in the fixed-length flight tube) being directly proportional to the root square of its m/z ratio.
4. The reflectron (an electrostatic mirror) corrects the initial kinetic energy dispersion, improving mass resolution and thus mass accuracy. (Isobaric ions with a larger kinetic energy will have a larger path in the *reflectron* and will be focalized with lower kinetic energy ions at the detector focal plane.)
5. Delayed or pulsed ion extraction corrects the initial spatial dispersion, thus improving again resolution and mass accuracy. (Isobaric ions spatially dispersed in the desorption plume will be *differentially* accelerated according to their position at the beginning of the ion extraction from the source to the analyzer.)

For more details concerning MALDI-TOF and the basics of MS, many good references are available. MALDI-TOF mass spectrometry is the technique of choice for PMF, thanks to several features:

1. Soft ionization technique (i.e., the peptide remains intact after the MALDI process).
2. Ionization method very tolerant to mixtures of different peptides and relatively tolerant to contaminants (buffer, salts, plasticizers, and so on).
3. Highly sensitive method (owing to both the discrete MALDI process and TOF mass analyzer).
4. Good resolution, currently exceeding 15,000 (full width, half mass [FWHM]), allowing

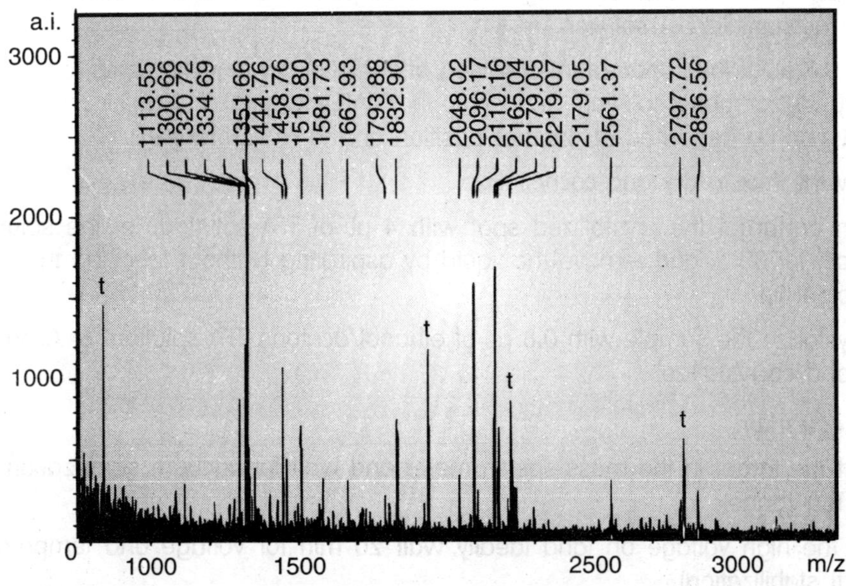

Figure 15.1: MALDI-TOF peptide mass fingerprint of an in-gel digest of a 2D spot from a Pisum sativum protein extract.

detection of monoisotopic peaks in the mass range of interest.

5. Excellent mass accuracy of the TOF analyzer (regularly in the range of 10-30 ppm).
6. Fast data accumulation.

On-Target Sample Deposition

Classical Dried Droplet Method

1. Prepare α-cyano-4-hydroxycinnamic acid matrix at half-saturation (approx 5 mg/ mL): dissolve (sonicate for 10 min) α-cyano-4-hydroxycinnamic acid (a small spoon tip) in 300 µL of 1:1 acetonitrile/TFA solution.

   Centrifuge for a few seconds to obtain a clear supernatant. Put 200 µL of the clear supernatant in a new microcentrifuge vial, and add the same volume (200 µL) of 1:1 acetonitrile/TFA solution to obtain half-saturation.
2. Rapidly (to avoid crystallization in the pipet tip) mix in a microcentrifuge vial 0.8 µL of digest solution with 0.8 µL of matrix and deposit rapidly on the MALDI target 0.8 µL of the mix. The remaining 0.8 µL can be deposited at another position on the MALDI target. Do not touch the target with the pipet tip. The droplet should be preformed at the pipet tip and deposited on the target by capillarity.
3. Allow the mix to dry and cocrystallize.

Dried Droplet Method on Prestructured Hydrophobic Target

1. Prepare a-cyano-4-hydroxycinnamic acid matrix: weigh 10 mg of a-cyano-4hydroxycinnamic acid, and dissolve (sonicate for 10 min) with 1 mL of 1:1 acetonitrile/TFA solution. In a new microcentrifuge vial, put 56 µL of that matrix solution and add 944 µL of 1:1 acetonitrile/TFA solution.
2. Take 0.8 µL of the supernatant obtained after the in-gel digestion step, and mix it rapidly with 0.8 µL of matrix solution. Deposit on the MALDI target 0.8 µL of the mix. The remaining 0.8µL can be deposited at another position on the MALDI target.
3. Allow the mix to dry and cocrystallize.
4. Wash on-target the crystallized spot with 4 µL of TFA solution, let the solution stay in contact for 30 s, and remove the liquid by aspirating (without touching the crystals with the pipet tip).
5. Recrystallize the sample with 0.8 µL of ethanol/acetone/TFA solution. 6. Allow the mix to dry and cocrystallize.

Spectrum Acquisition

1. Insert the target in the mass spectrometer and wait for vacuum stabilization (below the 10^{-6} Torr range).
2. Turn the high voltage on (and ideally wait 20 min for voltage and temperature [Joule effect] stabilization).
3. Tune laser power (attenuation) and target position to obtain a good signal-to noise ratio

and monoisotopic resolution over the mass range 700 to 4000 Th.

4. Fire 10 laser shots and discard them, as they are usually noisy in the low mass range (owing to matrix and/or salts clusters).
5. Acquire and sum 80 to 200 laser shots.
6. Calibrate the mass spectrum internally with autoproteolytic trypsin ions at 842.5099 and 2211.1046 Th.
7. Save spectrum.

Spectrum Annotation

Automatic monoisotopic mass assignment is possible, but we recommend a careful survey of the annotation.

Without spectrum smoothing, annotate (by centroiding at 70% height, or, better, by using advanced algorithms, e.g., SNAP from Bruker Daltonics) all first monoisotpic peaks apart from the main trypsin known autoproteolytic peaks.

Database Search Using MASCOT and Critical Review of the Search Result

The steps described below are for the MASCOT PMF search engine that can be licensed in-house or run on the distant London server.

However, the proposed search strategy is directly adaptable to other search engines.

First Search Step

The objective of this first step is to gain an overview of the data quality, to remove possible contaminants, and possibly to identify the plant protein directly.

1. Specify a global database (e.g., MSDB, NCBInr, or the cleaner Swiss-Prot), not a species-specific one, to have non-plant protein contaminants in the database.
2. Do not specify a species in the taxonomy field (i.e., "all entries"), to allow the identification of non-plant protein contaminants.
3. Allow one missed cleavage.
4. Set a large mass tolerance of 100 ppm, to evaluate the mass spectrum calibration. (The two trypsin autoproteolytic peptides used for calibrating the mass spectrum may have too low abundance and ion statistics to allow good calibration.)
5. Fixed modifications: carboxymethyl (C) or carbamidomethyl (C). For 2D-gel samples, cysteine residues are reduced and alkylated usually either with iodoacetic acid or iodoacetamide, giving carboxymethyl-cysteine or carboxyamidomethyl-cysteine (contracted in carbamidomethyl-cysteine), respectively.
6. Variable modification: none.

For good-quality data (*signal-to-noise* ratio, mass accuracy, low level of *contaminants*) and for protein referenced in the database searched, a *candidate* protein may appear.

Criteria to evaluate the candidate protein are as follows:

1. Score superior to the significant threshold value.
2. Large difference between the first-ranked protein and the following nonrelated one.
3. Protein from the studied species (or a close species with high *homology*) at the first rank.
4. Candidate protein MW and pI compatible with 2D-gel data.
5. Good mass accuracy (ideally) or bad mass accuracy (owing to *calibration* failure or *miscalibration* problem) but associated with a low and linear distribution of mass dispersion over the mass range.
6. A maximum of one miscleaved peptide for three matching peptides.
7. A minimum of five matching peptides.
8. Homogenous *localization* of the matching *peptides* in the protein sequence and sequence coverage.

Second and Following Search Steps

One or more steps may be required if the previous search failed to give a hit, or if too many peaks (e.g., more than 10) remain unmatched (which may indicate the presence of two or more proteins in the excised spot).

One or more search criteria can be adapted :

1. Remove peptide digest peaks from identified non-plant protein contaminants ("search unmatched" option button in Mascot).
2. Set the mass tolerance according to the real observed mass accuracy of the mass spectrum (e.g., as evaluated on non-plant protein contaminants or verified on trypsin autoproteolytic fragments). Typically, a good-quality search needs mass accuracy to be below 30 ppm (the lower value the better).
3. Allow the search with nonstoichiometric or unpredicted chemical modifications (mainly methionine oxidation, aspartic and glutamic acid methylesterification, N-terminal pyroglutamylation).
4. Restrain the taxon search range (e.g., to Viridiplantae).
5. Allow the search of a specific database.

The evaluation and validation criteria described elsewhere in this chapter remain necessary.

Identification of secondary (or minor) components in the mixture is important.

After successful identification of a protein by PMF, several peaks in the MALDI-TOF mass spectrum usually remain unidentified. These peaks can have different origins:

1. Peptide ion from other protein(s) comigrating in the electrophoresis gel.
2. Peptide ion from the same protein whose sequence differs slightly from the pro tein in the database (either by genome annotation mistakes or mutation).
3. Peptide ion from the same protein whose mass differs from the predicted peptide mass because of posttranslational modification (PTM).

4. Peptide ion from a contaminant protein (e.g., keratin or trypsin).
5. Nonpeptidic contaminant ion from a polymer residue or plasticizer. 6. Nonpeptidic contaminant ion with a chemical or biochemical origin.

MALDI-TOF/TOF Strategies

Proteins identified after PMF are only *candidates*, the scoring indicating the level of confidence. Thus, one should not forget that no *sequencing* step was *performed* to confirm this tentative identification.

Peptide sequencing on a MALDI mass *spectrometer* was difficult before the recently developed TOF/TOF tandem analyzers. In TOF/TOF analyzers, the first TOF allows the isolation of the peptide ion of interest; all other peptides from the path tube are discarded.

A collision cell after this first TOF breaks (by *collisionally induced dissociation*) the peptide into *sequence-specific* frag ments. The second TOF *separates* and *measures* the masses of these *sequencespecific* fragments.

Confirmation of Identification

To confirm an identification, one to three peptides identified from a single protein can be fragmented successively by TOF/TOF. The fragment ions must correlate with the predicted peptide sequence.

Identification of Unknown Proteins

In the case of an *unsuccessful* database search, some peptides (from two to five usually) can be *analyzed* by TOF/TOF. De novo sequencing and/or database search of the TOF/TOF fragments of these peptide ions may allow *identification*, either in the studied taxon or by cross-species homology (*see* also in other chapter elsewhere in this chapter for LC-MS/MS approaches).

Notes

1. Automation of the gel cutting (spot excision), in-gel digestion, and on-target sample deposition is possible with specific liquid handling automates. To increase throughput, large proteomic facilities are equipped with such technologies. *Automation* also limits sample handling and *contamination* possibilities by human keratins.
2. Cutting spots that are too large would be detrimental to protein *identification*. Indeed, it would increase *background* chemical noise, lower trypsin ability to penetrate the core of the spot, and reduce the *diffusion* of proteolytic peptides out of the gel.
3. In 2D-PAGE, proteins are reduced (typically with DTT or mercaptoethanol) and alkylated (usually with iodoacetamide or iodoacetic acid) during *equilibration* of the isoelectrofocusing gel prior to SDS-PAGE. This is to break the disulfide bridges between two cystein residues and thus prevent the secondary and tertiary structure of the protein from *interfering* with the migration of the protein in the polyacrylamide mesh. For 1D-PAGE separated proteins, reduction and alkylation has to be done before in-gel digestion.
4. Certain features of trypsin make it an appropriate *endoprotease* choice for MS *identification* of proteins after in-gel digestion:

a. Quite low molecular weight enzyme (approx 24 kDa), *penetrating* easily into the dehydrated gel with the rehydratation buffer.

b. Good specificity (cuts at the C-terminus of K and R, does not cut if the next amino acid is P, cuts less efficiently if acidic moieties of the lateral chains of D/E amino acid are present in the neighborhood, cuts less efficiently if the bulky F/W/Y amino acids are present in the neighborhood because of a steric effect of the lateral chains). Commercial porcine trypsin is treated with Ntosyl-L-phenylalanine chloromethyl ketone (TPCK) to avoid nonspecific chymotrysic activity

c. Occurrence of K and R generates most of the proteolytic peptides in the mass range 600 to 4000 Daltons, which is compatible with high mass accuracy MALDI-TOF-MS.

d. Commercial porcin trypsin lysine side chains are treated with reductive methylation to give few autoproteolytic ions, but two intense ions at 842.5099 Th and 2211.1046 Th can be used for internal mass spectra calibration.

e. Proteolytic peptides are easily protonated (and thus give good signal in mass spectra) because of their two basic sites: one at the N-terminus of the peptide and one on the lateral chain of the last C-terminus amino acid (K or R). Moreover, this property is very useful in ESI, giving mainly doubly charged species that generate informative b/y fragment ion series by MS/MS.

5. For a specific purpose (e.g., trying to increase the sequence coverage of the protein), other endoproteases can be used with the same protocol:

   a. Chrymotrypsin cuts after F/W/Y, generating more and thus shorter peptides.

   b. Endo Lys-C cuts after K (not R).

   c. Protease V8 cuts after DE or E (depending on the digestion buffer pH).

6. On-ice fresh trypsin preparation and trypsin in-gel penetration are recommended to minimize autoproteolysis.

7. Minimum digestion time is 3 h at 37°C, and overnight digestion should be considered as a maximum. *Overnight digestion* may be the most convenient. For long digestion time (e.g., overnight), digestion can take place at room *temperature* instead of 37°C. This reduces the amount of trypsin *autoproteolytic* signal.

8. In the dried droplet method, the goal is to obtain *homogeneous* and small crystals. Long *crystallization* will give large crystals, whereas rapid *crystallization* will give smaller crystals. To hold the crystallization process without changing the solvent composition, the target can be dried under vacuum or heated (3040°C) in an oven for few minutes.

9. Some MALDI-TOF mass spectrometer *manufacturers* have developed *prestructured hydrophobic/ lyophobic* targets (e.g., AnchorChip™ technology from Bruker Daltonics). These targets allow precise positioning of the sample and concentration on the target. Thus the sensitivity is increased by a factor 5 to 20. Careful cleaning of the target is required to avoid contaminant (mainly plasticizers) concentration on the anchor. Peptide digest can be spotted on the target directly after in-gel digestion, or after peptide extraction and concentration.

10. From the mass *spectrometry* point of view, two criteria are essential to allow reliable protein *identification*:
    a. Mass accuracy (ideally below 30 ppm, nowadays MALDI-TOF instruments being able to achieve 10-15 ppm with careful internal calibration of the spectra).
    b. Sensitivity to detect as many ions as possible and to increase sequence coverage.
11. Annotation of the spectrum is very often done *automatically* with good *efficiency* algorithms, but if the protein *identification* fails, we *recommend* surveying the *annotation* and limiting it to:
    a. The 40 most intense peaks as a maximum, assuming that the 2D-gel spot contains one or two *identifiable* proteins (i.e., with close *expression* level).
    b. A minimum signal-to-noise ratio of 3:1.
12. For intense mass spectra, annotation of low signal ions may be detrimental for protein identification. We *recommend* exclusion of signals whose intensity is below 2% of the most intense *protein-specific* peptide ion.
13. The main known *autoproteolytic* peaks from Promega modified *sequencing* grade porcine trypsin are: 842.51, 870.54, 1045.56, 1940.94, 2211.10, 2225.12, 2239.14, 2283.18, 2299.18, and 2807.31.

A complete list of less intense ions can be found at:

14. We recommend *annotating* even known *contaminants* (i.e., non-plant keratin peaks), performing the first database search step with these *contaminants*, and filtering (remove) them only after precise database *identification* of these *contaminants*. Indeed, keratin *proteolytic* peaks (e.g., 1475.74 Th) may belong to different keratin subclasses (e.g., keratin 10 and keratin 2), the other keratin *proteolytic* detected peaks being specific to one or both of these *subclasses*.
15. In-house licensing of the search engine (e.g., Mascot):
    a. Allows the installation of specific or *proprietary* databases. b. Preserves the *confidentiality* needed for some studies.
    c. Makes the search time independent of the network speed and the distant server busyness.
    d. Allows batch searches.
    e. Allows the archiving of all searches without time limit.
16. The sequence coverage *information* alone has limited sense, partly depending on the candidate protein predicted MW.

For a high MW protein candidate, a low sequence coverage may indicate that:

a. Some parts of the protein are less accessible to trypsin.
b. The protein contains little tryptic cleavage sites (e.g., PRP: proline-rich proteins), so tryptic peptide sizes are too big to be detected (out of the mass range).
c. Peptide ionization competition does not allow the ionization of some peptides.

d. The in-gel spot protein is only a fragment of this candidate protein.

e. The database does not contain exactly the sequence of the in-gel protein (mutation, or PTM, but more typically sequence polymorphism owing to cross-species identification).

On the other hand, for a low MW protein candidate, a high sequence *coverage* may solely indicate that the protein described in the database is too short: the sequence coverage may be *irrelevantly* too high. For these reasons, it is more *appropriate* to survey the *localization* of the matching peptides in the protein sequence than the *sequence* coverage: are the matching peptides localized all over the protein sequence (which may indicate a valid candidate)? Or, on the contrary, are they *localized* on a limited part of the protein (which may indicate the *identification* of a fragment *protein* if this correlates with in-gel MW and pI *measurements*)?

17. Multiple proteins are likely to be present in gel pieces excised from a monodimensional gel. This has to be taken into account to adapt database searches.
18. Widely opening the variable *modification* search parameter will increase both search time and nonspecific protein hits dramatically. If more variable *modification* are needed for the search, we recommend *constraining* the search (e.g., lower the mass tolerance, describe the taxon more precisely, set the number of missed cleavage sites to null).

16

Chapter

GENETIC CONTROL OF INSECTS

Genetic control refers to a variety of methods for suppresing insect pests through manipulation of the inheritance mechanisms of the population. The mass release of sterilized males, *sterile male release*, has proved most successful, but other techniques have been evaluated including the use of chemosterilants and the mass introduction of deleterious mutations (e.g., conditional lethals, cytoplasmic incompatibility, chromosomal translocations, and meiotic drive.

Sterile Male Release

The release of relatively large numbers of males, sterilized either by exposure to γ-radiation or by chemosterilants, into a population of virgin females has reduced the number of fertile female offspring, as shown in Table elsewhere in this chapter. The most effective use of sterile male release was in the eradication of the screwworm *fly Cochliomyia hominivorax* from the southeastern United States in 1959 by periodic release over 18 months of ca. 2×10^9 male flies, sterilized with γ-radiation from cobalt-60, over an area of 181000 km^2.

A similar program along the Texas–Mexico border has substantially controlled the screwworm over the past 25 years, through the annual release of as many as 10×10^9 sterile males. Dramatic population reduction of the Mediterranean fruit *fly Ceratitis capitata* was also achieved in California in 1981 when ca. 40×10^6 flies were released weekly.

Sterile male release has been used with limited success to suppress isolated populations of the codling moth *Laspeyresia pomonella* in the Pacific Northwest and is proposed as an integral part of a cotton boll weevil eradication scheme in the southeastern United States.

Success in sterile male release programs depends on several factors:

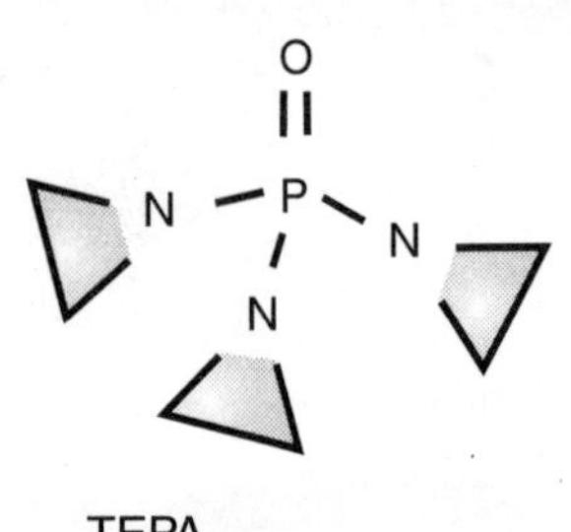

TEPA

1. method for rearing large numbers of insects,
2. adequate dispersion of released sterile males,
3. a sterilization procedure that does not decrease male vigor and competitiveness in mating,
4. a pest species that mates only once per lifetime or in which the sperm from sterile males compete with those of fertile males, and
5. a low pest population density or a means of reducing the population to a low level (e.g., by insecticide application).

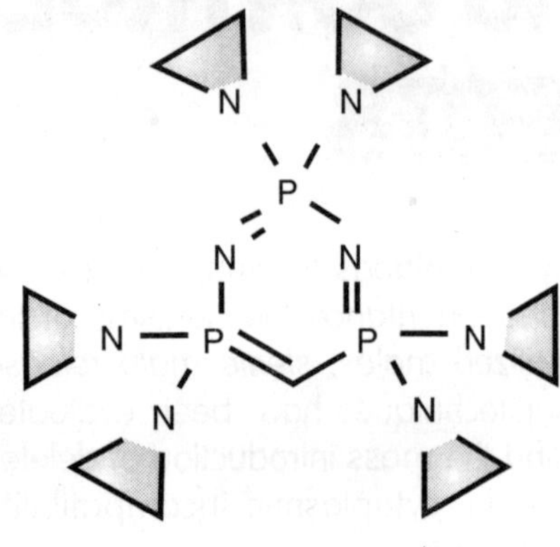

Many species of insects do not conform to these criteria, and the rearing and release of sterile males can be a costly procedure.

Chemosterilants

The availability of a variety of radiomimetic (i.e., mimic the effect of γ-radiation) and antimetabolite cancer suppressive drugs that can sterilize large segments of wild insect populations has led to many suggestions for their use in genetic control programs.

Aziridines have a highly reactive ethyleneimine group. They alkylate deoxyribonucleic acid and cause sterility in both sexes of house flies, mosquitoes, fruit flies, and other insects.

HEMPA

Compounds used experimentally include *TEPA* [57-39-6], triethylenephosphoramide, *mp* 41 °C, LD_{50} (rat, oral) 37 mg/kg, and its analogue *thio-TEPA*, triethylenethiophosphoramide, *mp* 51.5 °C.

Other radiomimetic compounds that have been investigated as chemosterilants include *apholate*, 2,2,4,4,6,6-hexakis(1-aziridinyl)-2,2,4,4,6,6-hexahydro-1,3,5,2,4,6-triazaphosphorine, *mp* 147.5 °C, LD_{50} (rat, oral) 98, 113 mg/kg; and *HEMPA*, hexamethylphosphoric triamide, bp 105–107 °C (1.5 kPa), d^{20}_{4} 1.03, LD_{50} (rat, oral) 2650, 3360 mg/kg. *Busulfan*, 1,4-butanediol dimethanesulfonate, $CH_3SO_2O(CH_2)_4OSO_2CH_3$, LD_{50} (rat, oral) 1.8 mg/kg, has been used in the mass production of sterile insects.

Unfortunately, all of these radiomimetic compounds are mutagens and carcinogens, and produce degenerative changes in human germ plasm at dosages of one-hundredth or less of the LD_{50}. Their casual and indiscriminate use in insect control is precluded.

Antimetabolites suggested for use as insect sterilants include *methotrexate*, amethopterin, N-{4-[(2,4diamino-6-pteridinyl)methyl]methylamino}benzoyl-L-glutamic acid, a folic acid antimetabolite; and *fluorouracil*, 5-fluoro-2,4-(1H,3H)-pyrimidinedione, a uracil antimetabolite.

HOST PLANT RESISTANCE

Breeding plants to incorporate genetic factors that deter insect pest attack is a most important technology for integrated pest management (IPM). Genetic host plant resistance is produced by antixenosis, antibiosis, or by tolerance. *Antixenosis* (nonpreference) results from the removal of kairomone attractants or alteration of morphological characters such as pubescence.

Table 16.1: Rate of Decrease in Wild Pest Population After Sustained Release of Sterile Males.

Generation	Wild Female Population Unsuppressed	Wild Female Population Suppressed	Sterile Male Population Released	Ratio of Sterile to Fertile Males	Number of Reproducing Females
Parental	1×10^6	1×10^6	9×10^6	9 : 1	100000
F_1	5×10^6	0.5×10^6	9×10^6	18: 1	26316
F_2	25×10^6	131580	9×10^6	68 : 1	1907
F_3	1.25×10^8	9535	9×10^6	942 : 1	10
F_4	1.25×10^8	50	9×10^6	180000 : 1	0

Antibiosis results from the production of deleterious allomones that deter insect growth and development. *Tolerance* is produced by enhanced plant vigor that results in rapid repair of insect injury. Genetic engineering provides opportunities to produce host plant resistance through transfer of genes for the production of antibiotic factors such as Bacillus *thuringiensis* (Bt) toxin into the genomes of crop plants such as tobacco, tomato, potato, corn, and soybean.

The insecticidal properties of the toxin should protect the crop plant from damage by insect herbivores. However, widespread utilization of this technology may also provide increased opportunities for the development of Bt resistance.

GENETIC ENGINEERING

Genetic engineering provides methods for isolating, selectively amplifying, and expressing genes encoding desirable traits. Because the majority of currently used plant-transformation techniques introduce genes into individual totipotent plant cells, it is important that an established tissue culture system, a working regeneration protocol, and an effective method for introducing genes must exist for successful deployment of plant genetic engineering.

Therefore genetic engineering of plants encompasses plant tissue culture, techniques of plant gene transfer, and plant molecular biology. Genes are often obtained from one organism and transferred into another organism to either produce large quantities of a gene product or to improve characteristics or traits of the transformed organism.

Interest in the transformation of agronomic plants with added traits of disease resistance is high and a major focus of plant biotechnology. Plant genetic engineering has provided the opportunity to not only develop disease-resistant transgenic plants but also to study structure and function of plant genes.

Recent advances in plant tissue culture, regeneration, and improvements of methods of introduction of foreign genes into plants have made it possible to overcome species barriers and introduce genes from virtually any source.

Plant transformation introduces genes into plants that become stably integrated into a plant chromosome and usually follow Mendelian patterns of inheritance and segregation. Transfer of genes by sexual means in conventional breeding often involves transfer of a block of undesirable genes linked with desirable genes. In contrast, molecular transfer of genes by genetic engineering involves transfer of a predetermined single gene or a few genes.

Methods of Gene Transfer in Plants

Genes can be transferred into plants by either of two commonly used approaches:

1) *Agrobacterium*-mediated gene transfer and

2) direct DNA uptake methods employing chemical, electrical, or mechanical means.

Agrobacterium-Mediated Transformation

Infection with *Agrobacterium* spp., armed with a natural gene transfer system, is the most commonly used method for gene transfer in plants. *Agrobacterium* tumefaciens and *A. rhizogenes* are characterized by the presence of large plasmids called Ti (tumor inducing) and Ri (root inducing) plasmids, respectively. Ti and Ri plasmids contain a segment of DNA, called the T-DNA (transfer DNA), that is transferred from bacterial cells to plant nuclei where it is integrated into chromosomal DNA.

The T-DNA contains oncogenes that produce tumors or hairy-roots when the oncogenes are transferred to host plants. The Ti plasmid contains a Vir (virulence) regulon consisting of several operons with coordinated regulation. A two component sensor–activator system, comprising of VirA and VirG proteins, senses environmental signals such as the cell wall degradation product acetosyringone or other phenolic compounds or five- or six-carbon sugars.

Interaction of signal compounds with VirA results in autophosphorylation of VirA. Subsequently, the phosphate moiety is transferred from VirA to VirG. The phosphorylated VirG acts as a sequence-specific transcriptional activator by binding to an 11-bp conserved vir-box sequence in the promoter of all *vir* genes.

The Vir proteins are responsible for formation and processing of the single-stranded T-strand (transfer strand) that is excised from the T-DNA of the Ti plasmid and transferred through the bacterial and plant cell membranes into the plant cell. Inside the plant cell, the T-strand is translocated to the nucleus, aided by nuclear-localizing signals present in $VirD_2$ and $VirE_2$ proteins that coat the T-strand. The T-strand is finally integrated into the plant chromosome via an illegitimate recombination process.

After integration, the T-strand becomes part of the plant chromosome and the genes present on the T-strand behave like any other plant gene. Although T-DNA contains several genes, none of these genes are expressed in *Agrobacterium;* consequently, these genes can be replaced with gene(s) of interest that are then transferred to plants. Exploiting this feature of T-DNA transfer, artificial binary vectors for plant transformation have been constructed. These wide host range replicon are made with one or more genes of interest, with appropriate promoters and terminators, flanked by T-DNA border sequences.

The binary plasmids are introduced into *Agrobacterium* containing a virulent, but disarmed Ti plasmid. Through the activation of the Ti plasmid genes from the binary vectors are transferred to plant cells and integrated into plant chromosomes, generating transgenic plants that carry one or more genes interest.

Until recently, most cereal crops were considered recalcitrant for *Agrobacterium*-mediated transformation. During the last five years, mostly from T. Komari and his group's efforts, protocols have been developed for *Agrobacterium*-mediated transformation of wheat, barley, maize, rice, millets, and a few other cereals. *Agrobacterium*-mediated transformation is considered the desired method of transformation because there are fewer gene copies transferred to the plants and the transgenic plants show less problems because of DNA rearrangements.

Direct DNA Transformation Methods

The chemical method of direct DNA uptake uses polyethylene glycol (PEG) to facilitate uptake of DNA into protoplasts by endocytosis. The transformed cells are then selected in the presence of an antibiotic and regenerated into whole plants.

Electroporation is used to introduced DNA into plant protoplasts through transient pores that are formed in the membrane bi-layer when exposed to high voltage electrical pulse. Plant cell walls are a barrier to direct DNA uptake by intact cells; hence, the cell wall is enzymatically removed to generate protoplasts prior to PEG treatment and electroporation.

Electroporation and PEG methods have been successfully used to transform rice and maize cells. However, until recently monocos transformation was achieved mostly by using the particle bombardment (biolistic) method. The biolistic process is defined as introduction of DNA into plant cells through the use of high-velocity microprojectiles.

High-density metals, such as tungsten or gold, are used to prepar spherical microprojectiles. DNA-coated microprojectile are accelerated by high-pressure helium gas so that the high-velocity particles can still breach cell wall and cell membranes and yet make nonlethal entry into cells. Apparently DNA from particles lodged inside the nucleus can stably integrate into chromosomes. Plants are selected and regenerated from transformed cells.

Soft penetrable tissues, such as embryos and suspension cells are most commonly used as targets for microprojectile bombardment. However, other types of tissues, such as leaves, axial buds, and flower organs have also been used. The first successful and reproducible monocot transformation employed biolistics for the development of transgenic wheat.

A slightly different approach, using electrical discharge to accelerate microprojectiles, has also been used to transform soybean, rice, and other plants. Biolistic transformation is achieved via nonhomologous integration of DNA into plant chromosomes, which results in a large number of transgene copies and various degrees of rearrangements.

Transformation efficiency on the biolistic method is lower than that obtained with *Agrobacterium*-mediated transformation in dicot plants. Nevertheless, as a universal gene delivery system the biolistic method can be used for many organisms. While the biolistic method has been used extensively for cereal transformation, the Agrobacterium-based system is currently considered as the method of choice for delivery of DNA for both dicot and monocot plants.

It is believed that the Agrobacterium-mediated approach provides precise, low copy, and stable incorporation of transgenes, features important for introduction of transgenic plants into commercial breeding programs and subsequent release of transgenic varieties.

Transgenic plants are characterized further to detemine the physical presence and active transcription of unselected transgene(s). Moreover, since many transgenes are of heterologous origin, it is necessary that functional expression be determined to ensure that the protein made in the plant is biologically active.

With the aid of newly acquired expertise in plant tranformation and improvement of gene isolation techniques, plants have been genetically engineered by using various approaches to obtain resistance against devastating dieases. There are numerous examples where the tools of biotechnology have been used to generate disease-resistant transgenic plants. For space constraints only a few are listed in this article.

Engineering Virus Resistance in Transgenic Plants

Since the landmark work of Roger Beachy in 1986 that demonstrated the feasibility of using a viral coat protein gene to obtain virus resistant plants, there have been intense efforts to engineer virus resistance in transgenic plants. Approaches that have been used to generate disease-resistant plants can broadly be categorized into two groups: 1) Pathogen-Derived Resistance (PDR) or 2) Pathogen-Targeted Resistance (PTR).

PDR is a term coined by R. Sanford in 1985 to describe an approach that uses a gene or DNA sequence from the pathogen itself to obtain resistance. Coat protein–mediated virus resistance is an example of PDR. In addition to coat protein genes other viral genes and viral cDNA sequences have been used for PDR. The resistance mechanism of PDR may be either protein- or RNA-based. In proteimediated resistance transgene mRNA is translated and there is a positive correlation between the level of protein accumulation and the degree of virus resistance.

In RNAmediated resistance there is little or no accumulation of transgene mRNA. Pathogen-derived resistance has been used extensively in the United States and abroad in many commercially released varieties that have provided excellent field resistance against many economically important virus diseases. This approach, however, has an important limitation, resistance is specific for the virus from which the transgene is obtained or closely related viral strains.

There are also some concerns about a potential for recombination between the host-resident transgene and a viral genome resulting in evolution of a more virulent virus, extended viral host-range, and vector specificity. Scientifically these possibilities are perhaps insignificant, as recombination is believed to occur routinely in natural viral infections.

Recent studies have indicated that in many cases PDR invokes an RNA-mediated silencing mechanism that is extremely efficient in eliminating viral RNA and host transgene RNA transcripts following viruf infection. This type of resistance has been termed homology-dependent resistance (HDR).

HDR mimics the well-documented posttranscriptional gene silencing or cosuppression mechanism in plants, which is dependent on homology between the transgene and the target gene. Pathogen-targeted resistance (PTR), on the other hand, uses known antiviral or antimicrobial systems to seletively inactivate pathogens following initial infection.

A recent work demonstrated the usefulness of PTR to obtain broad-spectrum resistance to economically impotant virus diseases. An interferon-induced mammalian antiviral pathway, called the 2-5A system, was intrduced into transgenic tobacco plants. A functional 2-5A system requires two enzymes, a 2-5A synthetase that produces 5′-phosphorylated, 2′,5′-linked oligoadenylates (2-5A) in response to double-stranded RNA, and the 2-5A-dependent RNase L. Infection of leaves of coexpressinf transgenic plants by several plant viruses resulted in the formation of necrotic lesions.

In contrast, leaves expresing only one of the transgenes, or control leaves, produced a typical systemic infection when inoculated. This antivral strategy has several advantages over other approaches currently in use as the 2-5A system is predicted to provide resistance to any plant virus that produces dsRNA in the infected cells.

Most plant viruses are positive sense, singlstranded RNA viruses that produce a dsRNA intermediate during their replicative process. Other PTR approaches involve use of dsRNA-degrading enzyme genes from yeast (pacl) or Escherichia coli (RNase III) and ribosome inativating protein (RIP) genes from pokeweed and other sources.

Engineering Bacterial Resistance in Transgenic Plants

Economically important phytopathogenic bacteria belong to one of five genera: *Pseudomonas, Xanthomonas, Erwinia, Clavibacter*, and *Agrobacterium*. Application of genetic engineering to combat bacterial diseases is still in its infancy. Sporadic uses of a few genes have been made to successfully control or reduce phytobacterial diseases in transgenic plants.

Most phytopathogenic bacteria thrive and multiply in the intercellular spaces in plants. Hencf, signal peptide fusions have been used to direct transgene products to intercellular spaces. A variety of strategies have been used to engineed bacterial resistance in transgenic plants, namely, expresion of nonplant antibacterial proteins, detoxification or inhibition of bacterial virulence factors, expression of antibacterial proteins of plant origin, and overexpression of cloned resistance genes in transgenic plants.

Expression of Non-Plant Antibacterial Proteins

Lytic peptides from insects and frogs are known bactericidal peptides that disrupt bacterial membrane integrity, eventually killing the cells. Several natural and synthetically designed lytic polypeptides have been expressed in transgenic plants. Expression of cecropin B, Shiva-1, attacin, and magainin in transgenic tobacco and potato had only limited enhanced resistance to *Ralstonia* (*Pseudomonas*) *solanacearum* and *Pseudomonas syringae pv. tabaci.*

Expression of a mammalian lactoferrin protein in transgenic tobacco plants demonstrated significant delays of bacterial wilt symptoms when inoculated with the bacterial pathogen *Ralstonia solanacearum.* Quantification of the expressed lactoferrin protein by enzyme-linked immunosorbent assay (ELISA) in transgenic plants indicated a significant positive relationship between lactoferin gene expression and levels of disease resistance.

Lactoferrin, a granule-associated glycoprotein, is a cationic protein with a high proportion of arginine and lysine at the N-terminal region, with two glycosylation and several iron-binding sites. Lactoferrin is highly antibacterial against both gram-positive and gram-negative bacteria at concentrations ranging from 3 to 50 μg/ml.

It is believed that these lethal effects are due to a direct interaction of lactoferrin with the cell surface and subsequent disruption of normal permeability functions of the membrane, a so-called dissipation of proton motive force action (23). Similarly, expression of an antimicrobial tachyplesin gene from Asian horseshoe crabs led to antibacterial activity against *Erwinia spp. in* transgenic potato.

Lysozyme is a well-known antibacterial protein, which exerts antimicrobial activity through the cleavage of the murein layer of bacterial peptidoglycan. As a result, the bacterial cell wall becomes weak, leading to cell lysis. Lysozyme genes from a number of sources, from T4 phage to plants, have been introduced into transgenic plants to obtain resistance to bacterial diseases.

Expression of lysozyme in transgenic potato plants demonstrated significantly reduced maceration of tuber pieces inoculated with *Erwinia carotovora.* A fungal glucose oxidase gene from *Aspergillus niger* has been expressed in transgenic potato plants to obtain resistance to *E. carotovora.* Expression of glucose oxidase in transgenic plants enhanced the levels of H_2O_2, which apparently activated cellular defense genes and triggered hypersensitive cell death.

Expression of Antibacterial Proteins from Plants

Expression of a barley thionin gene resulted in a significant decrease in the growth of Pseudomonas *syringae pv. syringae* and *P. syringae pv. tabaci.* Extracts of transgenic plant also exhibited antibacterial activity against *Clavibacter michiganensis pv. sepodonicus.*

Expression of Bacterial Genes

Toxins produced by some bacterial pathogens are responsible for the damage or disease caused by these pathogens. Because these toxins are deleterious to the pathogens, the bacteria produce compounds that detoxify these toxins or produce toxin-resistant target enzymes.

Phaseolotoxin, produced by the phytopathogenic bacteria *P. syringae pv. phaseolicola,* is cleaved by a plant endogenous peptidase to produce octicidin, an irreversible inhibitor of ornithine carbomyl-transferase (octase). Psedomonas *syringae pv. phaseolicola* produces two octase enzymes, one sensitive and the other resistant to the phaseolotoxin.

Expression of the pathogen-derived, toxin-resistant octase in transgenic tobacco plants confers resistance to *P. syringae pv. phaseolicola. Similarly,* expression of a tabtoxin-inactivating enzyme tabtoxin acetyl transferase (*ttr*) gene from *P. syringae pv. tabaci* enhanced resistance to the same bacterium in transgenic tobacco. Tabtoxin, a dipeptide, is an inhibitor of glutamine synthase and induces symptoms of wildfire disease of tobacco.

ENGINEERING FUNGAL RESISTANCE IN TRANSGENIC PLANTS

Transgenic plants with ectopic expression of various genes have been reported to have enhanced resistance against phytopathogenic fungi. Genes encoding hydrolytic enzymes such as chitinase, chitosanase, and glucanase, which can degrade fungal cell wall components, have been specifically used for generating fungal disease-resistant transgenic plants.

Some of the earliest work on developing fungal resistance in transgenic plants involved expression of a chitinase gene. Broglie et al. demonstrated for the first time that transgenic tobacco and canola plants expressing a chitinase gene were highly resistant to a root pathogen *Rhizoctonia solani* in greenhouse tests.

Chitin is a major component of fungal cell walls and degradation of chitin has been suggested to have protective effects against chitin-containing, plant-phytopathogenic fungi. Chitinases hydrolize the β-1,4 linkage of the N-acety-D-glucosamine polymer chitin.

S. Muthukrishnan's group has recently cloned several chitinase genes from insects and rice plants. They generated transgenic rice plants constitutively expressing these chitinase genes and reported that the transgenic plants showed a high degree of resistance to the sheath blight fungus, *R. solani*. Coexpression of chitinase and glucanase in transgenic crops heightened resistance to a number of fungal pathogens including *Alternaria* and *Cercospora spp.*.

A gene encoding a strong antifungal endochitinase from the mycoparasitic fungus *Trichoderma harzianum* provided broad-spectrum fungal resistance in transformed tobacco and potato plants. Transgenic plants were highly tolerant or completely resistant to *Alternaria alternate, A. solani, Botrytis cinera,* and *Rhizoctonia solani*. This work also indicated that biocontrol fungi might be a good source for antifungal genes.

Osmotin, a thaumatin-like protein isolated from plants, demonstrated antifungal activity against *Phytophthora infestans,* an extremely important pathogen of potato and other solanaceous plants. Nakajima et al. expressed a human lysozyme in tobacco to obtain resistance to the fungus *Erysiphe cichoracearum.*

Transgenic tobacco plants expressing a ribosome-inactivating protein (RIP) from barley were shown to be resistant to *Rhizoctonia solani.* RIPs inhibit eukaryotic protein synthesis via N-glycosidase modification of the *28S* RNA of ribosomes.

Interestingly these proteins show selective characteristics in that they are apparently inactive in the species from which they are isolated but can inactivate non-self ribosomes.

Constitutive high level expression of the pathogenesis-related protein PR-1a rendered transgenic tobacco resistant to two pathogenic oomycete fungi, *Peronospora parasitica* and *Phytophthora parasitica* var. *nicotianae.*

The exact nature of PR-1a-mediated disease resistance is not known but there might be an association with systemic acquired resistance (SAR), an effective long-distance defense response, via transmissible signals involving sali-cylic acid.

Jach et al. expressed three antifungal genes in tobacco individually or in tandem. Expression of a barley type II chitinase, a ,6-glucanase, and a type I RIP resulted in a synergistic disease resistance in transgenic tobacco inoculated with *R. solani.*

Transgenic tomato plants expressing a yeast Δ-9 desaturase gene have been shown to have enhanced resistance to powdery mildew fungus, *Erysiphe polygoni.* Resistance was attributed to the increased levels of 16 : 1 and 16 : 2 fatty acids in transgenic tomato.

Engineering Plant Diseases Resistance Genes

Disease resistance genes have been extensively used in classical plant breeding. Recent advances in biochemical, genetic and molecular bases of plant–pathogen interations have led to the identification and molecular cloning of a number of plant disease resistance genes (R genes).

These genes are either involved in detoxifying or neutraizing virulence factors produced by the pathogens, or they recognize pathogen over-gene-dependent ligands. Recognition of ligands triggers a cascade of signaling, involving multiple signal transduction pathways that eventually lead to the development of resistance by hyper-sensitive reactions (HR) or other means.

Hence many of the R genes are components of signal transduction pathways. Isolation and characterization of R genes from various sources have indicated that these genes have remarkable structural similarities, indicating that plants evolve and use similar signal transduction pathways to trigger resistance against a large number of unrelated pathogens.

The first plant resistance gene Hm1 was isolated from maize for resistance to the leaf spot fungus Cochliobolu*s carbonum*. The fungus produces a toxin that inhibits maize histone deacetylase activity, thereby killing plant cells.

The Hm1 gene product encodes an enzyme that neutralizes the toxin. The tomato *Pto* gene, a serine-threonine type of R gene that conditions a gene-for-gene type race–cultivar interaction, confers resistance to *Pseudomonas syringae pv. tomato*. Tobacco plants transformed with tomato *Pto* gene showed HR-based resistance to *Pseudomonas syringae pv. tabaci* pathogen carrying *avrPto*.

A barley R gene, *mlo*, provided remarkably durable field resistance to the powdery mildew fungus *Erysiphe graminis f. sp Hordei, presumably* through activation of the host defense response. Most of the other recently cloned R genes share a leucin-rich repeat (LRR) domain important for protein–protein interaction, although additional features are also found in many of the R genes.

In several tobacco species resistance to *Tobacco mosaic virus* (TMV) is mediated by a resistance gene N. The transposon tagging technique was used to clone the N gene from tobacco. Transfer of the N gene to nonN gene–containing transgenic plants conferred resistance to TMV. The N gene encodes a protein of 131-kDa with LRR domains, nucleotide binding sites (NBS), and similarities with the *Drosophila Toll* protein and the mammalian interlukin-1 receptors.

The *L6* rust resistance gene from flax was also cloned using transposon tagging. The L6 protein also contains NBS and LRR domains as in the N gene. An *Arabidopsis* RPP5 gene, specifying resistance to the downy mildew fungus *P. parasitica,* had similar protein features as the N and l6 genes.

Two nematode resistance genes, *Mi* cloned from tomato and *Cre3* from wheat, belong to this group. The rice *Xa21* gene that confers resistance to Xanthomonas *oryzae pv. oryzae* was isolated by positional cloning and overexpressed in transgenic rice plants.

The plants showed a high level of resistance to the pathogen, presumably by surface recognition of the pathogen and subsequent strong activation of an intracellular defense response. Interestingly this rice *R* gene contains an LRR domain and a kinase domain homologous to the *Pto* gene. A second *Xanthomonas* resistance gene, *Bs2,* from pepper induces an HR against *X. campestris pv. vesicatoria* expressing the *avrBs2* gene.

FUTURE OF GENETIC ENGINEERING FOR DISEASE RESISTANCE

Disease protection is a major goal of crop breeding. The tools of plant genetic engineering are invaluable for identification and introgression of disease resistance traits in crop plants. Plant biotechnology has already made significant progress towards development of disease resistant transgenic plants with several varieties being used commercially. Future designer plants will undoubtedly contain transgenes for protection from multiple diseases.

PLANTS GENETIC ENGINEERING

Several discoveries in the 1980s and 1990s permitted the transition of plant molecular biology from a fledgling science to commercial reality. These discoveries ranged from the identification of biologically important genes to the development of methods to introduce new genes into plants and regulate gene expression.

The former proces is commonly referred to as transformation. Nearly five dozen plant species have been transformed and the list of plant species subject to transformation include principal field crops such as corn, cotton (qv), rape, rice, soybean, and wheat. In addition, several horticultural species such as tomato, potato, petunia, chrysanthemum, apple, walnut, melons, etc., have been subject to transformation.

More than 500 field tests have been conducted and transgenic plants such as transgenic tomato, soybean, corn, rape, potato, petunia, melons, and cucumbers are in the advanced stages of commercial development and regulatory process. Four methods have been extensively investigated for the introduction of transferred deoxyribonucleic acid (T-DNA) into plants.

These include agrobacterium mediated T-DNA transfer, direct uptake of DNA by protoplasts, particle acceleration techniques such as electrostatic discharge or biolistics gun technology, and DNA uptake into partially digested immature embryos. By far the most commonly used method for gene introduction into dicotyledonous plants is the agrobacterium technology.

This bacterium delivers genes contained in the T-DNA region of the Ti plasmid to the nucleus of several dicotyledonous species. Within the nucleus, the T-DNA is randomly inserted into the chromosome of the recipient cell. The clonal progenies of the cell containing the inserted gene show a high degree of stability.

The gene is transmitted in a Mendelian fashion during sexual stages of cell division and development. Although *agrobacterium* mediated gene introduction into plants is highly efficient and routinely used, its primary limitation is that several plant species are recalcitrant to transformation via this *bacterium*. This is particularly so for monocotyledonous species such as corn, rice, and wheat. In these instances, particle gun technology is routinely used for the introduction of genes.

Whereas most genes introduced into plants via the gun technology appear to be nuclear localized, this technology also has been reported to be useful in transforming the chloroplast of plant cells. Several reviews documenting the progress in plant transformation during the early 1990s are available. Expression of genes that have been introduced into plants is regulated by promoters, although the extent of regulation of gene activity by the promoter is influenced at least to some extent by the insertion site of the gene within the chromosome. As of this writing methods for DNA transfer cause random insertion of the DNA into the chromosome.

Techniques for precise introduction of the transgene to specific sites with the plant genome are being developed. Numerous promoters have been used for gene expression in plants. The choice of promoters is dictated by the tissue and developmental specificity required for gene expression. By far the most commonly used promoter for constitutive gene expression in both mono- and dicotyledonous plants is the Cauliflower mosaic virus (CaMV) 35S promoter.

This promoter appears to be expressed in several plant organs and cell types; however, it is not truly constitutive in that it is not uniformly expressed in all plant tissues. DNA elements within the 35S promoter, which cause tissue specific expression of genes, have been described.

The activity of the 35S promoter may be enhanced by use of multiple copies of enhancer elements located within the 35S promoter. For tissue regulated gene expression, promoters have been described which are expressed in a tissue specific manner.

Examples of such promoters include the promoter for patatin which causes tuber specific expression of genes in potato, the 7S promoter of soybean, or the napin promoter of *Brassica*, which cause seed specific expression of genes, and the RB7 promoter which causes root specific expression of genes. These promoters may not only be spatially regulated in terms of cell and tissue specificity but may also be temporally regulated in that the promoters are active only at certain developmental stages of the cells and tissues in which the promoters are expressed.

In order to determine which plant cells have been transformed, selectable marker genes are introduced during transformation. These marker genes permit selective growth of transgenic cells on the medium used for tissue propagation whereas the nontransgenic cells are killed. Examples of selectable marker genes include antibiotic resistance genes such as neomycin phosphotransferase (NPT-II), hygromycin phosphotransferase, and chloram-phenicol acetyl transferase, as well as herbicide resistance genes such as phosphinothricin acetyl transferase, bromoxynil nitrilase, 2,4-D-oxygenase, etc.

Plant cells expressing the NPTII gene are able to survive kanamycin 8063-07-8 and addition of kanamycin permits selection of those cells receiving and expressing the NPTII gene during transformation (see ANTIBIOTICS; HERBICIDES). Herein two specific applications of plant biotechnology are discussed. The first is concerned with 5-enolpyruvylshikimate 3-phosphate synthase (EPSPS), the enzyme which is the target for the widely used herbicide glyphosate, $C_3H_8NO_5P$.

The second is directed toward a discussion of increasing starch biosynthesis in plants. The first application deals with a trait which directly impacts the farmer during the production phase of agriculture; the second application deals with a trait that impacts the consumer of agricultural products. These traits may be referred to as agronomic and quality traits, respectively.

A number of other agronomic and quality traits are being investigated. These include insect, virus, disease, and nematode resistance, fertilizer-use efficiency, ripening control, fruit firmness, etc. Of these traits the most advanced agronomic trait for bioengineering is insect resistance.

Insect resistant cotton and corn have been obtained by introduction and expression of a Bacillus *thurigiensis* kurastaki gene (BtK gene). The BtK protein encoded by this gene is selectively toxic to the lepidopteran pests, i.e., cotton boll worm, pink boll worm, and European corn borer, which attack these crops.

Insect-resistant potato has been obtained by expression of a BtT gene which encodes a protein, selectively toxic to the Colorado potato beetle, a principal pest of potato. This topic has been reviewed. Virus resistance, conferred by expression of the viral coat protein (CP) gene in transgenic plants, has also received considerable attention.

Products such as potato, squash, melons, etc., based on this technology are in advanced stages of development and commercialization. Both tomato fruit ripening and fruit firmness are among the advanced quality traits that are being investigated. A variety of approaches, based on inhibition of ethylene production are being pursued for enhancement of shelf life of tomato. For enhancing fruit firmness, cell wall hydrolytic enzymes such as polygalacturonidase and pectin methylesterase are being investigated.

BIOENGINEERING OF GLYPHOSATE TOLERANCE

The enzyme 5-enolpyruvylshikimate 3-phosphate (EPSP) synthase catalyzes the transfer of a carboxyvinyl moiety of phosphoenol pyruvate (PEP) to shikimate 3-phosphate (S3P), yielding inorganic phosphate and EPSP as reaction products. EPSPS has received considerable attention in recent years, in view of the demonstration that glyphosate (N-(phosphonomethyl)glycine 1071-83-6), the active ingredient of the herbicide Roundup, kills plants by inhibition of this enzyme.

EPSPS catalyzes the sixth reaction during aromatic amino acid biosynthesis via the shikimate pathway which exists only in plants and microorganisms. EPSP is the immediate precursor of chorismate, the first important branch point during aromatic amino acid and vitamin biosynthesis. Perhaps the most important inhibitor of the EPSPS reaction is glyphosate which inhibits the EPSPS reaction via formation of a ternary complex with either S3P or EPSP and enzyme.

Glyphosate is a competitive inhibitor with respect to PEP and an uncompetitive inhibitor with respect to S3P. Glyphosate, however, is not a structural analogue of PEP because the glyphosate does not inhibit any other PEP-dependent reaction. Whereas it has been suggested that glyphosate may be a transition-state analogue of the carbonium ion intermediate of PEP formed during catalysis, the bulk of the evidence suggests that this is unlikely.

Nevertheless, glyphosate inhibits a wide range of EPSPS enzymes of bacterial, fungal, and plant origin. A number of structural analogues of glyphosate have also been tested for the ability of inhibit EPSPS. Only a few such analogues, e.g., N-amino and N-hydroxy glyphosate, have been found to be inhibitors. Roundup is a nonselective, post-emergent herbicide having activity against a wide range of annual and perennial grasses as well as broadleaf weeds. Because

Roundup has no selectivity for weeds, use for weed control during active growth period of crops is fairly limited. Despite its nonselectivity, glyphosate, is extensively used in weed management because of broad-spectrum, systemic herbicidal activity; rapid inactivation in the soil (does not sterilize the soil); decomposition in the soil to the natural products, i.e., carbon dioxide (qv), ammonia (qv), and phosphate; no toxicity to animal, aquatic, and avian species; it binds tightly to soil and does not contaminate ground water; and its cost effectiveness in weed control.

In view of all the desirable features of glyphosate, the engineering of glyphosate tolerance in crop plants has the potential to open up new frontiers in weed management during cultivation. A substantial effort has been directed toward introducing Roundup tolerance to crop plants (35–38).

Engineering Roundup Tolerance

Knowing that the mode of herbicidal action of glyphosate is mediated via inhibition of EPSPS, at least two mechanisms can be considered for the introduction of Roundup tolerance to plants. The first option is to simply overproduce EPSPS so as to leave sufficient EPSPS enzymatic activity within the plant cells to satisfy the flux through the shikimate pathway.

Alternatively, a gene encoding a glyphosate tolerant EPSPS enzyme can be used so that the EPSPS reaction is unaffected even the presence of glyphosate. In addition, other approaches which are not related to the mode of action of glyphosate, such as glyphosate inactivation and inhibition of uptake, can be considered. These last are not discussed herein.

Overproduction of EPSPS

Overproduction of EPSPS has been demonstrated to confer glyphosate tolerance to both bacteria (39) and plant cells. Glyphosate tolerant plant cells have served as an excellent starting material for the isolation and purification of the EPSPS protein to homogeneity.

N-Terminal amino acid sequence of the resulting protein provided the requisite information for synthesis of oligoprobes which were used for screening a complementary DNA (cDNA) library of petunia cells tolerant to glyphosate. From the library, the cDNA encoding petunia EPSPS was isolated and sequenced.

The protein encoded by the cDNA had an N-terminal extension of 72 amino acids compared to the protein sequence obtained from the purified EPSPS enzyme. This N-terminal extension is necessary and sufficient to direct the EPSPS protein into the chloroplasts of plant cells (43). These studies also led to the conclusion that aromatic amino acid biosynthesis occurred primarily in the chloroplast of plant cells.

Whereas plant cells overproducing EPSPS could be generated by stepwise selection on glyphosate an the cells were glyphosate tolerant, these cells could not be regenerated into intact plants. Availability of the cDNA clone for EPSPS, however, provided a convenient tool for generating transgenic plants capable of overproducing EPSPS.

Using an *Agrobacterium tumefaciens* transformation system, the EPSPS gene wa e introduced into both petunia and tobacco plants. Petunia plants overproducing EPSPS were thus produced and shown to be tolerant to Roundup. However, the extent of tolerance was not adequate for commercial use.

Glyphosate-Tolerant EPSPS

Several groups have tried to introduce Roundup tolerance into plants using gene encoding glyphosate-tolerant EPSPS enzymes. A mutant glyphosate-tolerant EPSPS enzyme, fivefold less sensitive to glyphosate, was isolated from *Salmonella typhimurium*. The introduction of the *S. typhimurium* mutant EPSPS gene into tobacco plants resulted in expression of the mutant gene such that plants were tolerant to glyphosate, but the extent of tolerance was not commercial.

Other bacterial mutants, such as a mutant *Escherichia coli* enzyme tolerant to glyphosate, have been described. The *E. coli* mutant had a single amino acid change from the wild type, resulting in substitution of glycine 96 with alanine. An identical mutation was reported in glyphosate-tolerant *Klebsiella pneumoniae*. The nature of changes in the kinetic constants of the *K. pneumoniae* enzyme is similar to that of the *E. coli* enzyme.

The *E. coli* mutant EPSPS was fused to the chloroplast transit peptide (CTP) sequence of petunia EPSPS in order to target the bacterial protein to the chloroplast. *In vitro* uptake experiments confirmed that the bacteria enzyme could indeed be imported and processed to mature protein by chloroplast preparations. Introduction into petunia and tobacco plant cells led to regenerated plants expressing the bacterial gene either targeted to chloroplast or the cytosol.

Tobacco plants containing the *E. coli* mutant EPSPS targeted to the chloroplast had higher levels of Roundup tolerance compared to either plants overproducing wild-type EPSPS or the control nontransgenic plants, but the level of tolerance was not sufficient for commercial use. The level of Roundup tolerance of plants having the *E. coli* enzyme targeted to the cytosol was only slightly higher than that of control plants, suggesting that the cytosolic EPSPS reaction was unable to complement the chloroplastic deficiency of EPSPS.

The glycyl 96 (G96) and prolyl 101 (P101) residue occur in a conserved region of EPSPS which is present in bacterial, fungal, and plant EPSPS enzymes. Replacemen, of G96 with an amino acid other than alanine (A) result in an inactivation of the EPSPS activity of the protein However, the G96 to A mutation can be transferred to other bacterial and plant EPSPS enzymes, and in every case the alanyl enzyme has a higher glyphosate tolerance compared to the glycyl enzyme. This suggests that there is a high degree of conservation of the active site of EPSPS between bacterial, fungal, and plant enzymes.

Mutation of the conserved P101 to a serine residu, also results in glyphosate tolerance of the EPSPS enzyme. This mutation was introduced into petunia EPSPS by site-directed mutagenesis and the seryl enzyme was demonstrated to be glyphosate tolerant. Analogous to the *S. typhimurium* enzyme, this mutation confers only marginal (7-fold) glyphosate tolerance and no significant changes in the kinetic constants for the substrates.

The petunia cDNA containing the prolyl to seryl mutation and the targeting sequence was introduced into tobacco plants. The Roundup tolerance of the tobacco plants expressing the seryl mutant was intermediate between plants expressing the wild-type and alanyl mutant enzymes. Numerous bacteria which utilize glyphosate as a growth substrate were screened for the presence of EPSPS enzymes having binding constants for PEP close to those of the wild-type enzyme, but at least a 100–10,000fold increase in affinity for glyphosate.

The EPSPS from agrobacterium CP4 is perhaps the best studied. These enzymes are referred to as class II in order to distinguish them from the class I enzyme already described. The class II EPSPS enzymes have natural resistance to glyphosate and a low binding constant for PEP whereas the class I EPSPS enzymes are highly sensitive to glyphosate.

The agrobacterium CP4 enzyme has 28% identity to the *E. coli* enzyme and has the conserved glycyl to alanyl change. Antibodies which recognize the *E. coli* EPSPS recognize petunia EPSPS but do not recognize agrobacterium CP4 EPSPS. Similarly antibodies reacting with the agrobacterium CP4 EPSPS do not show immune reaction with either *E. coli* or petunia EPSPS.

Transgenic soybean plants expressing the CTP-CP4 EPSPS display commercial levels of Roundup tolerance. These results validate the importance of substrate kinetics of EPSPS in order to maintain adequate rates of aromatic biosynthesis. Furthermore, the fact that glyphosate tolerance can be obtained by expression of a glyphosate tolerant EPSPS illustrates that the herbicidal mode of action of glyphosate is related solely to inhibition of the EPSPS reaction.

As described earlier, translation of the EPSPS mRNA of plants results in the formation of a protein which has an N-terminal extension. The N-terminal extension, referred to as the chloroplast transit peptide, is necessary and sufficient for the import of the preprotein by the chloroplast. Once imported by the chloroplast, the transit peptide is cleaved releasing the mature enzyme.

As expected, introduction of the EPSPS transit peptide to other protein sequences results in the importation of the fusion protein by the chloroplast. The three-dimensional structure of EPSPS from *E. coli* has been established by crystallographic techniques. A number of amino acid residues have been modified to establish the necessity of these residues for enzymatic activity. At its N-terminus, the lysyl residue at position 22 of the *E. coli* enzyme has been shown to be highly reactive and essential for enzymatic activity.

It is likely that the lysyl residue is involved in substrate recognition. In addition to the lysyl residue at position 22, the arginyl residue at position 28 of EPSPS is conserved in all EPSPS enzymes studied to date. This arginyl residue is highly reactive, and its reaction with arginine reagents is inhibited by S3P and to a higher extent by a mixture of S3P and glyphosate.

By site-directed mutagenesis, the arginyl residue has been replaced *by lysyl*, histidinyl, and glutaminyl residues. The latter two replacements appear to be detrimental for EPSPS activity, whereas the lysyl enzyme retains substantial activity.

The roles of histidinyl, glutamyl, and cysteinyl residues of EPSPS have been probed by reaction with chemical modification reagents. These studies suggest that a glutamyl and histidinyl residue are critical for EPSPS activity. Similar studies with cystein modification suggest that cys-408 of *E. coli* EPSPS, although in a conserved region, is not essential for activity but is proximal to the active site.

Bioengineering of Increased Starch Content

The primary form of carbohydrate reserve in plants is starch (qv), entirely composed of the six-carbon sugar (qv) glucose. Starch typically is deposited in the form of water-insoluble granules, and is synthesized and stored in chloroplasts in photosynthetic tissues or in amyloplasts.

Starch is a generic term used to describe a very heterogeneous class of molecules which differ in size and structure between different plants, different tissues within a plant, and at different stages of plant development.

The heterogeneity of starch has proven useful in a number of different applications; for example, pea starch is widely used as a sizing agent in paper (qv) manufacture, and corn and potato starch are widely used to give viscosity, freeze–thaw tolerance, and body to a number of processed foods. The primary and likely sole pathway of starch biosynthesis is the adenosine diphosphate (ADP) glucose pathway.

In this pathway the first enzyme, ADPglucose pyrophosphorylase (ADPGPP), catalyzes the conversion of glucose-1-phosphate to ADPglucose. In plants, it has been proposed that sucrose synthase is involved in the production of the ADPglucose used in starch biosynthesis.

This model is not considered to be accurate given a number of mutants characterized affecting both starch and sucrose biosynthesis, and this topic has recently been reviewed. Another route for starch biosynthesis is through the action of starch phosphorylase. This enzyme is involved in the degradation of starch, forming glucose1-phosphate from successive removal of glucose units from the polymer.

The reaction is reversible *in vitro;* thus this enzyme potentially plays a role in the formation of starch. Through expression of antisense RNA, this enzyme has been eliminated in the amyloplast of potato tubers with no effect on starch content; thus any role in biosynthesis is proposed to be very minor.

Enzymes Involved in Starch Biosynthesis

Much of the early data dealing with starch biosynthesis in plants are derived from the study of various mutants. The shrunken-2 and brittle-2 mutants of maize have greatly reduced levels of ADPGPP activity owing to the absence of one of the two subunits of this enzyme, and result in a shrunken seed appearance. Mendel's early work on inheritance of traits was performed with a pea mutant deficient in branching enzyme activity. Mutations in plants affecting starch biosynthesis can have severe results to plant morphology and viability.

ADP Glucose Pyrophosphorylase

The rate-limiting reaction in both bacterial glycogen and plant starch biosynthesis is the first step, catalyzed by the enzyme ADPGPP. In bacteria the enzyme functions as a homotetramer subject to tight allosteric regulation by effector molecules that reflect the energy state of the cell, and is the only enzyme in the pathway of glycogen biosynthesis subject to such regulation.

The enzyme is activated by glycolytic intermediates and inhibited by adenosine monophosphate (AMP), ADP, and/or inorganic phosphate (Pi). Fructose 1,6-bisphosphate is typically the primary activator and AMP the primary inhibitor. The role of the activator is to increase the affinity of the enzyme for it substrates, adenosine triphosphate (ATP) and glucose-1-phosphate, and increasing amounts of the activator relieves inhibition caused by AMP, ADP, or Pi.

The allosteric regulation of this enzyme has been shown to regulate the flux of carbon through this pathway and control the level of glycogen that is produced. Much of this work has been performed with mutants of *E. coli* and *S. typhimurium* affected in their ability to accumulate glycogen.

The bacterial ADPGPP enzymes each have subunits that contain allosteric activator and inhibitor binding regions, substrate binding sites, and a site for binding Mg^{2+}. A series of chemical modification experiments lead to the elucidation of amino acid residues responsible for interacting with the various effector and substrate molecules.

The ADPGPP enzymes in plants function as heterotetramers consisting of two distinct subunits encoded by two different genes. These subunits differ in molecular weight, amino acid composition and sequence, and antigenic properties. Antibodies made against the large subunit only weakly react with the small subunit from a given plant (and vice-versa); but antibodies against the large (or small) subunit recognize the corresponding subunit from different plant species, i.e., certain sequences are conserved between widely divergent plant species. As in bacterial glycogen biosynthesis, ADPGPP catalyzes the rate-limiting step in starch biosynthesis. The levels of control are primarily via allosteric regulation, but regulation of gene expression also plays a role in controlling ADPGPP activity. The primary effector molecules differ from those in bacteria. For every plant system studied, the plant enzymes are activated by 3-phosphoglycerate (3-PGA) and inhibited by inorganic phosphate (Pi).

One possible exception is the wheat endosperm enzyme which appears not to be activated by 3-PGA. However, this enzyme is inhibited by Pi, and the presence of 3-PGA overcomes the inhibition. The importance of allosteric regulation to *in vivo* ADPGPP activity and starch content in plants has been demonstrated.

The gene encoding the ADPGPP enzyme from *E. coli* strain 618, which is relatively insensitive to allosteric control, was isolated and inserted into transgenic potato plants via *Agrobacterium tumefaciens* transformation. The gene was designed to express the active protein only in the potato tuber, and such expression resulted in a 25–50% increase in starch content.

In contrast, expression of the ADPGPP gene from a wild-*type E. coli* K12 strain, which encodes an enzyme subject to normal allosteric regulation, had little effect on starch content. These results showed the importance of allosteric control to ADPGPP activity, and circumvention of this control increases the flux of carbon through this pathway and results in an increase in starch biosynthesis and composition.

It is of interest to determine why the plant enzyme is composed of two distinct subunits and the bacterial enzymes only one. Because the enzyme must have binding sites for the allosteric activator and inhibitor, the substrates, and a catalytic site, it is possible that these sites are located on different subunits. The shrunken2 and brittle-2 mutants of maize endosperm lack the large and small subunits, respectively, of the ADPGPP enzyme.

These mutants have 12% and 17% of the wild-type ADPGPP activity and about 25% of wild-type levels of starch, demonstrating that both subunits are required for normal levels of enzyme activity and starch content, but that a single subunit by itself can form an active enzyme. This is supported by a starch-deficient mutant of *Arabidopsis* which lacks the large subunit, has about 5% wild-type levels of ADPGPP activity, and about 40% wild-type levels of starch.

In addition, elimination of one of the ADPGPP subunits in transgenic potato through expression of antisense RNA results in a reduction in ADPGPP activity to 1.5–17% of wild type, and starch content to 4–35% of wild type. These results suggest that allosteric, substrate, and catalytic sites reside on each of the subunit types. ADPGPP genes in plants are also controlled at the level of gene expression. In potato, the transcripts corresponding to the large and small subunits differ in their accumulation profiles in different organs.

The steady-state levels of transcripts corresponding to the large subunit of ADPGPP are highest in tubers and stolons and are inducible by sucrose. In contrast, the steady-state levels of transcripts corresponding to the small subunit of ADPGPP are relatively equivalent in tubers, stolons, and aerial portions of the plant and are not strongly influenced by carbohydrates. *Why* the gene encoding the large subunit of ADPGPP is more tightly regulated than that encoding the small subunit is unknown.

Starch Synthase

In contrast to the bacterial systems where a single synthase is responsible for the elongation of the glucose chain, in plants several synthases are involved in building the starch granule. These synthases are either soluble or granule-bound. The soluble synthases are divided into two forms, designated as Type I and Type II, distinguished by size, kinetic properties, and immunological properties.

These forms are encoded by separate genes which may show tissue and developmental regulation. Given these differences, the two types of enzymes likely play distinct roles in the formation of the starch granule, although this role is thought to be primarily involved in the synthesis of amylopectin, the branched form of starch. The granule-bound starch synthases are immunologically, physically, and kinetically distinct from the soluble synthases, and are encoded by one or more distinct genes.

In maize endosperm, two forms of granule-bound synthase have been identified, bringing the total number of synthases identified in this tissue up to four. Unlike the situation for the ADPGPP gene in potato, the potato granule-bound starch synthase gene has been shown to be regulated solely at the level of gene expression. The primary role of granule-bound starch synthase may be in the formation of amylose, the linear fraction of starch.

Waxy-like mutations which are devoid of amylose and granule-bound starch synthase have been characterized in a number of plant systems, including maize, rice, barley sorghum, and potato. The waxy mutation was obtained in transgenic potato through expression of antisense RNA to granule-bound starch synthase providing strong evidence that the waxy locus encodes the granule-bound starch synthase enzyme, and that this enzyme is responsible for the synthesis of amylose *in vivo*.

Branching Enzyme

Multiple forms of branching enzyme have been found in a number of plant species. These enzymes are all soluble and catalyze essentially the same reaction, but differ in physical, immunological, and kinetic properties, and like the synthases probably play different functional roles in the synthesis of the starch molecule. Branching enzymes are also encoded by multigene families which may show developmental and tissue-specific expression profiles.

The most detailed studies involve the isoforms from maize endosperm, where three different forms of branching enzyme have been purified and designated BEI, BEIIa, and BEIIb. Polyclonal antibodies against BEI do not react against either form of BEII, and vice-versa, but forms BEIIa and IIb appear to be closely related. Monoclonal antibodies have been produced which react with all three isoforms, showing that the enzymes share a few common epitopes but are otherwise divergent.

Each endosperm-branching enzyme has been highly purified and the branching characteristics studied. BEI was found to have high activity on amylose but little on amylopectin, and was found to preferentially transfer long chains. These chains would represent the B chains in the cluster model proposed for the structure of amylopectin. BEIIa and IIb were found to have low activity on amylose and high activity on amylopectin, and transferred preferentially short, or A chains.

Differences between these two isoforms in the types of branches produced were not noted, and these enzymes appear to be very similar. Branching enzymes have been characterized from a variety of other plant tissues. Only a single isoform has been detected in potato tubers. The gene for potato branching enzyme is regulated in a manner similar to the potato large subunit ADPGPP gene and is expressed most abundantly in the potato tuber.

Antisense RNA expression in transgenic potatoes has resulted in a 90% decrease in branching enzyme activity, but with no discernable effect on starch content or structure. This implies that either branching enzyme activity is present in vast excess, or a second enzyme indeed exists. The former seems to be the case.

In pea, the wrinkled seed phenotype has been linked to the locus and results in a 66–75% reduction in total starch, and an increase in amylose from 33% in wild-type pea up to 60–70% in the mutant. Branching enzyme activity is reduced to 14% of wild-type levels because of the complete absence of one isoform of branching enzyme. The decrease in total starch levels is caused by a similar mechanism as in bacteria lacking branching enzyme activity, i.e., as the glucose chain is elongated, it becomes a poorer substrate for the synthase enzyme.

One function of branching enzymes is to clip the elongating chain and provide additional substrate to the synthase enzymes. In this model, the synthase and branching enzyme work in concert, whereas the synthase elongates the chain, the branching enzyme cleaves, transfers a maltodextrin, and forms a new branch, which is then further elongated by the synthase.

The dependence of starch synthase on branching enzyme has been shown in *in vitro* systems where the activity of starch synthase is observed to be greatly enhanced by the addition of branching enzyme. This model of concerted activity also provides the rationale for the existence of multiple isoforms of starch synthase and branching enzyme in plants. Amylopectin is an asymmetric molecule formed of both short (12–42 residues) and long (>49 residues) glucose chains.

Synthesis of such an asymmetric structure requires starch synthases and branching enzymes having different specificities for elongation and for insertion of branch points at different distances along A- and B-chains. Further evidence for this comes from the study of a low starch mutant of *Chlamydomonas* that lacks soluble starch synthase II and shows a decrease in intermediate length chains in the amylopectin fraction.

Thus the structure of the starch granule can be influenced by the properties of both starch synthases and branching enzymes, and further controlled by regulation of gene expression in different tissues or during plant organ development.

GENETICALLY MODIFIED MICROORGANISMS (GMM) IN SOIL ENVIRONMENTS

The early 1970s witnessed a major breakthrough in molecular biology with the discovery of restriction enzymes. From that moment on, it has been possible to modify the genetic constitution of all organisms, and a new era in human history started. In the early years of genetic modification, the perspectives of these approaches were tremendous.

It was, for instance, thought that it would not take long before one would be capable of modifying organisms in such a way that they could easily cleanup spills of waste or polluting materials or successfully control pathogens. Also, in agricultural practice, expectations were high, for instance, considering the transfer of nitrogen fixation genes from bacterial hosts into plant genomes.

Application of such plants would enable developing countries to decrease their expenditures on chemical fertilizers. Also, genetic modifications would make it possible, at least in theory, to reduce crop diseases, improve nutrient availability and, thus, drastically increase agricultural yields. However, these expectations have been tempered ever since and have even led to an upheaval because of two reasons.

Firstly, genetically modified organisms might not be as well adapted to their natural environments as their unmodified ancestors and thus perform poorly after release. Secondly, public concern was raised about the infringement on nature by genetic modification (often called genetic "manipulation" by persistent opponents to the application of these techniques).

Maybe the first backlash was a result of the realization that genetically modified microorganisms (GMMs) would often indeed be outcompeted in their natural environments. In addition, the horizontal spread of recombinant genes was a second major concern because of the possible occurrence of unpredictable events due to lack of (ecological) knowledge about the effects of these new gene combinations on indigenous species.

However, it is currently well accepted that the application of GMMs to the environment needs a thorough understanding of the host strains carrying the genetic modification as well as the heterologous genes, with special emphasis on their regulating systems.

Also, it is essential to gain an understanding of the spread and potential dormancy of modified bacterial cells in the environment, in order to assess the potential risks involved in their use in nature. In this article, we focus on the requirements for the application of GMM strains in natural, unconstrained environments such as agricultural fields, and discus approaches for tracking released GMM strains as well as limiting their spread by the construction of so called self-containing mechanisms.

Construction of GMMs and Containment

For the construction of a genetically modified bacterial strain, developed to perform under field conditions, two considerations are important; 1) optimum survival and activity and 2) spatial and temporal containment. First, survival and activity of the introduced GMM often must be optimum; that is, metabolically active cells of the GMM strain should remain present in sufficient numbers for the required period of application.

On the other hand, the strain must ideally remain restricted only to the location and period for which heterologous gene activity is desired. Both considerations may be contradictory when optimum survival is necessary on the one hand, while on the other hand the GMM has to ultimately be eliminated in order to prevent its persistence or escape from the site of application.

However, sufficient survival as well as containment may be combined when different regulatory genes responding to intrinsic cellular or environmental signals are used. Regulatory genes necessary for controlling the heterologous genes inserted can be triggered under different conditions as genes necessary for containment of the GMM strain.

In this section, special attention is given to the regulation of the gene inserts, both to improve survival and to contain the introduced strains with respect to time and location. In general, GMM strains are ecologically less competent than their (unmodified) parent strains, as demonstrated by competition experiments in soil.

Because of the expression of the heterologous gene, the cellular energy expenditure is increased, and hence GMM cells may be deprived faster of their energy sources. Therefore, uncontrolled expression of inserted genes can be harmful to the released GMM; controlled gene regulation at the appropriate time and site of application would be preferable.

The use of regulatory genes specifically responding to cellular physiological status (e.g., exponential versus stationary phase) or specific environmental signals may help overcome cellular energy depletion. On the other hand, the signal used for beneficial gene expression can also be applied to regulate containment of gene expression via a "negative loop" construction.

A repressor gene can be placed under the control of a promoter regulated by the environmental signal, whereas the intended containment gene is constructed downstream of a promoter/operator region controlled by the repressor protein. The choice of different signals and their responding promoter sequences, with the aim to control heterologous gene expression, is an important first step.

Subsequently, a screening of appropriate promoter sequences should be carried out and these promoter constructs should be used for further application. In this section, a study is included in which two different promoters of *Pseudomonas fluorescens* that respond to signals from the soil environment were characterized.

Promoter probe reporter systems are excellent tools for investigating the expression of genes involved in the response of bacteria to soil conditions. Plant-induced reporter gene expression has been determined with transcriptional fusions using reporter genes such as lac*Z* [B-galactosidase], gusA [B-glucuronidase], lu*x* genes (light emitting genes from Vibrio fisheri), xyl*E* (xylene degradative gene), and *inaZ* (ice nucleation gene).

For microscopic detection of *in situ* induced reporter gene activity, specific fluorescence markers such as the green fluorescent protein (GFP) gene (*gfp*), which has been cloned from jellyfish (*Aequorea victoria*), can be used. Using promoterless *gfp* as a promoter probe reporter, specifically induced expression can be observed by epifluorescence UV microscopy in which intracellular GFP will emit a green light.

Although testing of induction of reporter genes in situ is relatively unexplored, progress has been made recently. Bacterial strains with reporters inserted in genes that show expression upon exposure to stress conditions such as phosphorus limitation have been studied *in vitro* and in soil, and the genes identified were shown to function under soil conditions.

Similarly, Kragelund and coworkers showed the expression of N and C responsive genes under soil related conditions. Responsiveness to root exudates and carbon limitation was investigated in soil, soil supplemented with nutrients, and rhizosphere soil. Reporter gene expression upon carbon limitation was observed in unamended bulk soil, indicating that the carbon limitation induced promoter can be used as a regulatory element to control gene expression in GMM strains intended for application in soil.

In situ root exudate controlled reporter gene activity under natural conditions has been reported by Vande Broek and coworkers and Brennerova and Crowley. Furthermore, *in situ* root colonization studies under gnotobiotic conditions have been performed with various different markers and bacterial species.

For example, de Weger and coworkers used naphthalene-inducible luxCDABE and constitutively induced luxCDABE and *luxAB* gene constructs for *in situ* root colonization studies with *P. fluorescens*. Reporter gene activity was detectable in the rhizosphere with the naphthalene-inducible luxCDABE and constitutively induced *luxAB* genes at a colonization level between 10^3 and 10^4 CFU/cm root.

Studies on the induction of reporter genes by soil or plant stimuli in natural soils are technically feasible. The promoter/reporter gene combinations can be used in the natural environment, for example, as biosensor systems or to determine the expression of genes of interest.

Reporter gene expression was observed near wheat roots in microcosms as well as in the fields; the promoter identified proved to be an excellent regulatory gene of the expression of recombinant genes in a new generation of GMM constructs.

BIOTIC AND ABIOTIC FACTORS INFLUENCING BACTERIAL SURVIVAL IN SOIL

For decades, bacteria have been isolated from soil and characterized in the laboratory with the aim to study and improve their performance as biopesticides, biofertilizers, or bioremediation agents. Symbiotic nitrogen-fixing bacteria such as rhizobia, free-living nitrogen-fixing bacteria such as *Azospirillum*, as well as plant growth-promoting rhizobacterial (PGPR) gram-negative species such as *Pseudomonas* and *Flavobacterium* or gram-positive species such as the *Bacillus* species are among the most frequently used soil inoculants.

However, many species do not survive well in soil, either because they are not indigenous to soil or because they survive well only in specific soil types. It should be noted that, in addition to the specific intrinsic properties, such bacteria have often been selected on the basis of the fast growth of colonies or broth cultures under laboratory conditions and their suitability for genetic modification.

Much effort has been spent on the isolation and identification of ecologically adapted bacteria from specific sites in soil, such as the rhizosphere, under the assumption that these organisms possess traits that helped them adapt to the conditions prevailing at these sites.

Copiotrophic gram-negative species such as those of the genus *Pseudomonas* are abundant in the rhizosphere and have been suggested to be the best soil inoculants because of their fast growth and alleged optimum survival under prevailing conditions in the rhizosphere as well as their genetic accessibility. However, upon introduction, bacterial cells are often subjected to the harsh conditions present in the rhizosphere and bulk soils, which can limit their survival. How the bacteria cope with these conditions depends on their genetic makeup and their physiological condition at the time of introduction into the soil.

For practical applications, bacterial cells should aptly colonize soil sites; optimum survival and activity is only ensured after successful occupation of such an ecological niche. It is clear that the site (i.e., rhizosphere or bulk soil) in which inoculant bacteria reside in soil greatly determines their physiological and metabolic status, which in turn determines the persistence and activity in soil.

Introduced Soil Populations

The main goals of field tests that have been conducted to date have been to monitor the fate of genetically marked bacteria as models for subsequent GMM releases with practical applications and to study the ecological effects of these inoculations.

Most of these studies assessed the putative risks involved in the release of GMMs into soil by measuring parameters such as survival, spread, and gene transfer, but not the fundamental properties of the introduced population, such as its physiological status or its response to environmental stresses. The latter items are important as they determine the efficacy of the application.

Knowledge of the behaviour and fate of indigenous soil bacteria upon their introduction into soil is of importance in all cases in which these bacteria are applied to practical purposes such as biological control or biore-mediation.

Survival of the introduced populations in soil has been studied for a wide range of organisms, and the generally observed progressive decline of bacterial numbers following introduction into soil has often hampered the effectiveness of bacterial inoculants. For instance, comparison of the population dynamics of fluorescent pseudomonas in different soils showed that all introduced populations declined to low numbers within time spans of several weeks to months.

Overall linear decay rates varied from approximately log 0.2 to log 1.1 colony forming units (CFU) per 10 days, and depended on the strain and the soil used. Similar responses have been found in other introduced bacteria such as *Salmonella typhimurium* and *Klebsiella pneumoniae* and *Flavobacterium* spp. and Alcaligenes spp..

Such responses might be characteristic of copiotrophic organisms exposed to soil conditions. Oligotrophic bacteria introduced into soil may also show progressive decrease in CFU numbers, albeit slower than copiotrophic bacteria. The slower decline of oligotrophic bacteria in bulk soil is possibly related to a better adaptation to the carbon-limited conditions of sof soil.

It is difficult to pinpoint a single dominating reason for the decline of introduced bacterial populations in soil. However, introduced bacteria, particularly shortly after release, are likely to be affected by the same adverse soil conditions that affect indigenous bacteria, such as limitation of a range of nutrient sources.

Upon introduction in soil, increasingly higher numbers of cells of an introduced *P. fluorescens* strain were detectable via specific immunofluorescence than via selective plating.

This suggested a possible conversion of part of the introduced population into nonculturable forms that, however, might be still viable. Such viable but nonculturable (VBNC) cells were also found in other studies; recently it was found that their presence does not enhance the persistence of introduced Pseudomonas populations.

The occurrence of nonculturable cells, in the indigenous as well as introduced bacterial populations, indicates that one or several major factors controlling the status of bacterial cells in soil are common to bacteria in soil.

Moreover, given the relative immobility of bacterial cells in soil, the localization of introduced bacteria following their introduction is likely to affect their ultimate fate, that is, localization determines

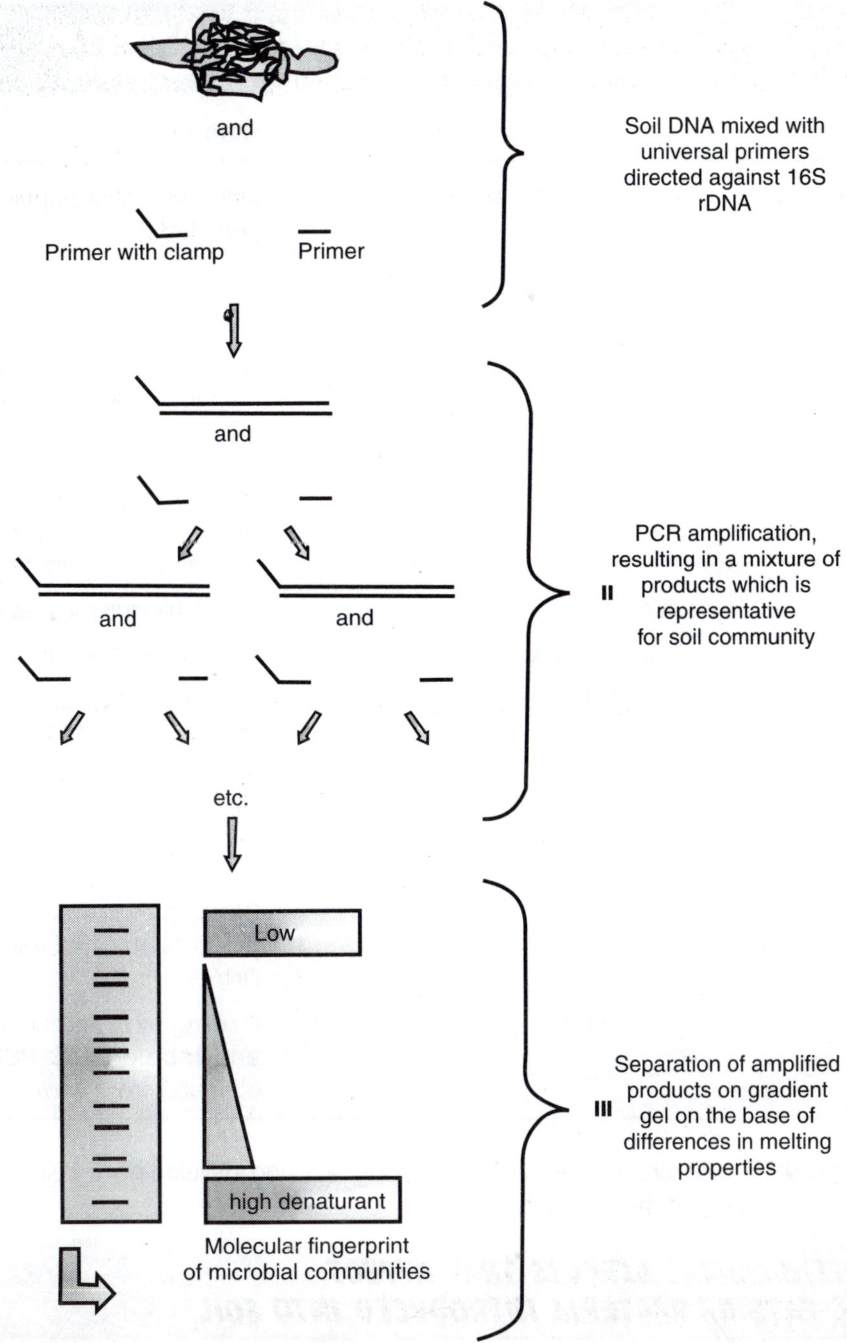

Figure 16.1: Schematic representation of the procedure for DNA fingerprinting °ngsoil habitats. Molecular analysis of soil communities is independent of preceding culturing and consists of the following steps: 1) extraction of soil DNA, 2) PCR amplification of 16S ribosomal DNA using a primer set with GC clamp (necessary for analysis of the products on denaturing gradient gel), and 3) separation of the PCR products on a gradient gel.

Table 16.1: Methods Used for Studying GMM Impact on Microbial Populations Indigenous to Soil Ecosystems.

Method		*Remarks*
Growth-based	Plating/colony formation on:	Only culturable populations assessed. Tedious.
	—General media	
	—Selective media	
	Biolog Physiological Community Level Profiling	Measurement of potential microbial community activity.
Microbial biomass assessments	Extraction of microbial compounds	Compounds related to specific groups of organisms such as muramic acid (bacteria) and phospholip-id fatty acid (fungi).
	Microscopy	Direct biomass assessment.
	Soil enzymes	Potential activity assessed.
Fingerprinting of PCR-amplified products from extracted nucleic acids	PCR/DGGE* or TGGE* and CSPD*	Profiling based on differential melting of 16S PCR products obtained from community DNA.
	PCR/ARDRA*	Profiling by restriction of 16S PCR products obtained from community DNA.
	PCR/T-RFLP*	Profiling by restriction of fluorescent end-labeled 16S PCR products obtained from community DNA.

whether inoculant cells are successful in colonizing soil and developing a cellular form resistant to soil stress or whether they will die out.

PHYSIOLOGICAL ASPECTS THAT AFFECT THE FATE OF BACTERIA INTRODUCED INTO SOIL

The metabolic activity and physiological state of bacterial populations will greatly affect their fate and efficacy after introduction into soil. Moreover, it is important to determine whether GMM strains are impaired in this adaptation due to the presence or expression of heterologous genes, resulting in an increased metabolic load. Different approaches have been taken to assess the metabolic and physiological conditions of cells in soil. Traditionally, the detection of specific strains in a natural environment has been based on CFU counts on selective plates or on immunofluorescence (IF) cell counts.

Using both techniques in combination, information on the presence of nonculturable cells can be obtained. However, these approaches provide little information about the activity of inoculated cells in soil. To asses the risks involved in the release of GMMs in soil, it is important to determine whether such nonculturable cells are alive (viable) and metabolizing, alive but in a metabolically arrested state, or dead.

To detect the presence of heterologous genes in bacterial cells in soil, including nonculturable ones, direct molecular detection techniques have been developed. Discrimination between viable and nonviable cells has often been performed using the direct viable count method (DVC, based on cell elongation by addition of nutrients in the presence of a cell division inhibiting compound) or redox dyes such as 2-(p-iodophenyl)-3-(p-nitrophenyl)-5phenyl tetrazolium chloride (INT) or 5-cyano-2,2-ditolyl tetrazolium chloride (CTC).

The viable cel numbers, determined by either or both techniques, often appeared to be intermediate between the total (IF) cel and selective CFU counts in different habitats such as plant surfaces and soil.

However, it remains to be determined whether viable but nonculturable populations are responsive to environmental stimuli, or if they are metabolically arrested.

At least in one study it was concluded that the viable but nonculturable state of introduced *P. fluorescens* cells does not represent a physiological stage in which cells are optimal adapted to the harsh conditions as present in soil.

Another criterion by which bacterial adaptation can be measured is their responsiveness to different, often stressful conditions that prevail in soil, such as carbon deprivation that limits growth.

In general, bacterial growth-limiting conditions provoke cellular responses, leading to an increased, overall, generalized, resistance. For instance, *Escherichia coli* cells that were limited in their growth after depletion of carbon sources showed an enhanced resistance to stress conditions such as high temperature, osmotic stress, and the presence of oxidative or noxious compounds.

A similar response was observed in a typical soil bacterium, *Pseudomonas putida,* and adaptation coincided with *de novo* synthesis of proteins upon growth limitation. Later, enhanced stress resistance was also demonstrated for *P. fluorescens* cells introduced into two texturally different soils.

Although the rates of cell decline upon introduction of this strain in the two soils were different, the time spans needed for complete adaptation (1–5 days) were the same.

This led to the conclusion that differences in survival time as a result of residence in different soil types are not related to differences in adaptation, but rather to other factors such as protection against predation. To address questions about bacterial responses that serve as signals for the onset of adaptation to soil, host cells carrying environmentally controlled reporter genes may help identify triggers to which these organisms will respond upon their release.

TRACKING GENETICALLY MODIFIED STRAINS IN SOIL ENVIRONMENTS

Sampling

Bacterial cells released into soil will not always remain at the site of application but may spread after their release. Such spread can be lateral (spurred by wind, rain, insects, or human activity) or vertical (induced by rain or burrowing worms). Also, cells may become nonculturable and die, whereas other cells become active and are mobilized.

The relative density of introduced strains may thus fluctuate in soil. In order to obtain a reliable overview of the presence of released GMM strains in the field, it is necessary to use an appropriate sampling strategy. Dispersal of a genetically modified *P. fluorescens* strain was monitored during a field release study in the Netherlands.

From the ninth day onward, detectable GMM CFUs were observed just outside the inoculated plot, with a maximal distance of about 2 m. However, CFU numbers in soil never exceeded 10^3 per g of dry soil, and the released GMM strain outside the plot could no longer be detected after 3 months, whereas in the inoculated part in the plot, the introduced strain survived for more than one year.

This example illustrates that the number and location of the samples drawn should carefully be taken into account. Thus, the sampling strategy chosen will depend on: 1) the mode of introduction, 2) the time span after release, and 3) factors influencing the mobility of inoculant cells such as plant growth, soil water flow, and spread by wind, animals, or human treatments.

Obviously, all considerations about the statistics of sampling play a role, as outlined by Van Elsas and Smalla. The mode of application of (GMM) strains will determine the location where cells reside. Cells can be applied either directly to soil in suspension via injection or spraying, or indirectly by immobilization in carriers using coatings (seed coating or root dipping) or slow-release carriers (e.g., in alginate or i-carrageenan beads).

Cells applied directly to soil will colonize the top soil layer, whereas immobilized cells tend to remain at the site of application. The way bacteria are introduced into the field (e.g., spraying versus application as immobilized cells) will influence the sampling strategy. Shortly after its release the effect of introduction is clearest, whereas, later, these effects may fade away because of dispersal in soil.

Also, it should be taken into account that the introduced strain may spread heterogeneously over the field. To establish statistically significant differences between treatments related to the release of GMM strains, enhanced numbers of replicates will be needed to overcome increasing standard deviations.

Vertical movement of introduced bacterial cells in soils has been observed under controlled conditions as well as in fields. Bacterial migration to deeper soil layers is mainly the result of water flow by heavy rainfall. However, in the same field experiment performed with a genetically modified *P. fluorescens* strain, vertical movement occurred without heavy precipitation.

Supposedly, bacterial cells migrated in substantial numbers to deeper layers by other means,

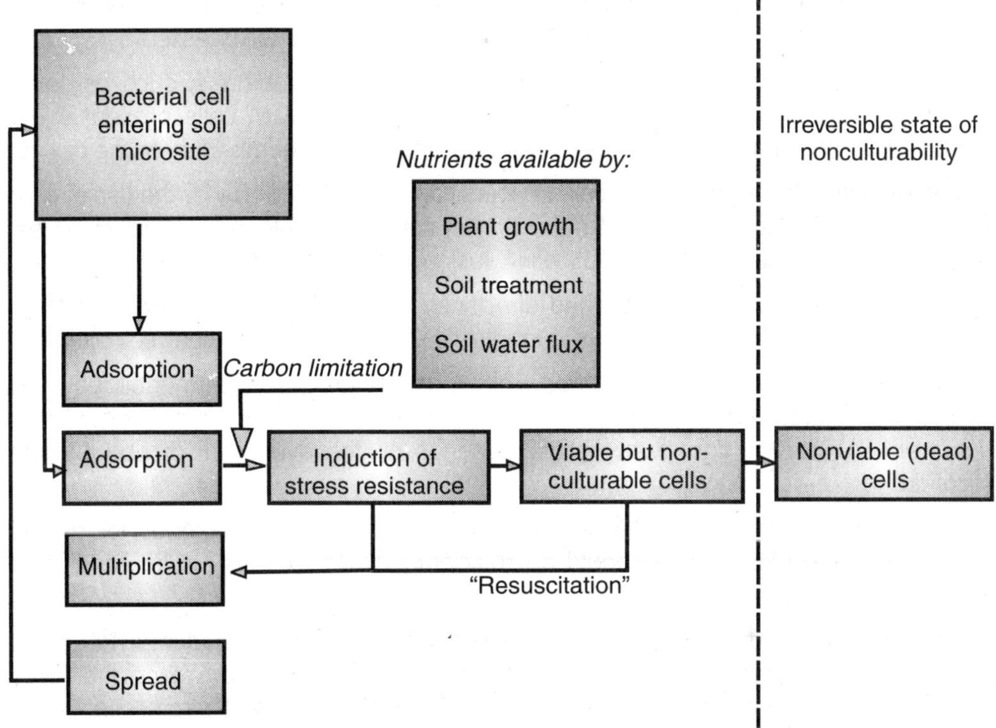

Figure 16.2: Concept of the sequence of events following the introduction of bacterial cells into soil and the specific factors influencing the physiological status of bacterial cells in soil.

for example, through soil cracks or by activity of soil animals such as earthworms. Plant growth is a major factor affecting the spread and activity of introduced bacterial cells. Bacterial growth near plant roots may result in increased ecological competence and thus competition with indigenous species residing near plant roots. Therefore, sampling of the rhizosphere of wild plant species near or in the test field is a prerequisite to determining the fate of the introduced GMM strain.

Methods for Sample Analysis

GMM strains in natural environments can be tracked with methods based on 1) their growth in selective media (cultivation-based methods), 2) intrinsic properties of the host (using antisera or probes targeting specific epitopes or nucleic acid sequences), and 3) the recombinant DNA (using probes or primers). In most cases, GMM strains, when released into soils, are monitored over time using cultivation-based methods.

Plating techniques are most convenient to use because of the selectable markers commonly present in GMM strains, as well as the possibility for quantification. Additionally, isolated colonies can be tested for phenotypic traits by carbon utilization testing [e.g., by using commercially available test systems such as API (Montalieu-Vercieu, France) or BIOLOG™ (Hayward, CA)] and molecular fingerprining [e.g., by using Repetitive Extragenic Palindromic (REP)-sequences, Enterobacterial Repetitive Intergenic Consensus (ERIC-) or BOX-PCR], and presence of the genetic insert can be tested by using molecular probes. However, bacterial growth is a prerequisite step for plant counting methods and, thus, only the culturable fraction of the populations is assessed.

Additional measurements by direct, cultivation-independent methods, such as seletive cell counts, will be necessary to determine the total number of GMM cells present in soil. The nonculturable fraction can then be deduced by subtracting the CFU numbers from the cell numbers.

The determination of the total GMM cell numbers can be accomplished by the use of fluorescent dyes conjugated to molecules targeted against host-specific properties such as epitopes (using antibodies) or nucleic acids, generally 16S rRNA (using molecular probes).

Cells extracted from soil can be stained directly by using specific antibodies conjugated to a fluorescent dye such as fluorescein isothiocyanate (FITC) (immunofluorescent cell staining).

However, when using molecular probes, cells must be fixed and permeabilized in order to allow entrance of the conjugated probes (whole-cell hybridization). Treated cells are visualized by their fluroscence under UV light (wavelength depends on the dye used) at about 1,000 times magnification.

In both cases, the targeted molecules present in bacterial hosts (eptopes for antisera and nucleic acids for molecular probes) should be sufficiently available, even under conditions of cell starvation.

Therefore, polyclonal antisera as well as probes directed against ribosomal 16S or 23S RNA are preferred, as the targets (multiple epitopes and ribosomal RNA) are expected to be most stable, albeit not constant. The inserted genes in the released GMM strain can be detected independent of cell growth. For that purpose, DNA extracted from soil can be amplified by PCR using primers directed against the heterologous gene.

Protocols for DNA recovery from soil are routinely applied in many research laboratories (68); they generally allow the detection of target molecules on the order of about 10^2 to 10^4 per g of soil. Quantification can be accomplished either by quantitative (Q)-PCR (quantification is based on the relationship between the number of target molecules initially present and the number of amplified products during several PCR cycles), competitive PCR (competition between amplification of target molecules and added target sequences), and most probable number (MPN) PCR (end point dilution of target DNA followed by PCR).

Moreover, PCR amplification in extracted and fixed cells (whole-cell PCR or *in situ* PCR) is a promising new tool for detection and quantification of released GMM strains. The side-by-side use of a range of different techniques for tracking GMM strains in soil is recommended in all field releases.

Only such a polyphasic approach allows the establishment of the possible occurrence of nonculturable cells, the number of cells that lost their heterologous DNA, as well as the occurrence of heterologous DNA without the presence of the original host cells.

Concerning the last point, heterologous DNA may be present in soil as 'naked' DNA (from lysed cells) or located in other hosts as a result of gene transfer (see later). Therefore, only a polyphasic detection strategy will allow an appropriate judgment of the presence and fate of GMM strains and their DNA in environmental samples.

ASSESSING THE EFFECTS OF GMMS ON POPULTIONS INDIGENOUS TO SOIL

Perhaps the most troublesome public concern about the release of GMM strains is the potential for concurrent undesirable effects on the indigenous microbial community upon introduction into the open field. Hence, most national governments currently require extensive data on such possible effects, prior to granting permission for field release of any GMM.

Preceding every field release, a protocol must be set up that includes measurements of expected effects and, if possible, consequences that may not be directly related to the application. These considerations can either be directly related to an effect of the inserted (heterologous) gene on the native organisms or related to changes in the occupation of different (micro) habitats of natural species caused by the application of inoculant strains.

Furthermore, an important potential consequence of the introduction of GMM strains into the environment is the transfer of recombinant DNA to indigenous species. The effects of GMM releases on natural populations can thus be considered by using two criteria; 1) an ecological and 2) a genetic criterion.

Ecological Effect of a GMM Release

In considering the impact of GMM releases, we deal only with the effects brought about by the novel genetic combination, thus disregarding effects that would have been caused by wild-type strains. Effects on natural populations can be expected from the expression of the inserted heterologous genes whose products are intended to target particular organisms, but which may, unintentionally, also affect nontarget, organisms.

For instance, constructs consisting of genes whose products are aimed at a reduction of plant damage caused by soil-born pathogens may also affect populations of nonpathogens. Examples are specific *Bacillus thuringiensis crystal* protein (*cry*) products targeted against larvae of dipteran insects and genes encoding antifungal compounds from *P. fluorescens* such as phenazines.

Both genes are not specific to just one particular pathogen, but can affect a broader range of organisms, although there is considerable specificity for limited insect groups among the various *cry* gene products. Assessment of effects that cannot directly be rationalized from the function of the heterologous gene is more complicated. In fact, all possible targets should theoretically be included in the risk assessment protocol, which is a truly daunting task.

Hence, we propose that, first, a reasonable assessment should be made as to which organisms might be possibly affected. Then, an appropriate technique that is capable of surveying the effects on all organisms of the selected groups must be chosen.

Recently, possible shifts in fungal community structures in field soil were studied upon release of a *P. putida* strain genetically modified by insertion of a gene responsible for the production of the antifungal compound phenazine-1-carboxylic acid, with the intention of a later application as a biocontrol agent. Fungal isolates were obtained from the rhizosphere of crop plants (*Triticum aestivum*) in different field plots treated with modified or parent strains and analyzed by Amplified Ribosomal DNA Restriction Analysis (ARDRA).

A transient change in fungal community structure was observed in wheat rhizospheres from plots inoculated with either strains, and the effect on fungal populations in rhizosphere soil with the GMM strains was different from that containing the unmodified strain.

Exposure of soil microbial populations to different carbon sources in BIOLOG™ microtiter plates is f commonly used approach to study differences or shifts in soil microbial compositions. Community Level Physiological Profiling (CLPP) using the 95 different carbon sources present in the Biolog microtiter plate helped determine the effect of a soil-inoculated GMM strain on indigenous populations.

As bacterial growth is not required, this approach can be considered as a measure of total activity of all bacterial population residing in the soil. Studying microbial populations by molecular techniques excludes biases caused by growth steps prior to analysis, whereas large numbers of species can be investigated in a single profile.

Nucleic acid extraction is a prerequisite, and methods for extraction and purification of DNA and RNA from soils are nowadays commonly available in many text books. In general, molecular fingerprinting of microbial communities is based on PCR amplification of hypervariable regions present on 16S rDNA genes using primers that target conserved regions directly adjacent to these. Ribosomal genes are "par excellence" suitable for molecular fingerprinting and different primers targeting 16S or 23S ribosomal genes have been developed.

Analyses of the amplified products are performed by Temperature or Denaturing Gradient Gel Electrophoresis (TGGE and DGGE) or Single Strand Conformational Polymorphism (SSCP). Both methods have successfully been applied by different laboratories to study microbial community changes in different habitats.

A fingerprint consisting of different bands, each representing individual organisms, reveals the complexity of the microbial populations in the environment, whereas, their activity can be determined by amplification from RNA. Individual bands can be sliced out from gels, cloned into vectors, and analyzed for DNA sequencing.

Database comparisons of the sequences, for instance, by BLAST searches on the Internet, helps establish the potential affiliation of the selected bands. Schwieger and Tebbe described a field release study in which molecular fingerprinting of microbial communities was applied.

Two *Sinorhizobium meliloti* strains chromosomally tagged with the luciferase gene (inserted for convenient recovery; one strain was mutated in a gene responsible for DNA repair after damage, i.e., *recA*) were introduced into field plots planted with alfalfa (*Medicago sativa*, the natural host for *S. meliloti*). Three months after introduction of both populations, the community structures near the alfalfa roots were clearly affected by the release, as demonstrated via SSCP.

The dominant *Pseudomonas* population was reduced in th rhizosphere of alfalfa plants treated with the modified strains. However, the rhizosphere community structure near a weed commonly observed in the field plots (*Chenopodium album*) was not affected by the introduced strains. Although the data obtained by molecular fingerprinting of inoculated soils show a great deal of detail, a major drawback of the method is the use of highly conserved ribosomal primers.

Only the most dominant species (over 0.1% of the total) can be visualized on gel. Therefore, development of primers for selected groups of organisms, including prokaryotes and archeal and fungal species, are in progress. Such systems would allow the in-depth study of specific groups of organisms. In conclusion, the development of molecular techniques, either or not in combination with classic technique such as culturing or enrichment steps, allows the study of changes in natural populations as a result of the introduction of GMM strains.

So far, different effects on natural populations caused either by the introduced strain or by expression of the inserted gene(s) have been shown. None of the effects shown are thought to pose a threat to ecosystem functioning. It is prudent to allow the testing of a wealth of novel microbial gene products on a case-by-case basis, building on the experience of no apparent risk gained so far.

Genetic Impact of GMM Releases

Transfer of genetic inserts from released GMM strains into ecologically competent species that reside in soil may have consequences for the persistence and expression of these genes in the environment. Therefore, special care has to be taken during construction of GMM strains with respect to the possibility of transfer of heterologous genes to natural species.

Most commonly, genes are inserted into the chromosome either by genetic crossing-over using DNA sequences homologous to the target site for insertion or by transposon mutagenesis. The advantage of chromosomally inserted constructs is: 1) the inserts are present in the cells as single copies, 2) the mutants are genetically stable, and 3) transfer is limited and only possible via interactions with mobilizing elements such as plasmids, conjugative transposons, insertion elements, and/or phages.

Such elements would first have to enter the inoculant cell from the natural microflora, a low-frequency occurrence in soil. A drawback of chromosomal insertion, however, may be the low copy number of the heterologous genes, possibly resulting in restrictions on expression.

Therefore, nonmobilizable plasmids carrying these constructs are favoured in some occasions. Movement of genes between organisms in soil is possible via the prokaryotic gene transfer mechanisms: transdution, transformation, and conjugation. Transduction, a phage-mediated DNA transfer, has been only incidentally described for soil environments but it cannot be ignored that it plays an important role in the movement of genes between bacterial species.

Transformation, the transfer of extracellular DNA by competent bacterial cells, and moblization, for example, plasmid-mediated transfer of DNA, are better known processes in soil environments. Tranformation between introduced strains in natural soils has been conclusively demonstrated by Lee and Stotzky. Nielsen and coworkers extended this work, showing that cells of the transformable bacterium *Acinetobacter* sp. BD413 can become naturally competent in soil and then effectively capture available DNA.

However, most information about bacterial gene transfer in soils has been obtained for conjugation using self-transmissible plasmids such as RP4 and pIPO2 and mobilizable plamids (IncP and IncQ plasmids) such as RSF1010 and pIE723.

Although transmissible plasmids are not recommended for use in GMMs intended for release, these studies have conclusively demonstrated that gene transfer does take place in natural soils. The use of mobiliable but non–self-transmissible plasmids (IncQ plasmids) may be a safer option, although the isolation of a cryptic mobilizable and self-transmissible plasmid from soil may imply that IncQ plasmids can be transferred into indigenous soil bacteria by retromobilization. As indicated, chromosomally inserted genes are transferable to other species via excision and insertion into incoming plasmids or conjugative transposons from the natural microflora, although the incidence of such an occurrence is likely to be very small. In general, gene transfer is an unwelcome event when releasing GMMs to the environment.

However, transfer of heterologous genes to indigenous species may sometimes be very advantageous as it may result in stable integration of these genes into species that are well adapted to the habitat and are, thus, ecologically more competent than the inoculant (host) strain.

This strategy has been applied under contained conditions in bioreactors, in which the xenobiotic compounds 3-chlorobenzoate and 3-chloro-aniline were degraded. Introduced GMM strains carrying the respective catabolic genes on mobilizable plasmids were transferred to indigenous strains and the transconjugants that appeared were successful in the degradation of noxious compounds. Although, up to now, this application is restricted to bioreactors, it may be extended to soils, in particular, in the degradation of xenobiotics in contaminated soils.

PROSPECTS FOR THE USE OF ENVIRONMENTALLY INDUCED PROMOTERS IN SOIL

Environmentally controlled promoter regions can certainly be applied to regulate the expression of biological control (control of plant diseases) or biological containment (containment of GMMs) of genes in soil inoculant strains. To achieve this goal, the reporter gene that enables the detection of the regulatory regions should be replaced by an appropriate biocontrol or biocontainment gene.

Suitable candidates for the construction of environmentally regulated biocontrol agents are bacterial genes involved in the synthesis of antifungal antibiotics, such as phenazine-1-carboxylic acid (103,104), and 2,4-diacetylphloroglucinol (105) from fluorescent *Pseudomonas* spp. or genes toxic to soilborne insect larvae, such as the *B. thuringiensis cryIVB* gene.

Genes of rhizosphere bacteria involved in antagonism toward plant-pathogenic fungi are of special interest and may be used as targets for cloning experiments. When these genes are brought under the control of a rhizosphere-induced promoter, the timing and extent of gene expression can be controlled depending on the promoter of choice.

This new strategy in the utilization of GMMs as biocontrol agents is promising because it can enhance and concentrate the beneficial action of the control agent to the site and time that it is most needed, for example, at the onset or peak of pathogen activity. However, in addition to the production of antibiotics, bacterial antagonism toward phytopathogenic microorganisms in the rhizosphere may also be affected by competition for nutrients, or it can be based on a combination of both actions.

Genes proposed for biological containment are the *E. coli gef/hok* (host killing) and *relF* genes and the *Serratia marcescens nuc* (nuclease) gene. Overexpression of *gef, hok,* and *relF* genes causes a collapse of the membrane potential, whereas *nuc* gene expression results in intracellular degradation of DNA. The result of both actions is severe damage to the host cell, resulting in cell death.

These genes can be brought under the control of a carbon starvation inducible promoter or a rhizosphere induced promoter region connected to the "negative loop" construction, as discussed before. Induction of the host killing genes by carbon limitation in bulk soil can result in cell death, thus confining the active GMM cells to the rhizosphere, which is the site whe expression of biocontrol genes is desired.

Thus, a GM strain can be constructed that expresses its biocontrol gene in the rhizosphere whereas, after escape from this site, expression of the biocontrol gene is switched off and the host strain will be killed by expression of the biocontainment gene.

CONCLUSION

The success of the application of GMM strains in soil environments depends on the selection of the bacterial host that should be ecologically competent in the selected habitat. Survival of the inoculant strain can be optimized by using appropriate carrier materials, whereas its activity can be regulated by selection of appropriate regulatory sequences used for the control of expression of the heterologous gene(s).

From a biotechnological point of view, there are no limitations in the optimization of suitable inoculants. However, more information is needed about the adaptation and survival of released GMMs, upon introduction into the target environment. Up to now, we lack fundamental knowledge of a full understanding of the adaptational processes with respect to the metabolic activity and dormancy of inoculant cells.

Unraveling these mechanisms will potentially allow an interference in the cellular properties and, thus, optimize survival under selected conditions. Restricting GMM strains to the site of application may decrease the putative risks involved in the release of these inoculants into the environment. The development of optimized modes of introduction, aimed to restrict released GMM strains to specific sites, as well as the tools necessary to track these strains and to determine fate, activity, and effect on natural microflora, is needed on a case-by-case basis.

Techniques for impact assessments now make it possible to pinpoint drastic effects on microbial fluctuations, but more development is necessary to enable detection of subtle changes in microbial community structure upon releases of GMMs into the environment.

Nevertheless, mankind should not lose its feeling of optimism about the great potential offered by the prospect of GMM releases in bioremediation and biocontrol. Compared with the prospects available only 10 years ago, major steps forward have been taken. It is very feasible that genetically modified biocontrol strains will replace chemical pesticides, or be used in an integrated fashion, to control soil-borne diseases in some of the key crop plants that feed the world.

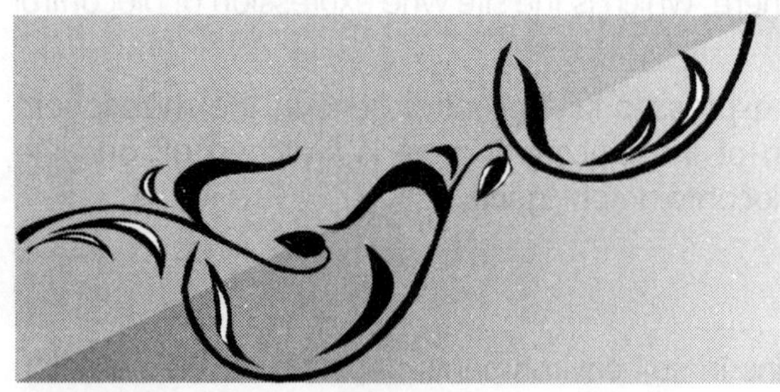

INDEX

A

B

C

D

E

F

G

H

I

J

K

L

M

N

O

P

Q

R

S

T

U

V

W

X

Y

Z